W9-BIY-553

PROCEEDINGS OF THE CONFERENCE ON INSTABILITY AND DISSIPATIVE STRUCTURES IN HYDRODYNAMICS

ADVANCES IN CHEMICAL PHYSICS

VOLUME XXXII

EDITORIAL BOARD

Advances in chemical physics
"

PROCEEDINGS OF THE CONFERENCE ON INSTABILITY AND DISSIPATIVE STRUCTURES IN HYDRODYNAMICS

EDITED BY

I. PRIGOGINE

University of Brussels,
Brussels, Belgium
and University of Texas,
Austin, Texas

AND

STUART A. RICE

Department of Chemistry
and
The James Frank Institute
The University of Chicago
Chicago, Illinois

VOLUME XXXII

AN INTERSCIENCE® PUBLICATION

JOHN WILEY AND SONS

NEW YORK · LONDON · SYDNEY · TORONTO

Library of Congress Cataloging in Publication Data:

Conference on Instability and Dissipative Structures in Hydrodynamics, Brussels, 1973.
Proceedings of the Conference on Instability and Dissipative Structures in Hydrodynamics.

(Advances in chemical physics; v. 32)
"An Interscience publication."
Includes bibliographical references.
1. Hydrodynamics—Congresses. 2. Stability—Congresses. 3. Heat—Convection—Congresses. I. Prigogine, Ilya. II. Rice, Stuart Alan, 1932–
III. Series.

QD453.A27 vol. 32 [QA911] 541'.08s [532'.5] 75-12717
ISBN 0-471-69934-9

Printed in the United States of America

10 9 8 7 6 5 4 3 2 1

INTRODUCTION

In many instances macroscopic systems, subject to appropriate boundary conditions, undergo spectacular morphological changes giving rise to patterns of spatial or temporal order. The characteristic parameters of these patterns are of macroscopic dimensions and differ therefore markedly from those arising in the familar phenomenon of phase transitions at equilibrium.

Recent progress in irreversible thermodynamics has revealed the necessary conditions for the emergence of these phenomena: nonlinear kinetic laws and a critical distance from the state of thermodynamic equilibrium. Stated differently: beyond a critical level dissipation can become an organizing factor, destabilize the disordered state constituting the extrapolation of near equilibrium behavior, and drive the system to an ordered configuration. Hence the term *dissipative structures*, introduced recently for this regime.

Meanwhile the development of mathematical analysis during recent decades had led to classes of nonlinear differential equations—ordinary or partial—whose solutions presented similar behavior. At the critical point of instability a branching, or *bifurcation* of two or more solutions, always occurs. Thus by the values of the bifurcation parameter—which, according to thermodynamics, can be the distance from equilibrium—the system can exhibit unique or multiple solutions, bistability, or even the highly interesting phenomenon of successive bifurcations.

During the last ten years chemical and biochemical kinetics have provided spectacular examples of spontaneous onset of a great variety of ordered patterns, such as sustained bulk oscillations of the limit cycle type, steady-state spatial structures, and standing or propagating concentration waves. In many cases the symmetry of the spatial domain was broken to give rise to polar patterns or to rotating waves. The importance of this type of behavior in biology is now widely recognized.

Theoretical models of nonlinear networks involving chemical reactions and diffusion have been developed, based both on the ideas of irreversible thermodynamics and on bifurcation theory. They have provided a first qualitative interpretation of the experimental data and have permitted a *classification* of the various types of organized states. An equally important area of research in this context has been fluctuation theory, which made it possible to understand the *origin* of as well as the *transitions* between different patterns.

Fluid dynamics, which was the first field to show, more than a century ago, the emergence of patterns of order, has long been developed independently of irreversible thermodynamics and fluctuations. On the other hand, it has always been the privileged field where new mathematical techniques and ideas were tested and applied. Unfortunately, the complexity of the Navier-Stokes equations is such that the origin and the mathematical mechanism of such basic phenomena as turbulence remain unsettled. For instance, the famous Landau-Hopf idea of successive bifurcations as a model of turbulence has not yet met with universal agreement.

Now, a lot of phenomena observed in the context of chemical kinetics have many aspects in common with fluid dynamics. The underlying mathematical treatment is much less complex than that for Navier-Stokes equations. Thus it is not unreasonable to expect that the development of nonlinear chemical kinetics will eventually lead to some new ideas applicable to fluid dynamics as well. In this respect it is interesting to note that many simple chemical networks can be shown rigorously to exhibit the phenomenon of Hopf (or successive) bifurcations which has never been established unambiguously in hydrodynamics.

An additional aspect which is likely to become increasingly important in many areas involving cooperative phenomena, refers to the role of fluctuations. In systems near thermodynamic equilibrium, fluctuations merely provide corrections to the average "observable" behavior. In most cases these corrections remain negligible. In contrast, the formation of a dissipative structure is entirely due to fluctuations. That is, instead of being a small correction, fluctuations determine the dynamics of such systems by driving the averages to a new regime.

The purpose of this volume is to present a number of problems involving hydrodynamic instabilities from the standpoint of irreversible thermodynamics of dissipative structures. We hope that the analogies with chemical kinetics and the existence of common underlying ideas in all phenomena involving emergence of order in a previously disordered medium will stimulate further research in these fascinating areas.

I. PRIGOGINE

CONTRIBUTORS TO VOLUME XXXII

M. S. BIOT, 117, Avenue Paul Hymans, Belgium

JEAN-PIERRE BOON, Faculte des Sciences, Universite Libre de Bruxelles, Bruxelles, Belgium

G. CHAVEPEYER, University of Mons, Faculty of Sciences, Mons, Belgium

P. GLANSDORFF, Faculte des Sciences, Universite Libre de Bruxelles, Belgium

ETIENNE GUYON, Physics Department, University of California, Los Angeles, California

D. T. J. HURLE, Royal Radar Establishment, Malvern, Worcester, England

E. JAKEMAN, Royal Radar Establishment, Malvern, Worcester, England

E. L. KOSCHMIEDER, College of Engineering and Center of Statistics & Thermodynamics, The University of Texas, Austin, Texas

D. R. MOORE, Department of Applied Mathematics and Theoretical Physics, University of Cambridge, England

J. C. MORGAN, Department of Mathematics, Hull College of Technology, England

G. NICOLIS, Faculte des Sciences, Universite Libre de Bruxelles, Belgium

J. PERDANG, Institut de Astrophysique, Cointe-Ougree, Belgium

PAWEL PIERANSKI, Physique des Solides, Universite Paris Sud, Orsay, France

J. K. PLATTEN, University of Mons, Faculty of Sciences, Mons, Belgium

I. PRIGOGINE, Faculte des Sciences, Universite Libre de Bruxelles, Belgium

P. H. Roberts, University of Newcastle upon Tyne, Great Britain

S. M. ROHDE, Engineering Mechanics Department, Research Laboratories, General Motors Technical Center, Warren, Michigan

F. SCHLÖGL, RWTH Aachen, Germany

G. THOMAES, Faculty of Applied Sciences, Free University of Brussels, Brussels, Belgium

RICHARD N. THOMAS, Institut d'Astrophysique, Paris, France

J. S. TURNER, Department of Applied Mathematics and Theoretical Physics, University of Cambridge, England

PIERRE R. VIAUD, Laboratoire de Aerothermique du C.N.R.S., Meudon, France

N. O. WEISS, Department of Applied Mathematics and Theoretical Physics, University of Cambridge, England

CONTENTS

ON THE MECHANISM OF INSTABILITIES IN NONLINEAR SYSTEMS

G. NICOLIS, I. PRIGOGINE,* AND P. GLANSDORFF

Faculté des Sciences,
Université Libre de Bruxelles, Belgium.

CONTENTS

I. INTRODUCTION

The existence of unstable transitions leading to new flow patterns has long been known and is considered by the hydrodynamicist to be a "natural" situation under certain conditions. Much more unexpected is the realization that *purely dissipative* systems having attained mechanical equilibrium, like chemical or biochemical reaction networks, may give rise to quite similar phenomena.[1] A major finding of the *generalized thermodynamics* developed during the last few years[1] is that, in spite of their extreme diversity, these instabilities share common features and have—in some sense—a common origin. Indeed, in all cases one deals with *nonlinear systems*. In the equations of fluid dynamics nonlinearity comes primarily from the convection term $\mathbf{u}\cdot\nabla$. In chemical kinetics, on the other hand, nonlinearity enters through the polynomial terms describing the reaction rates as a function of the concentrations.

A second important point is that the transition to instability is possible only when a certain set of parameters related to the *distance from equilibrium* exceed some critical value. In fluid dynamics these parameters may be, for example, a Rayleigh or a Reynolds number, whereas in chemistry

*Also at the center for Statistical Mechanics and Thermodynamics, The University of Texas at Austin, Texas.

they are always related to the overall affinity of the reaction sequence.

Finally, beyond instability the system appears to exhibit a completely new and unexpected type of behavior which is frequently associated with the appearance of *ordered patterns*, stationary or time-periodic or finally almost periodic.[1,2] The term *dissipative structures* has been introduced to describe these various spatiotemporal organizations emerging beyond a critical distance from equilibrium.

Two outstanding questions suggest themselves in this context:

1. Is it possible to analyze rigorously the *mathematical mechanisms* behind these transitions and, in particular, is it possible to provide a general classification of the situations which may emerge beyond instability?
2. What is the *microscopic origin* of these transitions? In different terms: how is the instability starting; is it a spontaneous phenomenon; and what are the time scales and ranges involved in the amplification of a mode in the neighborhood of the transition threshold?

These two questions are examined in the subsequent sections.

II. BIFURCATION ANALYSIS OF DISSIPATIVE STRUCTURES

In most problems of fluid dynamics of dissipative systems one is confronted with the solution of a nonlinear parabolic differential system. This introduces two major mathematical complications. The first one is quite general in nature: any nonlinear partial differential system implies, automatically, the existence of an *infinity of coupled degrees of freedom*. In contrast, the mathematician feels more at home with situations reducible to a finite dimensional phase space. In this case, one may apply the powerful tools of the qualitative analytical–topological theory initiated by Poincaré, continued by Andronov and his school,[3] and completed to perfection by Thom.[4] Thanks to these investigators one now has a *complete classification* of situations which may arise at the *bifurcation points* or, to use Thom's terminology, at the *catastrophe points*. Unfortunately there is not a comparable theory for the solutions of partial differential equations. What is possible, however, is bifurcation analyses and the derivation of analytic, usually approximate expressions for the solutions near the bifurcation points of certain branches of solutions (for a review of bifurcation theory see Ref. 2 or 5).*

*For a discussion of the relation between catastrophe theory and dissipative structures see the paper by Nicolis and Auchmuty.[6]

In this respect the second difficulty arises in connection with fluid dynamics. Indeed, it turns out that the structure of the nonlinear operator $\mathbf{u}\cdot\boldsymbol{\nabla}$ is such that the usual existence and uniqueness theories based on the properties of the Green's functions and, to a lesser extent, the explicit construction of approximate solutions is greatly complicated. As a result, most of the research in fluid dynamics is based on nonsystematic approximation schemes where the infinite hierarchy of equations for the normal modes is arbitrarily truncated to a finite order.

The situation appears to be more favorable in chemical kinetics for, in this case, the Green's function of the differential system can be derived straightforwardly. On the other hand, the similarity between chemical and hydrodynamic instabilities suggests that the preliminary classification of dissipative structures which turns out to be possible in chemistry and transport phenomena will be applicable, at least partly, in fluid dynamical situations as well.

More explicitly we will now discuss the results recently obtained[6] from a bifurcation analysis of a simple chemical network, in fact, the simplest one capable of exhibiting cooperative behavior. The chemical mechanism is represented schematically in Fig. 1.

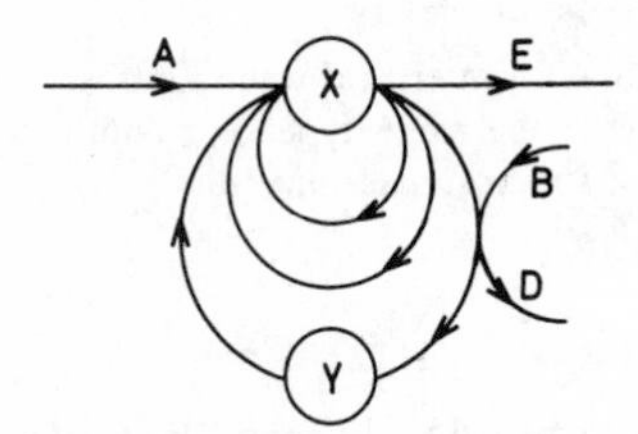

Fig. 1. A trimolecular reaction sequence involving the two intermediates X and Y, the initial products A and B, and the final products D and E. Each closed loop containing a given intermediate implies the presence of 1 mole of this intermediate in the corresponding reaction step.

The equations describing this mechanism are of the form:

$$\frac{\partial x}{\partial t} = f(x,y) + D_x \nabla^2 x$$
$$\frac{\partial y}{\partial t} = g(x,y) + D_y \nabla^2 y \tag{1}$$

where the functions f, g contain the cubic term x^2y but are otherwise linear.

A linear stability analysis of (1) reveals that the uniform steady-state solution may become unstable and that, under appropriate conditions, the system may evolve to a steady-state dissipative structure. Under other conditions one may have bifurcation of a standing or propagating concentration wave.

The surprising aspect of these transitions is that one observes quite different behavior depending on the *symmetry properties* of the critical mode. When the critical wave number is even, the system exhibits a *symmetry-breaking transition* to *two* possible new states, both of which are stable. The corresponding bifurcation diagram is shown in Fig. 2.

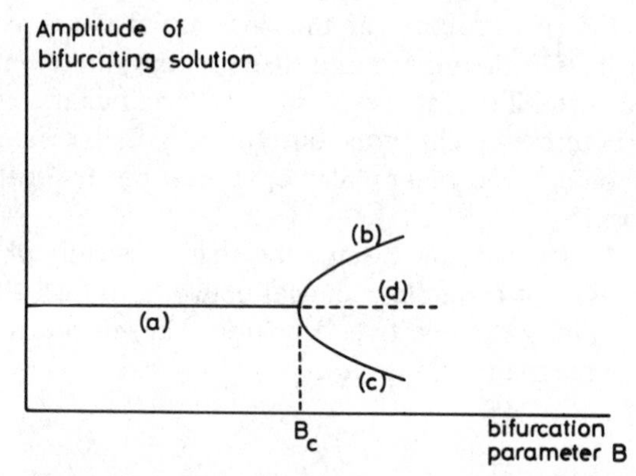

Fig. 2. Bifurcation diagram for scheme (1) corresponding to an even critical wave number. (*a*): stable uniform solution below the critical point B_c; (*d*): the same type of solution becoming unstable beyond the bifurcation point; (*b*) and (*c*): stable supercritical dissipative structures emerging beyond a symmetry-breaking instability.

In contrast, when the critical wave number is odd one observes *bistable behavior* combined with *hysteresis* as shown in Fig. 3.

For $B < B_c$ there may be two stable solutions, the uniform solution and a dissipative structure. For $B \geqslant B_c$ there are two dissipative structures, one of which requires an abrupt transition from the uniform solution. No symmetry-breaking in the strict sense of the term occurs in this case. Indeed, it is natural to conjecture that an adiabatic increase of B, when B is near B_c, will lead to a solution on branch (*b*). A sudden large increase of B, however, could lead with almost equal probability to solutions on either branches (*b*) or (*c*).

This phenomenon has striking similarities to recent observations and approximate calculations by Platten et al.[7] on the two-component Bénard problem with negative Soret coefficient and rigid boundary surfaces (see chapter by Platten and Chavepeyer in this volume). The situation is described on Fig. 4.

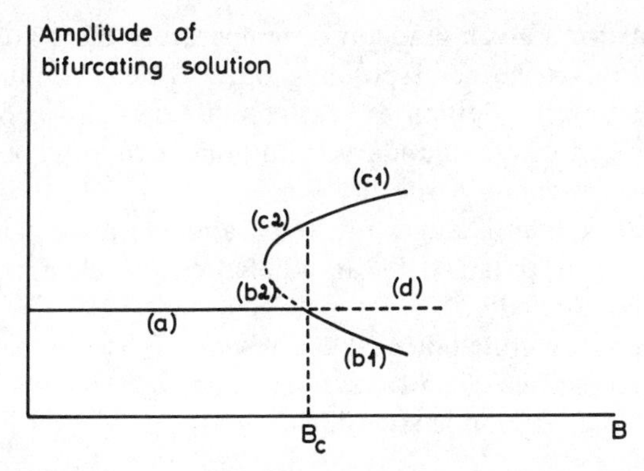

Fig. 3. Bifurcation diagram for scheme (1) corresponding to an odd critical wave number. (*a*) and (*d*) have the same meaning as in Fig. 2; (*b*1): supercritical stable dissipative structure arising continuously beyond bifurcation; (*b*2): subcritical unstable solution; (*C*1) and (*C*2): branches extending on both sides of the bifurcation point and separated from the uniform solution by a finite jump.

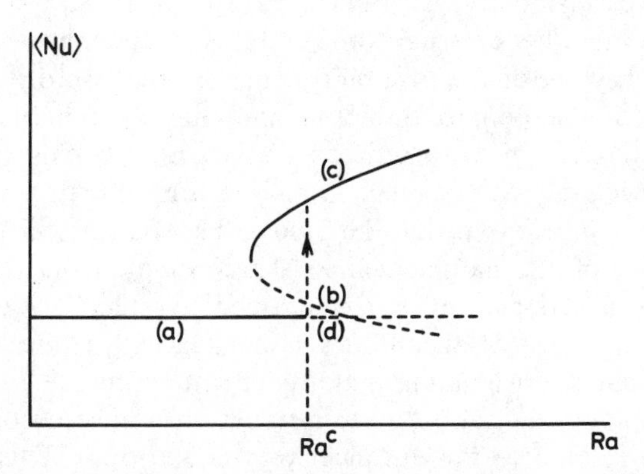

Fig. 4. Tentative bifurcation diagram for the two-component Bénard problem. ⟨Nu⟩: space averaged Nusselt number at the lower plate; Ra: Rayleigh number; (*a*) and (*d*): solutions corresponding to the state at rest, respectively, stable and unstable; (*b*): unstable branch of solutions; (*c*): stable branch corresponding to states with convective motion and emerging after an abrupt transition from branch (*a*).

It is quite remarkable that in a finite region below Ra^c there are three possible values of the Nusselt number (which measures the heat flow) for a given value of the Rayleigh number (which measures the gradient of temperature). The two extreme values correspond to stable flow patterns. The analogy with the well known Van der Waals situation is striking.

The fact that for a given gradient of temperature there are two different physical situations—one corresponding only to heat conduction and the other to convection in addition to heat conduction—shows how far we are from the situations where boundary conditions determine in a unique way the state of the system.

The fact that sudden changes in the parameters may lead to situations different from those obtained by an adiabatic slow change has been also observed by Koschmieder.[14]

We may therefore conclude that the results of Nicolis and Auchmuty,[6] while derived for simple chemical systems, contain elements which are also of interest for hydrodynamic situations.

III. THE ONSET OF INSTABILITIES

The foregoing examples strongly suggest an analogy between transitions leading to dissipative structures and phase transitions. One is surprised by the fact that, in spite of such strong analogies, the general belief in fluid dynamics is that the onset of convection, or of turbulence, or of any other similar kind of instability, is primarily triggered by the boundary conditions or by some other external stimulus. This attitude, however, masks the essential mechanism of the instability. Suppose that we drive a fluid very carefully up to the Bénard threshold and that by suitable isolation we avoid the influence of any spurious disturbance. Is this fluid going to evolve to the regime with convection or is it going to remain still?

The answer we believe is that the fluid will evolve and the reason for this is the existence of internal *fluctuations*. These spontaneous deviations from the mean are always present in a macroscopic system. They are known to be the primary causes of instabilities associated with phase transitions.

Now, in their overwhelming majority, fluctuations are *local* and *small*. On the other hand, the transition to instability is a macroscopic large-scale phenomenon related to the distance from equilibrium. This difference in scale strongly suggests a mechanism of *nucleation*, whereby fluctuations must first add up within a small spatial region of some critical size before being able to grow and drive the system to the new regime beyond instability.

At present it still remains difficult to substantiate this conjecture for fluid dynamical instabilities. The main reason is that the analysis of fluctuations is in this case very complex and requires a finer statistical mechanical treatment.[8]

Again, however, chemical kinetics turns out to be a more favorable field where the mechanism of nucleation and of the subsequent amplification process can be analyzed in detail.[9] The main feature of this analysis is the adoption of a stochastic description of fluctuations in the space of macroscopic variables. The interactions between particles appearing in the more detailed statistical mechanical description are then replaced by an average interaction between a "small" subvolume and the surrounding "big" system.

This description can therefore be regarded as intermediate between the "microscopic" method of statistical mechanics and the "macroscopic" method of continuum thermodynamics as illustrated below.

Consider a system described by a set of variables (the numbers of particles of various chemical constituents) whose macroscopic time evolution is given by the equation [see also (1)]:

$$\frac{d\bar{x}}{dt} = f(\bar{x}) \tag{2}$$

The system is assumed to remain macroscopically isothermal and homogeneous.

We consider a subvolume ΔV sufficiently small to permit a stochastic description of the dynamical processes therein in terms of the variable x only. In addition to the local processes, ΔV is coupled to the remaining part $V - \Delta V$ of the system, V. As we are interested here in chemical instabilities we neglect all processes responsible for this coupling other than the transport of matter across the surface surrounding ΔV. Let $\mathfrak{D}$ be the corresponding transport coefficient. One may then derive the following master equation for the probability $P(x,t)$ of having x particles at time t, within ΔV:

$$\begin{aligned}\frac{dP}{dt} &= R[x, P(x,t)] + \mathfrak{D}\langle x\rangle[P(x-1,t) - P(x,t)] \\ &\quad + \mathfrak{D}(x+1)P(x+1,t) - \mathfrak{D}xP(x,t)\end{aligned} \tag{3}$$

with

$$\langle x\rangle = \sum_{x=0}^{\infty} xP(x,t) \tag{4}$$

$R(x, P(x,t))$ stands for the contributions coming from the chemical reactions inside ΔV described on the average by (2). Note that in (3) the system surrounding ΔV has been taken into account in an average sense. The requirement of macroscopic homogeneity permits to relate this average to the mean value $\langle x\rangle$ of the subsystem itself. An important consequence is that the master equation (3) is *nonlinear* because of (4).

The moment equations generated by (3) can be derived straightforwardly. The first moment equation turns out to be independent of $\mathscr{D}$ [and approximately identical to (2)], in agreement with the requirement of spatial homogeneity. The second moment equation reads

$$\frac{d}{dt}\langle x^2\rangle = \sum_{x=0}^{\infty} x^2 R[x,P(x,t)] + 2\mathscr{D}\left[\langle x\rangle - \langle(\delta x)^2\rangle\right] \tag{5}$$

where $\langle(\delta x)^2\rangle$ is the mean square deviation. We note from (5) that the role of $\mathscr{D}$ in the evolution of fluctuations becomes more important as one deviates from the Poisson regime. In the limit $\mathscr{D}\to\infty$ the steady-state solution of (3), (5) reduces to a Poisson distribution.

Suppose now (3) or the corresponding moment equations predict an instability of the kind considered in the previous section. At the critical point, $\mathscr{D}$ will be related to the system's parameters contained in $f(\bar{x})$, such as rate constants. On the other hand $\mathscr{D}$ is also related to the size of the volume ΔV. A qualitative estimate yields:

$$\mathscr{D} \propto \frac{D}{l} \tag{6}$$

where D is Fick's diffusion coefficient and l a characteristic size parameter of ΔV. Then introducing (6) into the critical point relation one finds a relation between size of the subsystem and rate of growth of fluctuation. This calculation can be carried out on a number of chemical examples, including the model of the previous section. The results are shown in Fig. 5.

The meaning of this diagram is as follows. If the range of a disturbance, that is, the length over which it preserves a coherent character, acting at the vicinity of a point inside the subvolume ΔV belongs to the dashed region of Fig. 5, then the decay processes (roughly measured by $\mathscr{D}_c^{-1}$) take over and the disturbance dies out. For $k<k_0$ even if an infinite coherence length is imposed, the disturbance decays. For $k>k_0$ only those disturbances whose range exceeds l_c will be amplified and spread throughout the system.

The similarity between this picture and *first*-order phase transitions should be pointed out. In essence, Fig. 5 gives the values of an *entropy of activation* necessary to form an unstable nucleus under nonequilibrium conditions (for a review of the nucleation theory in the context of phase transitions, see Ref. 10).

A connection with the usual belief prevailing in fluid dynamics can now be established more clearly. If the critical coherence length l_c is macroscopic, then it would be very unlikely that thermal (and thus mostly local) fluctuations could trigger the instability. The latter would therefore require the presence of external distrubances. But, if l_c becomes comparable to some characteristic submacroscopic length scale, thermal noise would be able to create an unstable nucleus. Under certain conditions the latter could then propagate and "contaminate" the whole system.

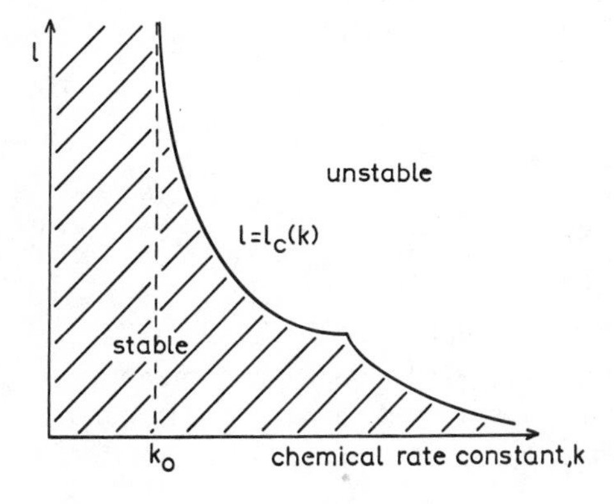

Fig. 5. Coherence length of a fluctuation versus a characteristic chemical rate constant at the critical point of instability.

IV. CONCLUDING REMARKS

The existence of an entropy of activation as we have discussed in the preceding section opens fascinating possibilities for a range of problems.

Obviously this question is beyond the macroscopic description as expressed in terms of the "chemical" equations (2). The situation is somewhat similar to that in the conventional nucleation theory. There also new surface properties must be introduced in addition to the bulk properties to account for phenomena such as overcooling or oversaturation. In fact, even for such problems our method provides an alternate approach which, in view of the incertitudes which plague the conventional methods, is worth exploring.

The basic equation (5) implies that the outside world always damps the fluctuations. This conclusion should be applicable to ecological models or social problems where equations similar to (2) are often a useful starting point. In fact the treatment outlined in (3) was inspired by the Prigogine-Herman theory for vehicular traffic flow.[11] The basic equation of this theory is the kinetic equation for the one-car velocity distribution $f(v)$,

$$\frac{\partial f}{\partial t} = -\frac{f-f_0}{T} + (1-P)c(\langle v\rangle - v)f \qquad (7)$$

Here f_0 is the "desired" velocity distribution function, T a relaxation time, P the probability of passing, c the concentration of drivers, and $\langle v \rangle$ the average speed

$$\langle v \rangle = \int_0^\infty dv\, v f \tag{8}$$

As is equation (3), the kinetic equation (7) is therefore a *nonlinear* integro-differential equation for the unknown distribution f. From (7) we obtain easily for time-independent conditions

$$\langle v \rangle = \langle v \rangle^0 - T(1-P)c\langle(\delta v^2)\rangle \tag{9}$$

where $\langle v \rangle^0$ is the "desired" average speed, and

$$\langle(\delta v^2)\rangle = \langle v^2 \rangle - \langle v \rangle^2 \geqslant 0 \tag{10}$$

Obviously (9) describes a "damping" in the average velocity as compared to $\langle v \rangle^0$ due to the overall effect of the population on the individual driver. In the case of traffic flow the theory necessarily has a phenomenological character. On the contrary the theory described in Section III may be justified by a more rigorous analysis[12] based on a phase space description of fluctuations.

As an example of the application of these ideas let us consider the relation between stability and complexity. A surprising outcome of the study of stability of multicomponent systems is that increased complexity (measured by the number of "links" between the components) tends to beget diminished stability. An excellent presentation of this problem is given in the recent monograph by May[13] on "Model Ecosystems." This seems to contradict the findings of ecologists that increased complexity leads instead to increased stability. Our approach permits the investigation of this important problem from a new point of view as it shows that the price of complexity may be metastability! The study of specific models from this point of view is at present in progress and we hope to report the results in the near future. For a more detailed presentation of the problem of stability of complex systems in the context of ecological and social systems, we refer to ref. 15 and 16.

References

1. P. Glansdorff and I. Prigogine, *Thermodynamics of Structure, Stability, and Fluctuations*, Wiley-Interscience, New York, 1971; French ed., Masson, Paris, 1971.
2. See e.g., *Nonlinear Problems in the Physical Sciences and Biology*, Springer-Verlag, Berlin, 1973.
3. A. A. Andronov, A. A. Vitt, and S. E. Khaikin, *Theory of Oscillators*, Pergamon Press, Oxford, 1966.

4. R. Thom, *Bull. Am. Math. Soc.*, **75**, 2 (1969); *Stabilité structurelle et Morphogénèse*, Benjamin, Reading, Mass., 1972.
5. D. Sattinger, *Topics in Stability and Bifurcation Theory*, Springer-Verlag, Berlin, 1973.
6. G. Nicolis and G. Auchmuty, *Proc. Natl. Acad. Sci. U.S.* **71**, 2748 (1974).
7. J. K. Platten and G. Chavepeyer, "A Hysteresis Loop in the Two Component Bénard Problem," Preprint, University of Mons, 1974.
8. Some early efforts are reported in G. Ludwig, *Physica*, **28**, 841 (1962). More recently H. Haken [*Phys. Lett.*, **46A**, 193 (1973)] has attempted a study of nonequilibrium fluctuations in fluid dynamics by adding Langevin forces to the conservation equations and subsequently deriving a Fokker-Planck type equation.
9. G. Nicolis, M. Malek Mansour, K. Kitahara, and A. Van Nypelseer, *Phys. Letters,* **48A** 217 (1974).
10. *Nucleation* A. C. Zettlemoyer, Ed., Dekker, New York, 1969.
11. I. Prigogine and R. Herman, *Kinetic Theory of Traffic Flow*, Elsevier, New York, 1971.
12. G. Nicolis and I. Prigogine, *PNAS*, **68**, 2102 (1971); G. Nicolis, P. Allen, and A. Van Nypelseer, *Prog. Theor. Phys. (Japan)*, **52**, 1481 (1974).
13. R. M. May, *Model Ecosystems*, Princeton University Press, 1973.
14. L. Koschmieder, *Advances in Chemical Physics*, Vol. 32, I. Prigogine and S. A. Rice, Eds., Interscience, New York, 1975, p. 000.
15. I. Prigogine, G. Nicolis, R. Herman, and T. Lam, *Collective Phenomena*, **2** (1974).
16. I. Prigogine, "*L'Ordre par fluctuation et le système social*", Preprint, University of Brussels, 1975.

ON A UNIFIED THERMODYNAMIC APPROACH TO A LARGE CLASS OF INSTABILITIES OF DISSIPATIVE CONTINUA

M. A. BIOT

117, Avenue Paul Hymans
Bruxelles, Belgium

As a contribution toward unification and interdisciplinary perspective it seems appropriate here to call attention to an extensive body of results established by another school regarding the thermodynamic aspects of instability *leading to dissipative structures*. The theory deals with the linear irreversible thermodynamics of deviations from a state of equilibrium which is unstable, as distinguished from unstable perturbations in the vicinity of a nonequilibrium steady state. However, as pointed out below, special cases of instability of steady-state flows may also be analyzed by similar methods.

This general approach provides a systematic thermodynamic analysis of continua which are initially stressed in a state of unstable equilibrium. Typical classical examples are elastic buckling in compression, and a viscous fluid in equilibrium in a gravity field with an unstable density gradient. In general perturbations of the unstable equilibrium obey the Lagrangian equations

$$\frac{d}{dt}\left(\frac{\partial \mathcal{T}}{\partial \dot{q}_i}\right)+\frac{\partial \mathcal{D}}{\partial \dot{q}_i}+\frac{\partial \mathcal{P}}{\partial q_i}=0 \qquad (1)$$

where $\mathcal{T}$ is the kinetic energy, $\mathcal{D}$ the dissipation function defined in terms of entropy production, and $\mathcal{P}$ is a new mixed mechanical and thermodynamic potential different from the classical thermodynamic potentials. The generalized coordinates q_i define the nonisothermal perturbations from the initial state of stressed equilibrium. Instability arises from the fact that $\mathcal{P}$ is not positive-definite.

The theory was developed over the last 20 years starting in 1954[1-3] with a Lagrangian-variational formulation of irreversible thermodynamics, followed by a sequence of applications and further developments, some of which are given in the reference section[4–17]. The theory of viscoelasticity of initially stressed solids and fluids including thermodynamic foundations was presented in systematic fashion in a book puplished in 1965.[18]

With the use of internal coordinates the thermodynamics embodied in (1) are readily applicable not only to viscous fluids but to viscoelastic solids with memory. This aspect of viscoelasticity based on internal coordinates as developed earlier,[1 to 3] was also presented in a book by Fung.[25]

Unstable solutions are proportional to $\exp(pt)$ where according to a basic theorem [Ref. 18, p. 441] p is real and positive. Thus for systems governed by (1) the incipient instability is *nonoscillatory*.

The theorem is applicable to viscoelastic solids if we assume that the hereditary behavior is due to a large number of unobserved internal coordinates included in the generalized coordinates q_i governed by (1). Note that the theorem assumes an initial state of equilibrium but an unstable one. It should be noted that it is still applicable to instability in the vicinity of a state of steady flow in cases were the system may be approximately represented by (1) as occurs in many problems of viscous buckling of solids. However other types of problems with steady heat flow may lead to oscillatory instability.[24]

It is not possible here to outline all the numerous applications and therefore only a few of the highlights will be mentioned. Stability in the thermodynamic context was discussed at a IUTAM colloquium in Madrid in 1955[3] followed by an application to the folding instability of an embedded viscoelastic layer in compression.[4] This brought out the important qualitative role played by thermodynamic principles and the appearance of dissipative structures embodied in the concept of *dominant wavelength*, thereby showing that such structures do not necessarily require a nonlinear thermodynamic behavior. The folding instability of a porous layer and its thermoelastic analogy were also analyzed[15] as an application of the general stability theory of porous dissipative solids.[13]

The instability of a multilayered viscous fluid in a steady state of compressive flow with finite strain subject at the same time to gravity forces was given a general and systematic treatment.[16,18,19] This theory considers small displacement perturbations superposed upon finite displacements which themselves are time dependent and constitute the initial unperturbed steady flow. A general theory of instability was also developed for a multilayered system including materials with couple-stresses and was applied to thinly laminated layers.[22]

Numerous applications have been made to problems of geological folding of layered rock under tectonic stresses and the results have been verified experimentally.[11,12] Such geological features provide another example of dissipative structures. A particularly interesting result is represented by a numerical evaluation of the actual time history of folding of a layered viscous medium in compression starting with a local bell-shaped deviation from a perfectly flat layer.[12,18] The gradual appearance of a dissipative structure in the form of folds may be followed as time goes on, and the wavelength of these folds turns out to be insensitive to the type of initial disturbance. It is also pointed out that a multiplicity of initial disturbances each generating its own structure similar to a sinusoidal wave packet may give rise to mutual interference patterns.

In the special case of destabilizing gravity forces the theory was applied to the formation of salt domes in geophysics, under transient conditions of gradual sedimentation and time dependence of thickness and compaction of the material overlying the salt layer.[20,21] A particularly interesting feature of the results is some kind of degeneracy exhibited by the cell pattern showing that triangular, hexagonal, circular, and many other patterns are equally unstable. Thus the appearance of any particular pattern should be very sensitive to boundary constraints, perturbations of these constraints, and initial conditions. It was shown that a system of localized irregular initial perturbations may produce a system of ring-shaped cells with a mutual interference pattern. Results are in good agreement with observed geophysical structures and geological time scales.

Finally it should be noted that the theory is not restricted to linear perturbations as exemplified by more recent work on nonlinear thermoelasticity, the thermodynamics of elastic instability, and postbuckling behavior.[23]

The Lagrangian approach outlined here provides not only deeper and unified physical insights but also powerful methods of approximate analysis formulated directly in terms of generalized coordinates of complex systems. This procedure does not require any preliminary knowledge of the differential field equations of the system. This is in contrast with the purely formal so-called projection methods based on abstract functional space theories.

References

1. M. A. Biot, "Theory of Stress-Strain Relations in Anisotropic Viscoelasticity and Relaxation Phenomena." *J. App. Phys.*, **25**, No. 11, 1385 (1954).
2. M. A. Biot, "Variational Principles in Irreversible Thermodynamics with application to Viscoelasticity." *Phys. Rev.*, **97**, No. 6, 1436 (1955).
3. M. A. Biot, "Variational and Lagrangian Methods in Viscoelasticity." *Deformation and Flow of Solids*, (IUTAM colloquium Madrid 1955), Springer–Verlag, Berlin, 1956, pp. 251–263.

4. M. A. Biot, "Folding Instability of a Layered Viscoelastic Medium under Compression," *Proc. Roy. Soc. (London), Ser. A*, **242**, 444 (1957).
5. M. A. Biot, "Linear Thermodynamics and the Mechanics of Solids." *Proc. U.S. Nat. Congr. Appl. Mech., 3rd New York*, 1958.
6. M. A. Biot, "The influence of Gravity on the Folding of a Layered Viscoelastic Medium under Compression." *J. Franklin Inst*., **268**, No. 3, 211 (1959).
7. M. A. Biot, "Folding of a Layered Viscoelastic Medium Derived from an Exact Stability Theory of a Continuum under Initial Stress." *Quar. Appl. Math.*, **17**, No. 2, 185 (1959).
8. M. A. Biot, "On the Instability and Folding Deformation of a Layered Viscoelastic Medium in Compression." *J. Appl. Mech., Ser. E.*, **26**, 393 (1959).
9. M. A. Biot, "Stability Problems of Inhomogeneous Viscoelastic Media," in *Non-homogeneity in Elasticity and Plasticity*, "Proceedings of IUTAM Symposium Warsaw, 1958" Pergamon Press, London, 1958.
10. M. A. Biot, "Instability of a Continuously Inhomogeneous Viscoelastic Half-Space under Initial Stress," *J. Franklin Inst*., **270**, No. 3, 190 (1960).
11. M. A. Biot, "Theory of Folding of Stratified Viscoelastic Media and its Implications in Tectonics and Orogenesis." *Geol. Soc. Am. Bull.*, **72**, 1595 (1961).
12. M. A. Biot, H. Ode, and W. L. Roever, "Experimental Verification of the Theory of Folding of Stratified Viscoelastic Media," *Geol. Soc. Am. Bull.*, **72**, 1621 (1961).
13. M. A. Biot, "Theory of Stability and Consolidation of a Porous Medium under Initial Stress," *J. Math. Mech..*, **12**, No. 4, 521 (1963).
14. M. A. Biot, "Stability of Multilayered Continua Including the Effect of Gravity and Viscoelasticity," *J. Franklin Inst.* **276**, No. 3, 231 (1963).
15. M. A. Biot, "Theory of Buckling of a Porous Slab and its Thermoelastic Analogy." *J. Appl. Mech., Ser. E.*, **31**, No. 2, 194 (1964).
16. M. A. Biot, "Theory of Viscous Buckling of Multilayered Fluids Undergoing Finite Strain," *Phys. Fluids*, **7**, No. 6, 855 (1964).
17. M. A. Biot, "Internal Instability of Anisotropic Viscous and Viscoelastic Media under Initial Stress," *J. Franklin Inst.* **279**, No. 2, 65 (1965).
18. M. A. Biot, *Mechanics of Incremental Deformations*, Wiley, New York, 1965, 504 pp.
19. M. A. Biot, "Theory of Viscous Buckling and Gravity Instability of Multilayers with Large Deformation." *Geo. Soc. Am. Bull.*, **76**, 371 (1965).
20. M. A. Biot and H. Ode, "Theory of Gravity Instability with Variable Overburden and Compaction," *Geophysics*, **30**, No. 2, 213 (1965).
21. M. A. Biot, "Three-Dimensional Gravity Instability Derived from Two-dimensional Solutions," *Geophysics*, **31**, No. 1, 153 (1966).
22. M. A. Biot, "Rheological Stability with Couple-Stresses and Its Application to Geological Folding," *Proc. Roy. Soc., (London) Ser. A*., **298**, 402 (1967).
23. M. A. Biot, "Non-linear Thermoelasticity, Irreversible Thermodynamics, and Elastic Instability," *Indiana Univ. Math. J.* **23**, 309 (1973).
24. J. C. Legros, "Influence de la thermodiffusion sur la stabilité d'une lame liquide horizontale," *Bull. Acad. Roy. Belgique (classe Sci. 5e série*., **59**, 382 (1973–1974).
25. Y. C. Fung, "Foundations of Solid Mechanics," Prentice-Hall, Englewood Cliffs, New Jersey, 1965, 525 pp.

CONCEPTS IN HYDRODYNAMIC STABILITY THEORY

P. H. ROBERTS

University of Newcastle upon Tyne, Great Britain

Abstract

The present state of hydrodynamic stability theory is reviewed, and is illustrated principally by one of its clearest examples, Bénard convection.

CONTENTS

I. INTRODUCTION

It is a common experience that the physical laws governing a given system admit more than one solution. The system can, of course, adopt only one of these, and stability theory seeks to decide which it will be.

A situation often encountered in the stability theory of continuous systems concerns a one-parameter family whose members differ essentially from one another by only one externally imposed condition, for example, in the case of stability of pipe flow, by the applied pressure gradient. For all values of this parameter, R (for example), a solution of a particularly simple type, which we will call "the basic state," exists: for example, in the case of pipe flow it is a simple laminar unidirectional motion down the pipe. While the basic state may be the only solution for $R < R_c$ (for example), two (or more) new solutions may come into being when $R > R_c$. It is frequently observed that, as R is increased through R_c, the system forsakes its basic state in favor of the new family. In these circumstances, the basic solution is said to be "unstable" and the new solution "stable."

The nature of the transition at R_c depends very much on the system under consideration. It may, as in pipe flow, be of the "snap-through" variety. That is, the new state to which the system moves at R_c differs by a finite amount from the corresponding basic state, that is, the families are disjoint at R_c. In such cases it frequently happens that R_c itself increases with decreasing amplitude of the initial disturbance to which the system is subject, but that no amplitude is sufficient to create instability at sufficiently small R. On the other hand, the new family may be continuous with the old at $R = R_c$. A simple example of such "bifurcation" is provided by Bénard convection:

A horizontal layer of fluid is heated from below and cooled from above. A one-parameter family of static basic states exists which may be conveniently labeled by the Rayleigh number

$$R = \frac{g\alpha d^3 \Delta T}{\nu\kappa} \tag{1}$$

a dimensionless measure of the temperature contrast, ΔT, applied across the layer. Here d is the fluid depth, g is the acceleration due to gravity, α is the coefficient of volume expansion of the fluid, ν is its kinematic viscosity, and κ is its thermal diffusivity.* The behavior of the system is completely determined by R and by a second dimensionless quantity, the Prandtl number $P = \nu/\kappa$, which is less easy to vary continuously, since it is intrinsic to the working fluid selected, and which we will usually assume is fixed.

A critical value of R exists which is independent of P and which, when the bounding surfaces are fixed perfect conductors, is about 1707.76. For $R < R_c$, only the basic state is possible in which the heat is conducted upwards in obeyance to Fourier's law, and the fluid rests in hydrostatic equilibrium. As R increases through R_c, bifurcation occurs and the static state becomes unstable to a new family of solutions. Each of these consists of an interlocking pattern of steadily overturning cells, which can, under properly controlled conditions, be strikingly regular (see Fig. 1). The fluids in the cells draws heat from the lower boundary and, from its resulting expansion, can rise by bouyancy to the upper boundary, where it gives up its heat, contracts, and sinks to replenish its supply. These convective motions substantially enhance the efficiency of the layer in transporting heat, a quantity usually measured by the Nusselt number, N, defined as the heat flux transmitted through a unit area of the layer divided by the heat flux which would be caused by the basic conductive solution at the same value of R. (Clearly $N = 1$, for $R \leqslant R_c$.)

*It is simplest to adopt the theoretician's idealization of the Boussinesq fluid, defined as the limiting case in which α and the compressibility of the fluid tend to zero, while g approaches infinity in such a way that $g\alpha$ remains finite. This limit removes all the inessential complications of the mathematics while preserving the physically important effects.[1] Sufficiently thin layers of actual fluids obey the Boussinesq theory with good accuracy.

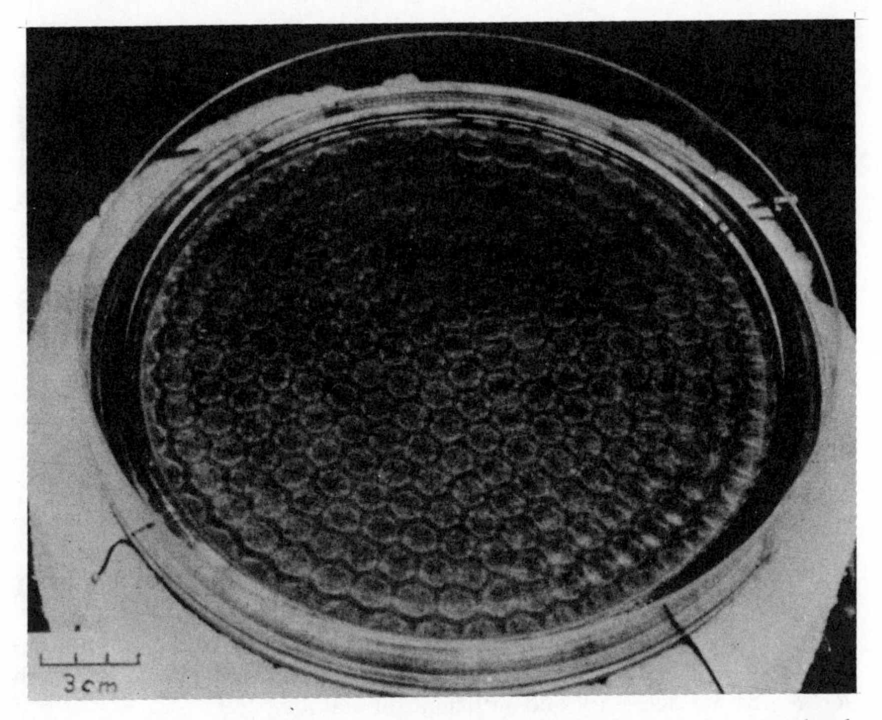

Fig. 1. Bénard convection: a famous photograph by E. L. Koschmieder, here reproduced, with permission of Chicago University Press, from Chapter 10 "Non-linear hydrodynamic stability theory and its applications to thermal convection and curved flows" by L. A. Segel in *Non-equilibrium Thermodynamics, Variational Techniques and Stability* (R. J. Donnelly, R. Herman and I. Prigogine, Eds.,) © 1966 by The University of Chicago. All rights reserved. Published 1966.

As R is increased, N and the amplitude of motion grow. Ultimately, however, irregularities develop and these give way, at even larger values of R, to a fully developed turbulence whose energy density, like N, increases without limit as $R \to \infty$.

A number of problems face the theoretician. For instance, can the stability ($R < R_c$), and ultimately instability ($R > R_c$), of the basic state be understood? One must first define what is meant by the words "stability" and "instability" and this, it transpires, is not as easy to decide unequivocally in a continuous system as one's experience of the stability of the discrete constructs of classical mechanics might lead him to expect.[2] We attempt to short circuit these difficulties here, and merely state that, through linear stability theory (see Section II), the theoretician has met this challenge with considerable success for Bénard convection, and most other examples of bifurcating systems. Analytic[3] and computational[4] difficulties can occur, as in the Orr-Sommerfeld problem governing the linear stability theory of parallel flows, but these have not proved insurmountable. It is often found that the instability that arises at bifurcation is not time independent, a phenomenon generally known as "overstability," following the terminology introduced by Eddington[5] in the first known example of this behavior (the pulsations of a variable star). When the transition is effected through steady states, as for the Bénard problem, it is often said that "the principle of the exchange of stabilities" holds.

It must be admitted that linear stability analyses are of little practical interest for the snap-through instabilities. For example, no bifurcation point at all has been located for circular Poiseuille flow or for plane Couette flow; indeed, probably none exists.[6] Nevertheless, snap-through instability is observed experimentally at sufficiently large Reynolds numbers. Again, the difference between the R_c of the linear theory and the observed R for the onset of turbulence is alarmingly great for plane Poiseuille flow. It appears that the bifurcation point, R_c, of linear stability theory can only be confidently applied as a necessary condition ($R < R_c$) for stability. This difficulty does not arise for Bénard convection, for which $R < R_c$ is both necessary and sufficient for stability.

Next, although it is not strictly a stability problem, the theoretician may be asked whether he can explain the structure of the new family of solutions arising for $R > R_c$. He will immediately find two avenues attractive, both of which are perturbation techniques. He may adopt (regular) perturbation in $R - R_c$ to obtain the equations governing small amplitude disturbances, or he may adopt (singular) perturbation methods for large R to obtain the structure of large-amplitude laminar flows. Analyses of Bénard convection in the large R limit have been completed by Roberts.[7] Even though the analytic attractions of such studies appeal to many theoreticians, and even though the results obtained throw light on processes at work at large R, they have the grave disadvantage of conflicting with experience: the flows at large R are turbulent and not laminar. One must conclude that the laminar solutions at large R are themselves unstable. We return to them briefly in Section IV.

At intermediate values of R, only approximate and numerical avenues are possible. The latter involves the solution of several coupled partial differential equations, and is not to be undertaken lightly even with generous support from a large modern computer. This is, of course, particularly the case when three-dimensional solutions are sought.

Moreover, there are usually several dimensionless numbers, in addition to R, defining the system (for example, P, in the Bénard situation) and a complete survey of solutions in this parameter space is generally prohibitively expensive. These difficulties have not deterred a number of investigators from undertaking valuable numerical work (for Bénard solutions, see Ref. 8).

Turning next to approximate methods, we should mention bounding techniques and we should also discuss the application of the generalizations which have been made of successful methods of nonequilibrium thermodynamics. In both cases, the aim of obtaining detailed information about the structure of the solution is to some extent relaxed. Instead, reliable information about its global properties are sought; for example, the Nusselt number, N, of Bénard convection provides an illuminating, and often adequately informative, portmanteau description of the vigor of the motions and their ability to transmit heat. The two methods are rather different in approach.

In the bounding technique, a quantity such as N is expressed in terms of integrals of the solution over the domain of interest. The solution is, of course, unknown but, drawing arbitrarily from the class from which it belongs (e.g., functions obeying the same continuity and boundary conditions, and possibly other integral constraints as well), it is found that the quantity of interest is bounded. The bounds themselves may be obtained by the techniques of the calculus of variations, but there is no suggestion that the minimizing or maximizing functions actually represent the solution precisely. Nevertheless, the results obtained are rigorous and useful.[9]

The use of variational methods above is reminiscent of nonequilibrium thermodynamics, a theory that contains a principle of minimum entropy production[10] that controls precisely the flux of heat through a stationary layer. The success of this theorem has naturally led to a search for similar extremum principles for continua in motion. For example, Malkus[11] has postulated a principle of maximum Nusselt number for the Bénard layer. This proposal has generated much excitement and has led to some satisfactory explanations of convective behaviors. Nevertheless, it has not yet been given formal justification and should therefore perhaps still be regarded as heuristic. Indeed, no such simple extremum principle has been discovered to date which leads to the exact solution. The only such principle that has been found is that of Glansdorff and Prigogine, and it is by no means simple. In application, it can provide the best approximate solution from a given class, but the values of quantities (such as the Nusselt number) so obtained are rigorously neither upper nor lower bounds. We return to their method in Section IV.

II. LINEAR STABILITY

Let $\mathbf{f}$ be a one-dimensional array of the physical variables governing the behavior of the system according to equations of the form

$$\frac{\partial \mathbf{f}}{\partial t} = \mathbf{F}(\mathbf{f}) \tag{2}$$

where $\mathbf{F}$ is a nonlinear differential operator which depends parametrically on R and perhaps other dimensionless numbers. The basic state, $\mathbf{f}_0$, existing, for example, at all values of R, obeys

$$0 = \mathbf{F}(\mathbf{f}_0) \equiv \mathbf{F}_0, \text{ (say)} \tag{3}$$

It is, of course, supposed that $\mathbf{f}_0$ like $\mathbf{f}$ obeys the boundary requirements. The linear stability of $\mathbf{f}_0$ is examined by writing

$$\mathbf{f} = \mathbf{f}_0 + \mathbf{f}' \tag{4}$$

and supposing that $\mathbf{f}'$ is so small that its squares and higher powers are negligible. The equation obtained by substituting (4) into (2), namely,

$$\frac{\partial f_i'}{\partial t} = P_{0ij} f_j' \quad \left[P_{ij}(\mathbf{x}) = \frac{\partial F_i}{\partial f_j} \right] \tag{5}$$

is therefore necessarily linear and, by (3), homogeneous. This is also true of the boundary conditions. The operator $P_{0ij}(\mathbf{x}) = \partial F_{0i} / \partial f_{0j}$ acting on f_j' now generally contains $\mathbf{x}$ both differentially and, through $\mathbf{f}_0$, parametrically.

Since the original problem for $\mathbf{f}_0$ is properly posed, the equations (5) and boundary conditions governing $\mathbf{f}'$ now completely define an initial value problem, which may be solved by, for example, Laplace transformation from t to s. Although there are instances in which branch points and cuts in the complex s-plane cannot be ignored in the subsequent inversion, it generally happens that stability is decided by the poles, that is, by the normal mode solutions,

$$\mathbf{f}'(\mathbf{x}, t) = \mathbf{f}''(\mathbf{x}) e^{st} \tag{6}$$

of (5). On substitution of (6) into (5) we have

$$P_{0ij} f_j'' = s f_i'' \tag{7}$$

a linear equation which, together with the boundary conditions, now defines an eigenvalue problem for s. Unless dissipative effects are ignored, the spectrum $\{s_\alpha\}$ of s is discrete and denumerable with limit point at $s=-\infty$ (but no finite point of accumulation), and the corresponding eigenfunctions $\{f'_\alpha\}$ form a complete orthonormal set in terms of which the initial value problem can be completed solved:

$$\mathbf{f}' = \sum_{\alpha=1}^{\infty} \mathbf{f}''_\alpha e^{s_\alpha t} \tag{8}$$

For reasons made obvious by (8) the solution $\mathbf{f}_0$ is said to be *stable* if $\Re(s_\alpha)<0$ for all α, and *unstable* if $\Re(s_\alpha)>0$ for any α. If $\Re(s_\alpha)=0$ for the single mode, $\alpha=1$ (for example) (or conceivably, but less frequently, for several modes), and $\Re(s_\alpha)<0$ for all other modes α, the solution $\mathbf{f}_0$ is said to be *marginally stable*. In our formulation, $\mathbf{f}_0$ passes continuously from stability to instability as R increases through the marginal state ($R=R_c$), and usually

$$s_\alpha \sim i\omega + \sigma(R-R_c) \qquad (R\to R_c) \tag{9}$$

where ω is real and $\Re(\sigma)>0$. If $\omega\neq 0$, the state $\mathbf{f}_0$ is overstable for $R>R_c$; if (as for Bénard convection) $\omega=0$, the principle of the exchange of stabilities holds. In the latter case, R can usually be efficiently located by setting $s=0$ in (7), and regarding that equation as an eigenvalue problem for R_c.

In many situations of interest, $\mathbf{f}_0$ depends on only one spatial coordinate, for example, the vertical height, z, in the Bénard layer. In this case, the operator $\mathbf{P}_0$ in (7) involves only that coordinate parametrically, and the differentiations with respect to the remaining coordinates can usually be made algebraic by Fourier decomposition. For example, in a Bénard layer in a horizontally rectangular dish, we would write

$$\mathbf{f}''(\mathbf{x}) = \mathbf{f}'''_{l,m} e^{i(lx+my)} \tag{10}$$

and would then generally find that a discrete set $\{l_i, m_i\}$ of l and m would exist satisfying suitably chosen side wall conditions, and that (7) would reduce to an ordinary differential equation in z. In the case of a layer of small depth compared with its horizontal dimensions, it is found that the set $\{l_i, m_i\}$ is dense in the sense that, although at $R=R_c$ there is only one pair, $\{l_c, m_c\}$, for example, for which σ vanishes, a minute increase in R above R_c results in the appearance of a second unstable mode. Indeed, in the case of a horizontally infinite layer, the modes become continuous and, for all positive $R-R_c$, there is a range $k_1<k<k_2$ of $k=(l^2+m^2)^{1/2}$, containing k_c, in which unstable modes exist [see Fig. 2, in which curves of

equal s are sketched in the (k,R) plane]. It proves simplest to avoid the inessential complications of the discreteness by confining attention to this case.

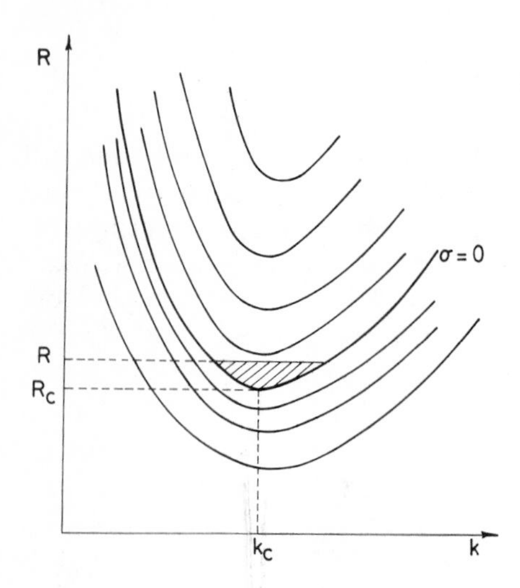

Fig. 2. Schematic diagram of the growth rates, s, as a function of wave number, k, and Rayleigh number, R, for Bénard convection.

Although it may appear that some of these remarks are quite special to the Bénard problem, they have in fact a wide application. Perhaps the least typical features are first the vanishing of ω in (9), and second the horizontal isotropy of the system which makes k, but not l/m, relevant in the dispersion relationship. The second of these, results in significant departures from the general case now to be described.

Including all $\{l,m\}$ in the neighborhood of the critical mode $\{l_c,m_c\}$, (9) becomes

$$s \sim i\omega_c + \sigma(R-R_c) - iU(l-l_c) - iV(m-m_c)$$
$$-\left[a(l-l_c)^2 + b(m-m_c)^2 + 2c(l-l_c)(m-m_c)\right] \tag{11}$$

Here ω_c, U and V are real, and the real parts of a, b, and c define a positive quadratic form, so that $\Re(s)$ is correctly maximal at $\{l,m\} = \{l_c,m_c\}$.

In preparation for the next section, it is worth emphasizing that, when we follow the evolution of an arbitrary initial state by the decomposition (8) or more precisely by the generalization

$$\mathbf{f}' = \int\int \mathbf{f}'''_{l,m}(z)e^{i(lx+my)+st}\,dl\,dm \tag{12}$$

to the continuous case, we find that the contributions from all modes except those near $\{l_c, m_c\}$ become increasingly negligible in comparison with the contributions made by those modes. In other words, the disturbance develops into a wave packet, and the integral (12) may be approximated by the steepest descent method. In fact, introducing stretched length and time scales by

$$X=(x-Ut)|R-R_c|^{1/2}, \qquad Y=(y-Vt)|R-R_c|^{1/2}, \qquad T=t|R-R_c| \tag{13}$$

we obtain, from (11) and (12),

$$\mathbf{f}'=A(X,Y,T)\mathbf{f}'''_{l_c,m_c}(z)E \tag{14}$$

where

$$E=\exp[i(l_c x+m_c y+\omega_c t)] \tag{15}$$

For general initial $\mathbf{f}'$, it is found that

$$A(X,Y,T)\propto T^{-1}\exp\left[-\frac{(\alpha X^2+\beta Y^2+2\gamma XY)}{4T}\pm\sigma T\right] \tag{16}$$

where α, β, and γ define the inverse matrix associated with a, b, and c, and the sign before σ is that of $(R-R_c)$. It may be seen that (16) is the fundamental solution of the equation

$$\frac{\partial A}{\partial T}=\pm\sigma A+a\frac{\partial^2 A}{\partial X^2}+b\frac{\partial^2 A}{\partial Y^2}+2c\frac{\partial^2 A}{\partial X\partial Y} \tag{17}$$

which bears an obvious relation to (11).

According to (14), after a time, which is large compared with the pertinent diffusion times, the solution approaches the product of two parts. The first is defined by the critical eigenfunction and varies over a short length-scale, and possibly ($\omega_c\neq 0$) over a short time-scale also. The second represents a modulation of this pattern which, in the frame moving with the horizontal group velocity (U, V) of the critical mode, is slowly varying both in space and time. The typical length and time scales are respectively proportional to $|R-R_c|^{-1/2}$ and $|R-R_c|^{-1}$.

The situation is reminiscent of the slow fluctuations with long-range order that are associated with a second-order phase transition, and Zaĭtsev and Shliomis[12] have discussed the linear Bénard problem in these terms. According to (14), the spatial variations disappear for fixed X and Y as $T\to\infty$, leading to a completely regular pattern of critical modes. And it is interesting to observe in Fig. 1 that, although the interlocking pattern of hexagons strongly suggests long-range order, there are some deviations suggesting that t is still not large compared with the diffusion times defined by the radius of the dish. Both Snyder (from his Couette flow experiments[13]) and Krishnamurti (for Bénard convection[14]) commented on the enormous times necessary to achieve complete regularity of the instability patterns.

Because of the high degeneracy resulting from the horizontal isotropy, $s=s(k)$, of the dispersion relation for Bénard convection, it is not strictly correct to apply (16) or (17). For the particular case of convection rolls in the y-direction, in which $\{l_c, m_c\}=\{k_c, 0\}$, we obtain, by expansion about the critical mode,

$$s \sim \sigma(R-R_c) - a\left[(l-k_c) - \frac{m^2}{2k_c}\right]^2 \tag{18}$$

which leads in place of (17) to

$$\frac{\partial A}{\partial T} = \pm\sigma A + a\left[\frac{\partial}{\partial X} - \frac{i}{k_c}\frac{\partial^2}{\partial Y^2}\right]^2 A \tag{19}$$

with a Y-scaling $[Y=y|R-R_c|^{1/4}]$ that differs from (13) (see Ref. 15).

A grave disadvantage of the linear theory of this section is apparent from (16): it becomes self-contradictory for $R>R_c$, since A then grows without limit. For this reason we discuss nonlinear effects in the next section.

III. SMALL AMPLITUDE SOLUTIONS AND THEIR STABILITY

If $R-R_c$ is small and positive, we may expect that the steady amplitude, A, of the new solution, as we have called it in Section I, will be small. Landau[16] seems to have been the first to realize that A would be of order $|R-R_c|^{1/2}$ as $R \to R_c^{+}$. Landau directed his argument towards the case of plane Poiseuille flow, but it has an almost general validity. The first formal justification for Bénard convection was provided by Sorokin.[17] The result may be established by the straightforward process of iteration starting from the critical eigenfunction (14)* of the linear stability analysis, or more

*When multiple degeneracy occurs, as in the Bénard case, the final equation for A depends on the precise combination of critical modes selected in (14). The results for the single roll (14) will, for example, differ from that for the three equally inclined rolls of equal amplitude that define the hexagonal planform.

precisely from

$$\mathbf{f}' = A\mathbf{f}'''_{l_c,m_c}(z)E + \text{c.c.} \tag{20}$$

where c.c. stands for the complex conjugate of the preceding term. The arguments of Section II suggest the use of two scale methods,[18] and so, following (13), we write

$$\begin{gathered}\frac{\partial}{\partial x} \longrightarrow \frac{\partial}{\partial x} + \epsilon\frac{\partial}{\partial X}, \qquad \frac{\partial}{\partial y} \longrightarrow \frac{\partial}{\partial y} + \epsilon\frac{\partial}{\partial Y} \\ \frac{\partial}{\partial t} \longrightarrow \frac{\partial}{\partial t} - \epsilon U\frac{\partial}{\partial X} - \epsilon V\frac{\partial}{\partial Y} + \epsilon^2\frac{\partial}{\partial T}\end{gathered} \tag{21}$$

where, as we will soon find, ϵ is again of the order $|R-R_c|^{1/2}$. A new term arises in (17) because of the nonlinearities. This is most readily found by first ignoring the spatial modulation of X and Y, supposing that A in (20) depends only on T.

To higher order in $\mathbf{f}'$, (5′) is replaced by

$$\frac{\partial f_i'}{\partial t} = P_{c,ij}f_j' + Q_{c,ijk}f_j'f_k' + S_{c,ijkl}f_j'f_k'f_l' + (R-R_c)W_{c,ij}f_j' \tag{22}$$

where

$$P_{ij} = \frac{\partial F_i}{\partial f_j}, \qquad Q_{ijh} = \frac{1}{2!}\,\frac{\partial^2 F_i}{\partial f_j\,\partial f_h}, \qquad S_{ijkl} = \frac{1}{3!}\,\frac{\partial^3 F_i}{\partial f_j\,\partial f_k\,\partial f_l}, \qquad W_{ij} = \frac{\partial^2 F_i}{\partial R\,\partial f_j} \tag{23}$$

and the suffix c is added to emphasize that these operators are evaluated using the basic state of the marginal case $R = R_c$.

Consideration of the iteration process starting from (14) suggests that the expansion of $\mathbf{f}'$ in powers of A takes the form

$$\begin{aligned}\mathbf{f}' = {} & (A\mathbf{f}'''_{11}E + \text{c.c.}) + (|A|^2\mathbf{f}'_{20}{}'' + A^2\mathbf{f}'''_{22}E^2 + \text{c.c.}) \\ & + (|A|^2A\mathbf{f}'''_{31}E + A^3\mathbf{f}'''_{33}E^3 + \text{c.c.}) + \cdots\end{aligned} \tag{24}$$

Substitute (22) into (23) and equate like powers of A and like dependence on E. To order A, we recover (5) and its solution (14) for $\mathbf{f}'''_{11}$. The equations for order A^2 pose no problems. In the third order, using (21), we often* obtain an equation of the form

$$
\begin{aligned}
|A|^2 A(P_{c,ij} - i\omega_c \delta_{ij}) f'''_{31j} E = \epsilon^2 \frac{\partial A}{\partial T} f'''_{11i} E + (R - R_c) W_{c,ij} A f'''_{11j} E \\
- |A|^2 A \left[Q_{c,ijk} \left(f'''_{20j} f'''_{11k} + f'''_{22j} f'''_{11k}{}^* \right) + 6 S_{c,ijkl} f'''_{11j} f'''_{11k} f'''_{11l}{}^* \right] E
\end{aligned} \tag{25}
$$

It is important to notice that (25) cannot be solved for a general right-hand side, since the operator on the left annihilates the nontrivial solution $\mathbf{f}'''_{11}E$ (see Ref. 7). It can be solved only if a certain consistency condition is obeyed. This may be obtained by scalarly multiplying (25) by $\bar{\mathbf{f}}'''_{l_c m_c}(z) E^{-1}$ (where the bar denotes the adjoint of the critical eigenfunction) and integrating over the entire $\mathbf{x}$-domain. In this way, we obtain from the right-hand side an equation of the form

$$
\frac{\partial A}{\partial T} = \frac{\sigma(R - R_c)}{\epsilon^2} A - \frac{K}{\epsilon^2} |A|^2 A \tag{26}
$$

where the constant K is obtained from the final $\mathbf{f}'''$ products in square brackets in (25). Since K is generally nonzero, self-consistency requires that ϵ and A are both $O(|R - R_c|^{1/2})$.

If σ is real and K is positive, as for Bénard convection, steady finite amplitude solutions $|A|^2 = \sigma(R - R_c)/k$ exist, to which an arbitrary initial state will tend. If σ and K are complex, but $\Re(K) > 0$, the solution approaches a steady oscillation with a frequency which differs from ω_c by order $(R - R_c)$. If σ is real and K is negative, A increases catastrophically; after a time (T) of order 1, (26) will fail to govern it. In this case subcritical instabilities occur, and the criterion $R < R_c$ of the linear theory is necessary but not sufficient for stability. If $R < R_c$, small amplitude perturbations will decay, but those for which $|A|^2 > \sigma(R_c - R)/(-K)$ will grow indefinitely. In the case of complex σ and K, with $\Re(K) < 0$, the situation is similar; the subcritical constant amplitude oscillation is unstable to either an increase or a decrease in $|A|$. These cases are summarized in Fig. 3. In Fig. 3*a*, the (stable) steady-state value of $|A|^2$ for the finite amplitude solutions is plotted against R for the case $\Re(K) > 0$. The arrows indicate the direction in which the system will evolve if, for the given R, a different value of $|A|^2$ is chosen initially. A typical situation for $\Re(K) < 0$ is shown

*Perhaps more frequently,[19] the mean flow does not, in leading order, assume the simple form $|A|^2 \mathbf{f}'''_{20}$ shown in (24), but requires the addition of a second function involving a new amplitude factor B. This appears on the right-hand side of (25), and gives rise to a new term in AB on the right of (26) and (27). These equations must then be supplemented by an additional evolutionary equation for B.

in Fig. 3*b*. For $R < R_c$, two steady values of $|A|^2$ are possible, one (dotted) corresponding to an unstable state and the other (solid line) to a stable situation. The small amplitude theory described here is only competent to locate the less interesting unstable state.

The different cases just described, and some additional ones of great interest, also arise when (26) is generalized to include spatial modulation. It is clear, on comparison with (17), that we then must replace (26) by

$$\frac{\partial A}{\partial T} = \frac{\sigma(R-R_c)}{\epsilon^2} A + a\frac{\partial^2 A}{\partial X^2} + b\frac{\partial^2 A}{\partial Y^2} + 2c\frac{\partial^2 A}{\partial X \partial Y} - \frac{K}{\epsilon^2}|A|^2 A \quad (27)$$

and similarly (19) gives

$$\frac{\partial A}{\partial T} = \frac{\sigma(R-R_c)}{\epsilon^2} A + a\left[\frac{\partial}{\partial X} - \frac{i}{h_c}\frac{\partial^2}{\partial Y^2}\right]^2 A - \frac{K}{\epsilon^2}|A|^2 A \quad (28)$$

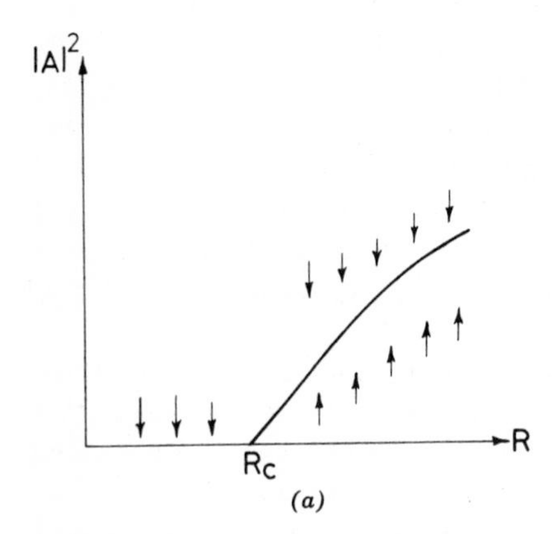

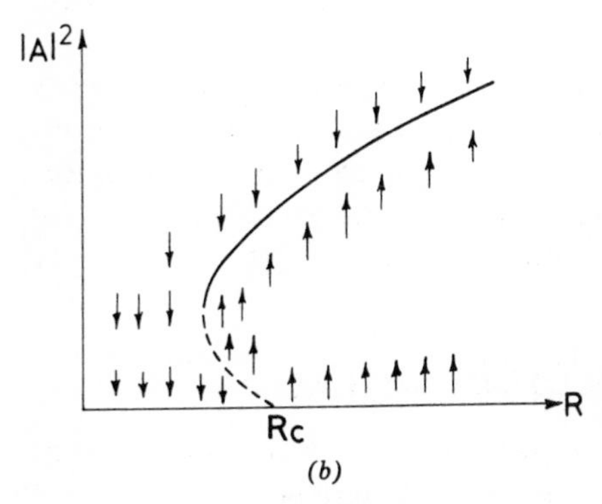

Fig. 3. Schematic diagrams illustrating the variation with R of $|A|^2$, where A is the amplitude of the finite amplitude states. Two cases are shown (*a*) $\Re(K) > 0$, (*b*) $\Re(K) < 0$.

Many cases have been examined in the context of parallel flow. It has been shown[20] that, since $\mathcal{R}(K)<0$ for these instabilities, the disturbance may generate an "instability burst," that is, a general initial perturbation will tend to amplify preferentially in a restricted XY-area. Indeed, according to (27), it focuses on an infinite peak at one point in a finite time, T_0. Although clearly (27) is not competent to follow the solution to such an extremity, it has been shown in the particular case of parallel flows that it does so up to times T, of order ϵ less than T_0. The theoretical existence[20] of these bursts, for both positive and negative $R-R_c$, appears to be related to the experimentally observed turbulent spots in parallel shear flows, that prelude transition to turbulence.

It appears that the approach described here complements (and may eventually provide an alternative for) the earlier "preferred mode" theory of the stability of the finite amplitude motions developed by Schlüter, Lortz, and Busse.[21] They demonstrated the linear instability of finite amplitude Bénard convection in cells of all planforms except the roll. They showed that the rolls are stable to infinitesimal perturbation by rolls of a slightly different size only if their wavelengths, L, lie within a certain band [$L^2<\pi^2(R-R_c)/8R_c$ for free boundaries], which is itself contained within the band ($L^2<3\pi^2(R-R_c)/8R_c$) of all possible finite amplitude rolls. They could not decide which roll was preferred within this band. It was recently shown by Newell, Lange, and Aucoin,[21] essentially by following the general (Y-independent) solution of (28) to large times, that only one roll size remains, namely, one with wave number k_c, to leading order in $R-R_c$.

We should note that Busse[21] extended the analysis he had performed with Schlüter and Lortz to examine the linear stability of rolls when P is infinite but $R-R_c$ is not small. His approach was numerical, and it located the famous balloon-diagram shown in Fig. 4. There is no known practical way at the present time of extending the work of Newell, Lange, and Aucoin into this balloon to determine the preferred mode which is presumed to exist at every value of $R-R_c$ within it.

One may wonder why Fig. 1 should show hexagonal cells when theory so persistently predicts roll-like structures. It transpires that deviations from the Boussinesq approximation, such as variations of thermal conductivity with temperature, may cause the hexagons to be preferred. Also, for R close to R_c, the shape of the side wall boundaries have a long-range effect on the patterns.

IV. GENERAL EVOLUTION CRITERIA

The detail and intricate arguments intrinsic to analyses of the type described in Sections II and III should not be underestimated. It is natural

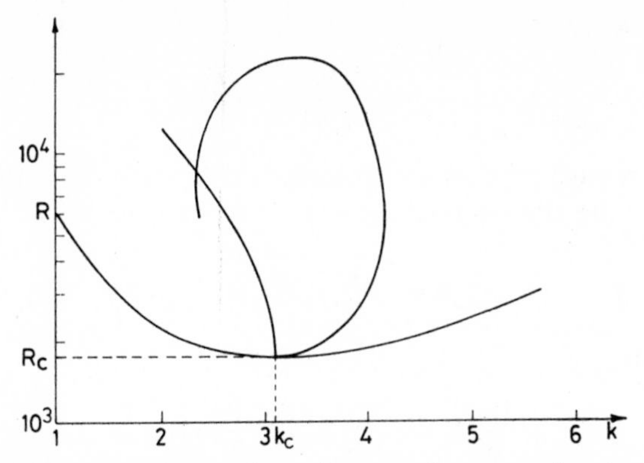

Fig. 4. Busse's balloon. The diagram shows the range of wave numbers in which convection in the form of rolls is stable to perturbation by other rolls of different wave number.

to seek simple, but fundamental, principles by which they can be understood. Moreover, were such principles available, we would also hope that they would provide practical methods by which nonlinear stability theory could be comprehensively attacked, perhaps not achieving complete mathematical precision but at least avoiding serious physical error. The task, then, is in two parts: first the principles must be found, and then their practical usefulness must be assessed by applications.

A general thermodynamic theory governing continua in motion has been given by Glansdorff and Prigogine.[22] Their work has three main aims: first, to determine the general evolution of the system under study; second, to decide whether it is stable or not; and third, to assign probabilities to fluctuations about the mean state. In view of the debate[23] that has arisen, particularly over the way in which they have met the third of these aims, it is worth emphasizing that their answers to the first two objectives are independent of their answer to the third.

In essence, the Glansdorff-Prigogine theory starts from the Gibbs relation for a system in complete thermodynamic equilibrium, namely,

$$dS = \sum_i P_i \, dX_i \tag{29}$$

where S is the entropy, X_i represents the extensive variables, and P_i represents the corresponding intensive variables. [For continua, the summation in (29) over variables at different points takes the limiting integral form.] Assuming stable local thermodynamic equilibrium, they can suppose that the same relation holds far from thermodynamic equilibrium, so that the matrix

$$g_{ij} = \frac{\partial^2 S}{\partial X_i \partial X_j} = \frac{\partial P_i}{\partial X_j} \tag{30}$$

is negative definite. The excess entropy $\delta_2 S$ created by a fluctuation δX_i from the evolving state is then

$$\delta_2 S = \frac{1}{2} \sum_{i,j} g_{ij} \delta X_i \delta X_j < 0 \tag{31}$$

It follows that

$$\frac{\partial}{\partial t} \delta_2 S = \sum_{i,j} g_{ij} \delta X_i \frac{\partial}{\partial t} \delta X_j \geqslant 0 \tag{32}$$

for all choices of δX_i. Glansdorff and Prigogine call this quantity, $\partial \delta_2 S / \partial t$, "the excess entropy production." The first-order part of the inequality provides the evolutionary criterion for the mean state, and the second-order part yields a sufficient criterion for its stability.

In the notation of Sections II and III the evolutionary criterion requires that

$$\int \delta f_i \left[\frac{\partial f_i}{\partial t} - F_i(\mathbf{f}) \right] w \, dV = 0 \tag{33}$$

for all $\delta \mathbf{f}$ satisfying the boundary conditions; w is a weighting factor which would depend on i. Clearly, when $\delta \mathbf{f}$ is general, (33) implies (2), and the converse is also true, that is, the evolutionary criterion is completely equivalent to the problem posed. In practical applications, $\mathbf{f}$ will be drawn from a suitable class of trial functions, and (33) will be used to select the "best" $\mathbf{f}$ from that class. As a variational method, it is generally not self-adjoint; indeed, it is not usually linear.

Several authors have noted the relationship of (33) to the well-known Galerkin approximation. It seems possible that, particularly in view of its convergence properties, the w implied by the Glansdorff-Prigogine method may give it a privileged position among the Galerkin methods. In this connection, the discussion of the heat conduction problem given by Glansdorff and Prigogine in Chapter 10 of their book[10] deserves notice. Nevertheless, it is not often practical to study the convergence of the solution in the complex applications mentioned above; indeed, it is usually not easy to proceed beyond the leading term. It is then difficult to claim any superiority for the results compared with any other Galerkin method.

As for the Galerkin approximation, the quality of the results is no better than that of the trial function from which they are derived.

A number of applications of the theory were presented by Glansdorff and Prigogine in Part II of their book[10]; for more recent studies, see Ref. 24. It was shown by the present author[25] how the mean field equations[26] of Bénard convection could be obtained readily from the evolutionary criterion, when the convection was in rolls; for convection with self-interacting planforms, such as hexagons, additional terms appeared in the mean field equations. An asymptotic analysis by Stewartson showed that, in the limit $R \to \infty$, $N \propto R^{1/3}$ for laminar flow in rolls contained between stress-free boundaries, while $N \propto (R \log R)^{1/5}$ for rigid walls. As shown in unpublished results by Gough, Spiegel, and Toomre, the same asymptotic forms (with different coefficients) follow in the case of laminar hexagonal convection. These results should be compared with the exact asymptotic results obtained by Roberts[7] for laminar convection in rolls at infinite Prandtl numbers; he obtained $N \propto R^{1/3}$ and $N \propto R^{1/5}$ for rolls between stress-free boundaries and rigid walls, respectively.

The present author also used the evolutionary criterion to study the stability of the nonlinear solutions and concluded, independently of (but in agreement with) Schlüter, Lortz, and Busse,[21] that the rolls were stable and the hexagons were not.

A further vindication of the method was provided by convection in a layer containing heat sources.[27] Tritton and Zarraga found experimentally that the horizontal dimensions of the convection cells increased dramatically with R, in contradiction with the findings of the parallel theoretical study by the present author, a discrepancy they attributed to the choice of trial function used in the theory. In order to examine this criticism, Thirlby undertook the full numerical integration of the partial differential equations. He obtained results in essential agreement with the approximate theory of the present author. Further experiments by Hooper did not support the findings of Tritton and Zarraga: the dimensions of the cells were found to increase only slowly with R (see also de la Cruz Reyna, and McKenzie, Roberts and Weiss[27]).

We should observe that Graham[23] has recently suggested an alternative method of obtaining a generalized thermodynamics for continua in motion, and has set up a functional equation of the Fokker-Planck type for the probability of different states in Bénard convection at small $R - R_c$. He has shown that its only time-independent solution is that governed by (28).

Acknowledgments

I am grateful to Professor K. Stewartson of University College London, Professor W. H. Reid of Chicago University, and my colleagues Dr. A. Davey and Dr. A. M. Soward for their criticism of the first draft of this paper. I also wish to thank the British Council for their support.

References

1. E. A. Spiegel, and G. Veronis, *Astrophys. J.*, **131**, 442 (1960); see also G. Veronis, *Astrophys. J.*, **135**, 655 (1962).
2. R. J. Knops and E. W. Wilkes, in *Handbuch der Physik*, Vol. VIa/3, C. Truesdell Ed., Springer-Verlag, Berlin, 1973, p. 125.
3. W. H. Reid, *Studies Appl. Math.*, **51**, 341 (1972).
4. A. Davey, *Quart. J. Mech. Appl. Math.*, **26**, 401 (1973).
5. A. S. Eddington, *The Internal Constitution of the Stars J. Mech. Appl. Math.*, **26**, 401 (1973), Cambridge, University Press (Second Impression), 1930, p. 201.
6. A. Davey, *J. Fluid Mech.*, **57**, 369 (1973).
7. Roberts, G. O., Unpublished work; 1968; see also J. L. Robinson, *J. Fluid Mech.*, **30**, 579 (1967).
8. J. W. Deardorff, *J. Atmos. Sci.*, **21**, 419 (1964). J. W. Deardorff, *J. Atmos. Sci.*, **22**, 419 (1965). J. E. Fromm, *Phys. Fluids*, **8** 1757 (1965). T. D. Foster, *J. Fluid Mech.*, **37**, 81 (1969). G. Veronis, *J. Fluid Mech.*, **26**, 49 (1966). F. B. Lipps, and R. C. J. Somerville, *Phys. Fluids*, **14**, 759 (1971).
9. L. N. Howard, *J. Fluid Mech.*, **17**, 405 (1963). L. N. Howard, *Ann. Rev. Fluid Mech.* **4**, 473 (1972). See also F. H. Busse, *Z. Angew. Math. Phys.* **20**, 1 (1969a); F. H. Busse, *J. Fluid Mech.*, **37**, 457 (1969b); F. H. Busse, *J. Fluid Mech.*, **41**, 219 (1970); F. H. Busse, *Symp. Math. Inst. Naz. Alt. Mat.*, **9**, 493 (1972).
10. I. Prigogine, *Bull. Classe Sci., Acad. Roy. Belg.*, **31**, 600 (1945). See also Sect. 3.4 of P. Glansdorff, and I. Prigogine, *Thermodynamic Theory of Structure, Stability and Fluctuations*, Wiley, London, 1971.
11. W. Malkus, *Proc. Roy. Soc. (London), Ser. A*, **225**, 185 (1954); see also F. H. Busse, *J. Fluid Mech.* **30**, 625 (1967); A. C. Newell, C. G. Lange, and P. J. Aucoin, *J. Fluid Mech.*, **40**, 513 (1970).
12. V. M. Zaǐtsev, and M. I. Shliomis, *Soviet Phys. JETP (Engl. Transl.)*, **32**, 866 (1970); see also Newell et al.[11]
13. H. A. Snyder, *J. Fluid Mech.*, **35**, 273 (1969).
14. R. Krishnamurti, *J. Fluid Mech.*, **42**, 295, 309 (1969). For other experimental work, see, for example, E. L. Koschmieder, *Beitr. Phys. Atmos.*, **39**, 1 (1966); M. M. Chen, and J. A. Whitehead, *J. Fluid Mech.*, **31**, 1 (1968); H. T. Rossby, *J. Fluid Mech.*, **36**, 309 (1969); F. H. Busse, and J. A. Whitehead, *J. Fluid Mech.*, **47**, 305 (1971); **66**, **67** (1974).
15. A. C. Newell, and J. A. Whitehead, *J. Fluid Mech.*, **38**, 279 (1969); L. A. Segel, *J. Fluid Mech.*, **21**, 345 (1965); L. A. Segel, *J. Fluid Mech.*, **38**, 279 (1969).
16. L. D. Landau, *C. R. Acad. Sci., U.R.S.S.*, **44**, 311 (1944).
17. V. S. Sorokin, *Prikl. Mat. Mekh.*, **18**, 197 (1954).
18. N. N. Bogoliubov, and Y. A. Mitropolsky, *Asymptotic Methods in the Theory of Oscillations*, Hindustan Publ. Co., Delhi, 1961.
19. J. Pedlosky, *J. Atmos. Sci.*, **29**, 680 (1972); A. Davey, L. M. Hocking, and K. Stewartson, *J. Fluid Mech.* **63**, 529 (1974); A. Davey and K. Stewartson, *Proc. Roy. Soc. London Ser. A.*, **338**, 101 (1974).
20. K. Stewartson, and J. T. Stuart, *J. Fluid Mech.*, **48**, 529 (1971); L. M. Hocking, K. Stewartson, and J. T. Stuart, *J. Fluid Mech.*, **51**, 705 (1972). L. M. Hocking, and K. Stewartson, *Proc. Roy. Soc., London Ser. A*, **326**, 289 (1972). L. M. Hocking, and K. Stewartson, *Mathematika*, **18**, 219 (1971).

21. F. H. Busse, Ph.D. Thesis, Translated by S. H. Davis Rand Corp., Santa Monica, Calif., 1962. A. Schluter, D. Lortz, and F. H. Busse, *J. Fluid Mech.*, **23**, 129 (1967). F. H. Busse, *J. Math. Phys.*, **46**, 140 (1967). A. C. Newell, C. G. Lange, and P. J. Aucoin, *J. Fluid Mech.*, **40**, 513 (1970). F. H. Busse, *Fluid Mech.*, **52**, 97 (1972). R. M. Clever and F. H. Busse, *J. Fluid Mech.* **65**, 625 (1974).
22. P. Glansdorff, and I. Prigogine, *Physica*, **30**, 351 (1964). I. Prigogine, and P. Glansdorff, *Physica*, **31**, 1242 (1965). P. Glansdorff, and I. Prigogine, *Physica*, **46**, 344 (1970). See also P. Glansdorff and I. Prigogine, Ref. 10.
23. R. Graham, in Coherence and Quantum Optics L. Mandel and E. Wolf Eds., Plenum Press, New York, 1973, p. 851. See also R. Graham, "A Generalized Thermodynamic Potential for the Convection Instability," to appear.
24. W. Unno, *Publ. Astron. Soc. Japan*, **20**, 356 (1968); W. Unno, *Publ. Astron. Soc. Japan*, **21**, 240 (1969); R. van der Borght, and B. E. Walters, *Proc. Astron. Soc. Aus.*, **2**, 47 (1971); R. van der Borght, J. O. Murphy, and E. A. Spiegel, *Aust. J. Phys.*, **25**, 703 (1972); R. van der Borght, and J. O. Murphy, *Aust. J. Phys.*, **26**, 341 & 617 (1973). R. Van der Borght, *Aust. J. Phys.*, **27**, 471 (1974); J. C. Morgan, *J. Fluid Mech.*, **57**, 433 (1973).
25. P. H. Roberts, in *Non equilibrium Thermodynamics, Variational Techniques, and Stability*, R. J. Donnelly, R. Herman, and I. Prigogine Eds., University Press, Chicago, 125, 299, 1965. See also J. O. Murphy, *Proc. Astronom. Soc. Aust.*, **1**, 381 (1970); R. van der Borght, and B. E. Walters, *Proc. Astronom. Soc. Aust.*, **1**, 382 (1971); J. O. Murphy, *Proc. Astron. Soc. Aust.*, **2**, 51, 53 (1971); E. A. Spiegel, *Ann. Rev. Astron. Astrophys.*, **9**, 323 (1971); E. A. Spiegel, *Ann. Rev. Astron. Astrophys.*, **10**, 261 (1972); E. A. Spiegel, *Ann. Rev. Astron. Astrophys.*, to appear (1975); G. O. Gough, Y. Toomre, and E. A. Spiegel, *J. Fluid Mech.*, to appear (1975).
26. J. R. Herring, *J. Atmos. Sci.*, **20**, 325 (1963). J. R. Herring, *J. Atmos. Sci.*, **21**, 277 (1964). See also J. W. Elder, *J. Fluid Mech.*, **35**, 417 (1969).
27. D. J. Tritton, and M. N. Zarraga, *J. Fluid Mech.*, **30**, 21 (1967). P. H. Roberts, *J. Fluid Mech.*, **30**, 33 (1967). R. Thirlby, *J. Fluid Mech.*, **44**, 673 (1970). T. Hooper, Ph.D. material (to be published) S. De la Cruz Rayna, *Geofis. Internaz.*, **10**, 49 (1970). D. P. McKenzie, J. M. Roberts, and N. O. Weiss, *J. Fluid Mech.*, **62**, 465 (1974).

SOME REMARKS ON VARIATIONAL METHODS, THE LOCAL POTENTIAL, AND FINITE ELEMENT METHODS WITH APPLICATION TO CERTAIN CONTINUUM MECHANICS PROBLEMS

S. M. ROHDE

Engineering Mechanics Department,
Research Laboratories,
General Motors Technical Center,
Warren, Michigan

CONTENTS

Abstract

This paper is a discussion of classical and nonclassical variational methods. Emphasis is placed on the associated approximation schemes and, in particular, the finite element method. The local potential formulation as conceived of by Glansdorff and Prigogine is connected with these more modern approximation schemes. Examples are given which arise from certain nonlinear fluid and solid mechanics problems. The extension of the local potential–finite

element method to hydrodynamic, chemical, and hydromagnetic stability problems is outlined. Variational inequalities are discussed briefly. Examples of the variational formulation and construction of approximate solutions to free-boundary problems arising in hydrodynamic and elasticity theory are discussed in this context. The extension of these ideas to nonclassical potentials and, in particular, the local potential is outlined.

I. INTRODUCTION

In this paper certain variational methods will be discussed. Recently, much interest has again been directed toward these methods from both the mathematical and the physical points of view. A large portion of this renewed interest is closely connected with modern developments in approximation theory, particularly finite element theory. It is the intent of this author to connect certain variational methods, particularly the local potential, with some of newer approximation concepts. In particular we are here concerned with certain nonlinear elliptic boundary value problems arising from situations of physical and engineering interest. Currently, we are also considering the variational formulation of certain free-boundary problems and the construction of approximate solutions to such problems. These are likewise discussed as is the extension of these ideas to classes of hydrodynamic and hydromagnetic stability problems.

Sections II and III concern both classical and nonclassical variational formulations and the resulting approximation schemes associated with these, respectively. These sections are in a sense pedagogical, but are included to serve as a basis for the introduction of the finite element method in Section IV. Sections V and VI illustrate the application of some of these methods, that is, the local potential iterative scheme coupled with finite element methods, to the solution of certain nonlinear fluid mechanics problems. These problems are the slow viscous flow of a compressible fluid between finite permeable eccentric cylinders (Section V) and the flow of a certain non-Newtonian fluid between nonparallel plates (Section VI). In Section VII the variational formulation of certain free-boundary problems is discussed. Section VIII gives applications of these formulation to certain fluids problems, elasticity problems, and elastic-plastic torsion problems. In Section IX we discuss the application of finite element methods to hydrodynamic stability problems.

II. VARIATIONAL FORMULATIONS

This section is a brief general outline of what is meant by a variational formulation or embedding corresponding to a given boundary value problem. In this regard we distinguish between "classical variational" principles and the more modern concepts of a variational statement. Let us begin with the former. Consider a solution to the Dirichlet problem

$$-\nabla^2\phi = f, \qquad x \in \Omega \subseteq R^n,$$
$$\phi|_{\partial\Omega} = 0. \tag{1}$$

If $\partial\Omega$ and f are sufficiently smooth, a solution to (1) may be characterized as minimizing the quadratic functional (Dirichlet integral or potental)

$$J[\phi] = \int_\Omega \left[(\nabla\phi)^2 - 2f\phi \right] d\Omega \tag{2}$$

among all ϕ belonging to some Banach (Sobolev) space. We remark that a function minimizing (2) may not be a classical solution to (1) (i.e., be twice $[C^2(\Omega) \cap C^0(\bar{\Omega})]$ differentiable). Conversely a solution to (1) may *not* have a finite Dirichlet integral. For discussions of regularity the reader is referred to Refs. 1–3. We will not dwell on such technical points here but rather develop an overview regarding the variational formulation and approximate solution of certain linear and nonlinear partial differential equations of physical interest.

As is well known,[4] any positive definite symmetric operator can be embedded in a classical variational form as was done above. For non-self-adjoint and most nonlinear problems, such classical variational principles do *not* exist. Nevertheless certain variational formulations, based in particular on thermodynamic considerations, have been discovered and applied with great success to obtaining approximate solutions to classes of non-self-adjoint and/or nonlinear elliptic boundary value problems, evolutionary equations, and eigenvalue problems. In particular we refer to the local potential formulation as discovered by Glansdorff and Prigogine;[5–7] some of the above-mentioned applications of the local potential may be found in Refs. 8 to 10. A mathematical discussion of the local potential may be found in Ref. 11 where convergence for a class of boundary value problems is proved. Here we remark that such nonclassical variational formulations usually involve a construction involving twice as many *dependent* variables as in the original problem. Stationarizing, rather than extremizing, that functional yield a set of Euler-Lagrange equations which contain the physical equations of interest. In this regard the local potential formulation has a distinct advantage in that a "minimality" property is present which leads naturally to an iterative scheme as discussed in Ref. 11; and which will be used in subsequent sections.

III. APPROXIMATE METHODS BASED ON VARIATIONAL PRINCIPLES

Variational principles were used by mathematicians, such as Dirichlet, Hilbert, and Riemann, to prove existence and uniqueness theorems for differential equations. On the other hand, approximate methods based on these variational principles are what scientists and engineers such as Rayleigh, Ritz, and Galerkin devised. More recently, we have the approximate methods devised by Glansdorff and Prigogine. Let us consider the philosophy involved in these methods as this will be our point of departure in discussing the more recent application and connection with modern approximation theory and, in particular, finite-element methods.

For the linear, self-adjoint case, we seek a function ϕ belonging to a certain space of function H such that ϕ minimizes a quadratic functional $J[\phi]$ corresponding to our original problem as discussed previously.[4] The Rayleigh-Ritz procedure then consists of choosing a sequence of finite dimensional spaces $H_0 \subseteq H_1 \subseteq H_n \subseteq H$ such that any element in H can be approximated arbitrarily closely in the appropriate norm by elements of H_n for sufficiently large n. Thus, letting $\{\phi_i\}_{i=1}^n$ be a basis for H_n the problem of minimizing J over H is replaced by the simpler problem of minimizing J over H_n; that is,

$$J[\tilde{\phi}^n] = \min_{\phi^n \in H_n} J[\phi^n] = \min_{\bar{\alpha} \in R^n} J(\bar{\alpha}) \tag{3}$$

where $\phi^n = \sum_{i=1}^n \alpha_i \phi_i$. Clearly, this is the problem of minimizing a quadratic form $J(\bar{\alpha})$ which leads to a nonsingular *linear* system for the vector $\bar{\alpha}$. It is fairly simple (in some cases) to prove that $\tilde{\phi}^n \to \tilde{\phi}$ in the appropriate norm. Significant use is made of the fact that we have an extremum.

Galerkin type methods, on the other hand, are based on the concept of orthogonality of the residual or error to each of the subspaces H_n. For example, if we wish to solve the equation (L linear or nonlinear)

$$L\phi = f \tag{4a}$$

we set $\phi^n = \sum_{i=1}^n \alpha_i \phi_i$ and determine α_i such that

$$(L\phi^n, \phi_i) = (f, \phi_i) \qquad 1 \leqslant i \leqslant n \tag{4b}$$

This, of course, leads to a linear (in fact identical to the Rayleigh-Ritz system if L is self-adjoint) or a nonlinear system of algebraic equations. Here (,) denotes an inner product; for example,

$$(a,b) = \int_\Omega ab\rho(x)\, d\Omega$$

where $\rho(x)>0$, $x\in\Omega$. In the case of evolutionary problems, a set of ordinary differential equations are obtained. Note that the Galerkin system is exactly what is obtained when one applies the Rayleigh-Ritz method to the nonclassical variational statement

$$J[\phi,\tilde{\phi}]=\int_{\Omega}(L\phi-f)\tilde{\phi}\,d\Omega \tag{5}$$

when we set $\phi^n=\sum_{i=1}^{n}\alpha_i\phi_i$, $\tilde{\phi}^n=\sum_{i=1}^{n}\tilde{\alpha}_i\phi_i$ and stationarize $J(\alpha,\tilde{\alpha})$ with respect to $\tilde{\alpha}$. Other classical approximate schemes include the least square method which is again becoming popular.[4,12] The reader is referred to Ref. 4 for a detailed discussion of classical variational methods.

Approximation schemes, based on the local-potential formulation, can be divided into two basic categories as first conceived of by Glansdorff and Prigogine: the self-consistent method and the local-potential iterative scheme. The first of these schemes can be shown, in general, to result in a "Galerkin" set of equations as discussed in Ref. 11 (although the structure involved in this formulation like that involved in a classical potential is of theoretical importance in showing convergence as Glansdorff[13] has done).

The second scheme, used not nearly as much as the self-consistent method, hinges heavily in the actual approximation upon the structure of the local potential, that is, the minimality property involved.[6,11] Let us now outline that scheme.

Our local potential $J[\phi,\tilde{\phi}]$ (sufficiently extended) was shown to have the following property:[5,11] for every $\tilde{\phi}\in H(\tilde{\phi}\in H_n)$, $J[\phi,\tilde{\phi}]$ is a quadratic functional corresponding to a positive definite operator (form) on $H(R^n)$ (more generally a strictly convex functional). Hence, we obtain a sequence $\overset{m}{\tilde{\phi}}\,(\overset{m}{\tilde{\phi}}{}^{n})$ defined by

$$J\left[\overset{m}{\tilde{\phi}},\overset{m-1}{\tilde{\phi}}\right]=\min_{\phi\in H}J\left[\phi,\overset{m-1}{\tilde{\phi}}\right]$$

$$J\left(\overset{m}{\tilde{\alpha}},\overset{m-1}{\tilde{\alpha}}\right)=\min_{\alpha\in R^n}J\left(\alpha,\overset{m-1}{\tilde{\alpha}}\right) \tag{6}$$

where $\overset{m}{\phi}{}^{n}=\sum\overset{m}{\tilde{\alpha}}_i\phi^i$. Under certain conditions, we have $\overset{m}{\tilde{\phi}}\xrightarrow{m\to\infty}\overset{*}{\tilde{\phi}}$ and more important $\overset{m}{\tilde{\phi}}{}^{n}\xrightarrow{m,n\to\infty}\overset{*}{\tilde{\phi}}$ where $\overset{*}{\tilde{\phi}}$ is the required solution. We do not present convergence proofs here but will remark that the author's own experience shows that these methods are considerably more general than would be presumed by the sufficient conditions for convergence which one may obtain.

IV. THE FINITE ELEMENT METHOD (FEM)

As mentioned in the Introduction in recent years considerable emphasis has been placed on variational formulations and the approximations these formulations lead to in certain finite dimensional spaces, particularly the piecewise polynomial spaces. The origin of these methods goes back to Courant[14] (and Synge[15]). Courant suggested that the domain over which a solution to a partial differential equation is sought be triangulated (decomposed into a number of triangles) and the Ritz method be applied with trial functions which are assumed to be piecewise linear over such a triangulation. In the last 10 years or so, this idea of decomposing a region into a finite number of "elements," approximating the field variables we wish to determine over each element by polynomials which satisfy certain continuity conditions at element boundaries, and then using a variational method to determine the best approximation over the entire solution region, has been used extensively by engineers to solve complex problems in structural mechanics.[16,17] More recently, many applications to other branches of continuum mechanics have been found.[18,19] In particular classes of fluid mechanics problems have been effectively handled using the finite element method.[20,21] Thus variational formulations for both particular and classes of problems are again in "vogue."

Figure 1*a* shows a region in R^2 which we see triangulated in Fig. 1*b*. To obtain a piecewise linear approximation, we assume that our field variable is linear (C^1) over each element and continuous along element boundaries. A basis $\{N_i\}$ for this approximation may be constructed as follows. At the ith node P_i (vertex of a triangle) consider the "star," S_{P_i} of P_i, that is, the set of all triangles connected to P_i (Fig. 2*a*). Let $N_i(x,y)$ be that "tent function" which is linear over each of these triangles, has the value 1 at P_i and 0 at the other two nodes of each of these triangles, and is identically 0 outside of S_{P_i} (Fig. 2*b*). Clearly any piecewise linear approximation to ϕ may be written as

$$\phi^n(x,y) = \sum_{i=1}^{n} \phi_i N_i(x,y) \tag{7}$$

The connection with other approximate methods derived from variational principle is now clear—the finite element method can be considered a special case of the Ritz-Galerkin method. Its advantage is its ability to handle complex geometries, boundary conditions, and discontinuities in coefficients (e.g., interface problems) effectively. Moreover higher order element methods yield extremely accurate approximate solutions[22] with relatively few degrees of freedom.

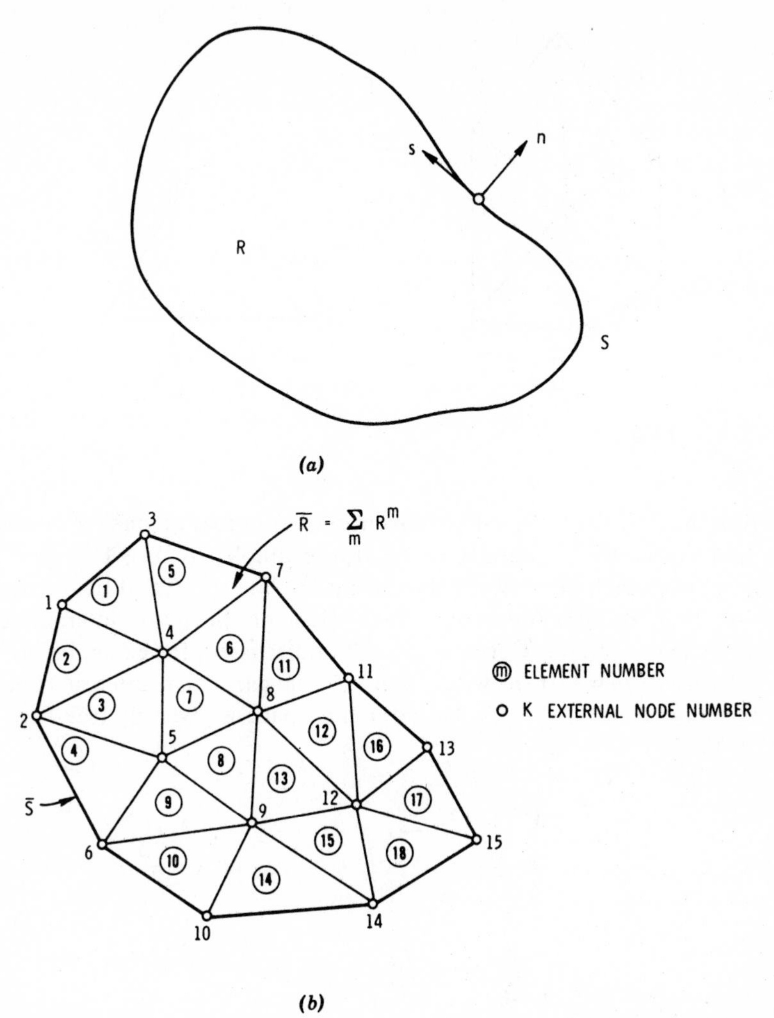

Fig. 1. Finite element formulation example: (*a*) original region; (*b*) finite element dicretiza-tion.

Before presenting particular results, let us look at some of these higher-order spaces. We will first restrict ourselves here to one-dimensional problems; numerous hydrodynamic and hydromagnetic stability problems are, of course, in this class. The simplest example of such an approximation is again the piecewise linear approximation, a basis for which is given in Fig. 3*a*. The piecewise Hermite spaces, which we use subsequently are examples of some higher-order piecewise polynomial spaces. Here the

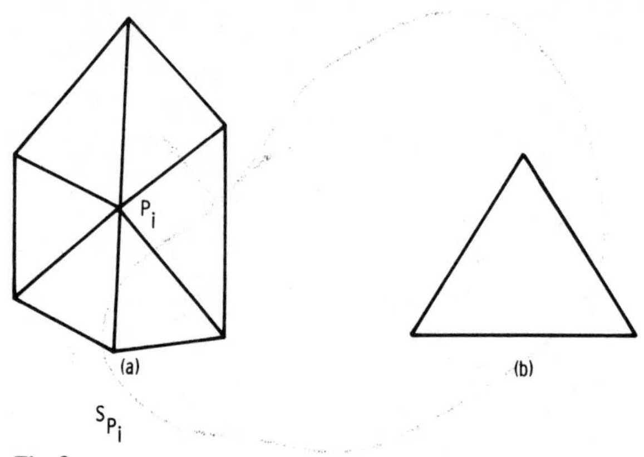

Fig. 2.

solution interval say $0 \leqslant x \leqslant 1$ is broken into a number of smaller intervals, the field variable is assumed to be, for example, a $2m+1$ degree polynomial over each interval which is to interpolate the actual field variable and its first m derivatives over each element. Such an interpolate is $C^m(0,1)$. These smooth Hermite spaces are discussed, for example, in Ref. 23. The case $m=1$ (piecewise cubic) is of particular interest. Letting $0=x_0<x_1<x_1\cdots<x_n=1$, a basis for this space is given by (see Figs. 3*b* and 3*c*)

$$\phi_i=\begin{cases}-3z^{+2}+2z^{+3}+1, & x_i\leqslant x\leqslant x_{i+1}\\ 3z^{-2}-2z^{-3}, & x_{i-1}\leqslant x\leqslant x_i\end{cases}$$

$$\xi_i=\begin{cases}\dfrac{z^+}{\Delta_i}(z^+-1)^2, & x_i\leqslant x\leqslant x_{i+1}\\ \dfrac{(z^--1)}{\Delta_{i-1}}z^{-2}, & x_{i-1}\leqslant x\leqslant x_i\end{cases}\tag{8}$$

where $z^+=(x-x_i)/\Delta_i$, $z^-=(x-x_{i-1})/\Delta_{i-1}$, and $\Delta_i=x_{i+1}-x_i$. If we are considering, for example, a two-point boundary value problem with $u(0)=u(1)=0$, then we represent our approximate u_n by

$$u^{2n}(x)=\sum_{i=1}^{n-1}u_i\phi_i(x)+\sum_{i=0}^{n}u_i'\xi_i(x)\tag{9}$$

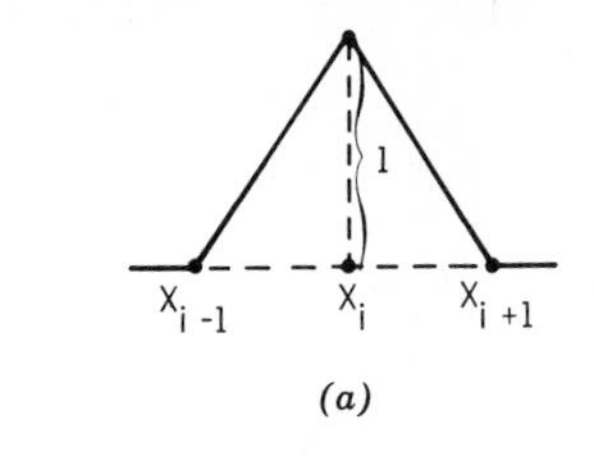

(a)

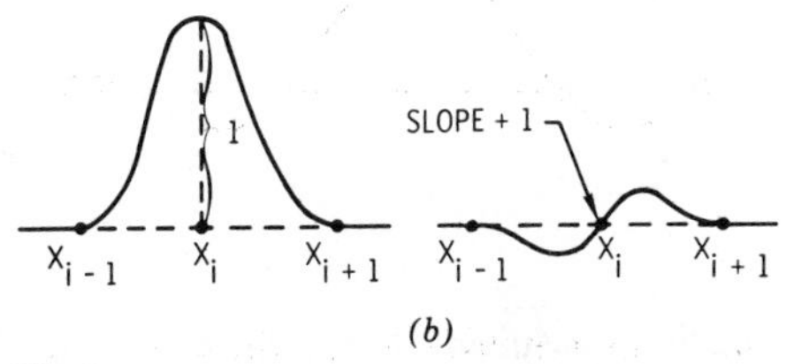

(b)

Fig. 3.

Our functional then becomes a function of the $2n$ variables $\{u_i\}_{i=1}^{n-1}$ and $\{u_i'\}_{i=0}^{n}$ which represent the approximate nodal value of u and its first derivative, respectively.

Another important class of piecewise polynomial approximating functions are the cubic splines.[24,25] A cubic spline $S(x)$ on $(0,1)$ interpolating u at x_i, $0 \leqslant i \leqslant N$ is a piecewise cubic function, cubic on each interval $x_i \leqslant x \leqslant x_{i+1}$ which interpolates u at x_i and is $C^2[0,1]$. The latter continuity requirement leads to a set of $n-1$ equations for $S'(x_i)$. Spline bases may be constructed. For example, splines satisfying the boundary conditions $S'(0)=S'(1)=S(0)=S(1)=0$ may be written as

$$u^{n=1}(x)=S(x)=\sum_{i=1}^{n-1} u_i C_i(x)$$

where the C_i are the cardinal splines[25] (Fig. 4) which satisfy:

$$C_i(x_j)=\begin{cases} 0 & j\neq i \\ 1 & j=1 \end{cases} \qquad 0 \leqslant i \leqslant n$$

$$C_i'(x_0)=0$$

$$C_i'(x_n)=0$$

The choice of a spline basis can lead to more stable approximations to certain problems as the author found for a class of highly nonlinear

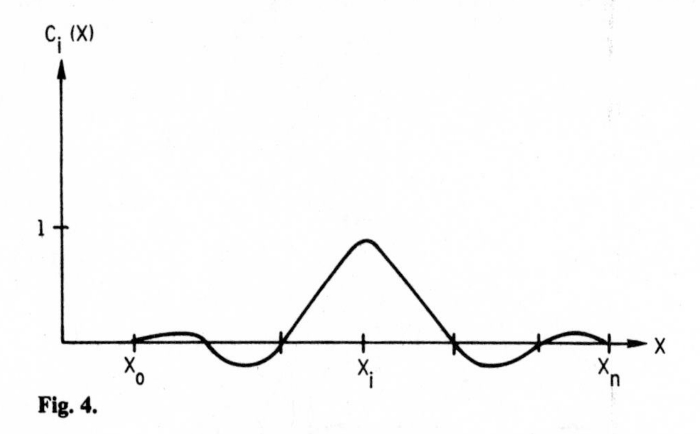

Fig. 4.

integrodifferential equations resulting from the coupling of hydrodynamic and elastic systems in certain applications.[26]

V. APPLICATION OF THE LOCAL POTENTIAL FINITE ELEMENT ITERATIVE SCHEME TO THE SLOW VISCOUS FLOW OF A COMPRESSIBLE FLUID BETWEEN FINITE ECCENTRIC PERMEABLE CYLINDERS

A problem which is of practical interest is the problem of the slow viscous flow of a compressible fluid between eccentric closely spaced finite cylinders; one or both of the cylinders possibly being permeable (Fig. 5). Neglecting curvature and inertial effects the momentum equations for this system are

$$\begin{aligned} \frac{\partial p}{\partial x} &= \frac{\partial}{\partial z}\mu\frac{\partial u}{\partial z} \\ \frac{\partial p}{\partial y} &= \frac{\partial}{\partial z}\mu\frac{\partial v}{\partial z} \\ \frac{\partial p}{\partial z} &\equiv 0 \end{aligned} \tag{10}$$

where $x = R\theta$. Typical velocity boundary conditions are

$$\begin{aligned} u(x,y,0) &= 0 \\ u[x,y,h(x,y)] &= R\omega \\ v(x,y,0) &= v[x,y,h(x,y)] = 0 \\ w(x,y,0) &= w[x,y,h(x,y)] = 0 \end{aligned} \tag{11}$$

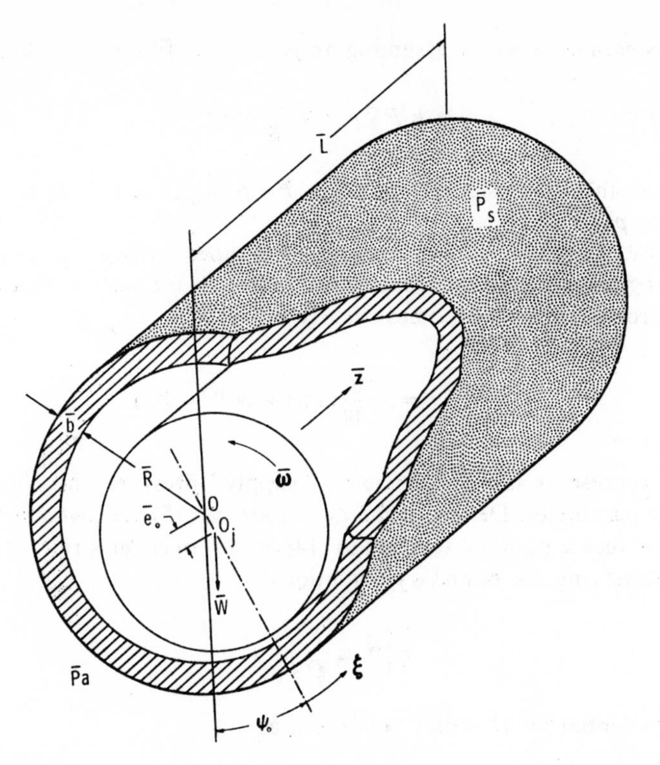

Fig. 5.

where

$$h(x,y)=1+\varepsilon\cos\frac{x}{R} \tag{12}$$

is the gap distance up to $O[(R_2-R_1)/R_2]$ [where $(R_2-R_1)/R_2=C/R$ is typically 10^{-4}]. The momentum equations (10) can be thought of as stemming from the boundary layer equations when the latter are expanded in a modified Reynolds number ($R_e^*=R_eC/R$). These equations are then the $R_e^*=0$ approximation. Equations (10) subject to the boundary conditions (11) may be integrated to obtain the velocities as functions of the pressure field (combined Poiseuille-Couette flow). These are then substituted into the continuity equation:

$$\nabla\cdot(\rho\bar{V})=0 \tag{13}$$

Equation (13) is then integrated across the gap to yield the so-called

Reynolds equation, which assuming an isothermal film ($\rho \propto P$) becomes:

$$\nabla \cdot h^3 P \nabla P = \Lambda \frac{\partial}{\partial \theta}(Ph) \tag{14}$$

Here P is the dimensionless pressure $P = p/p_{\text{amb}}$ and Λ is a constant $[\Lambda = (6\mu\omega/p_a)(R/c)^2]$.

In those cases where, for example, the inner cylinder is porous, by accounting for the flow (unidimensional) through the wall of that cylinder using Darcy's law, (14) becomes slightly modified:

$$\nabla \cdot h^3 P \nabla P = \Lambda \frac{\partial}{\partial \theta}(Ph) + \gamma(P^2 - P_S^2) \tag{15}$$

Here P_s represents the dimensionless "supply" pressure and γ is a permeability parameter. Details of the derivation of (15) are given in Ref. 27.

We thus seek a periodic solution to (14) or (15) over the strip $-L/D < y < L/D$ satisfying the boundary conditions

$$P\left(\theta, \pm \frac{L}{D}\right) = 1 \tag{16}$$

A local potential for (15) may be written as

$$J[P, P_0] = \int_\Omega \left\{ \frac{1}{2} h^3 P_0 (\nabla P)^2 + \left[\Lambda \frac{\partial (P_0 h)}{\partial \theta} + \gamma (P_0^2 - P_s^2) \right] P \right\} d\Omega \tag{17}$$

following (11). Alternately a local potential may be derived from more fundamental considerations as was done for (14) in Ref. 28.

The finite element "space" we chose in this example was the bicubic Hermite space which has recently been shown to have significant advantages over more classical solution techniques for problems of this type,[22] that is, for the linear problem involved at each step of the iteration. This space interpolates the unknown function, its first partial derivatives, and the second mixed derivatives between nodes. The rectangular region was decomposed as shown in Fig. 6. A basis given by the Cartesian product of the ϕ_i's and ξ_i's discussed in Section IV was used. Thus, at each interior node, we have four degrees of freedom (P, $P_{,x}$, $P_{,y}$, and $P_{,xy}$). At the upper and lower boundaries, P is known and $P_{,x} \equiv 0$ (actually symmetry about $y = 0$ was used to cut the number of unknowns in about half).

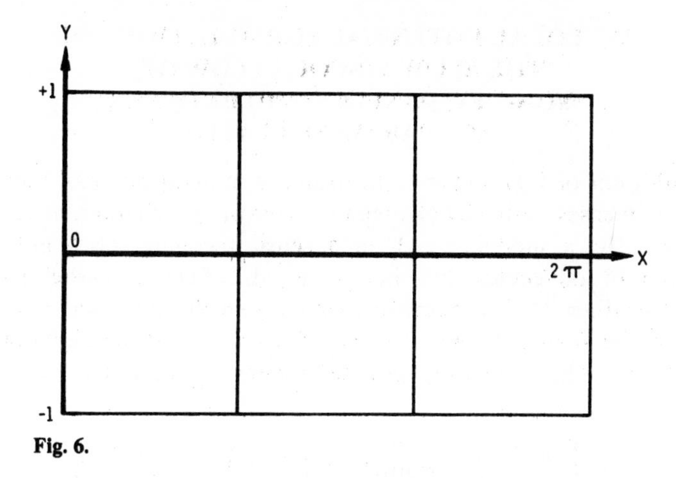

Fig. 6.

The pressure was initially assumed to be identically equal to 1 and the local potential iterative scheme was used. Figure 7 shows a typical converged solution. We remark that with only 12 degrees of freedom ($n = 3, m = 1$) more accurate results were obtained than with over 100 degrees of freedom and a lower order method! Extensive results and convergence studies for this problem are reported in Ref. 29.

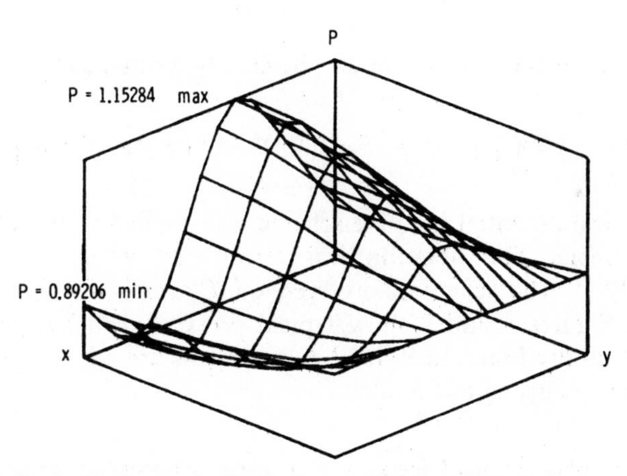

Fig. 7. Three-dimensional pressure distribution; $\epsilon = 0.2$, $\Lambda = 2.0$, $\gamma = 0$.

VI. LOCAL POTENTIAL FORMULATION FOR THE SLOW VISCOUS FLOW OF NON-NEWTONIAN FLUID BETWEEN NONPARALLEL PLATES

The addition of highpolymers to ordinary mineral oils has been found to, in certain cases, radically change the macroscopic characteristics of that flow. Recently, a model, which in a sense accounts for a nonrandom orientation of molecules, has been studied.[30] Such a model has been alluded to earlier.[31]* The pointwise viscosity in the fluid was assumed to depend on the angle β between the velocity vector and the direction of the viscous forces. The momentum equations were as (10) and were integrated to yield:

$$\nabla\cdot\left[\frac{F_2F_0-F_1^2}{F_0}\nabla P\right]=\frac{\partial}{\partial x}\left(\frac{F_1}{F_0}V\right), \qquad P|_{\partial\Omega}=0 \tag{18}$$

where

$$F_i=\int_0^{h(x)}\frac{\xi^i\,d\xi}{\mu(x,y,\xi)}, \qquad h(x)=2-x, \qquad \Omega=(0,1)x(-.5,.5) \tag{19}$$

$$\mu=\mu_M(1-a\cos^2\beta) \tag{20}$$

$$\cos\beta=\frac{\nabla p\cdot\bar{v}}{|\nabla p|\cdot|\bar{v}|} \tag{21}$$

A local potential for this system can be clearly written as

$$J[p,p_0]=\int_\Omega\left[a(x,y,p_0,\nabla p_0)(\nabla p)^2-2b(x,y,p_0,\nabla p_0)p\right]d\Omega \tag{22}$$

Again the local potential iterative scheme was applied using a linear finite element scheme. The iteration was terminated when $\|p^{m+1}-p^m\|^\dagger/\|p^{m+1}\|<10^{-4}$. The scheme converged quite rapidly for $a<\cdot 8$($\sim$3 to 5 iterations). Figures 8 and 9 show typical pressure distributions obtained. Further details are found in Ref. 30. Experiments are now being conducted to verify the validity of this model.

*The author wishes to thank Professor P. H. Roberts, University of Newcastle-Upon-Tyne for informing him of similar models which are discussed in Ref. 47.

†Here $\|\ \ \|$ denotes the usual Euclidean metric.

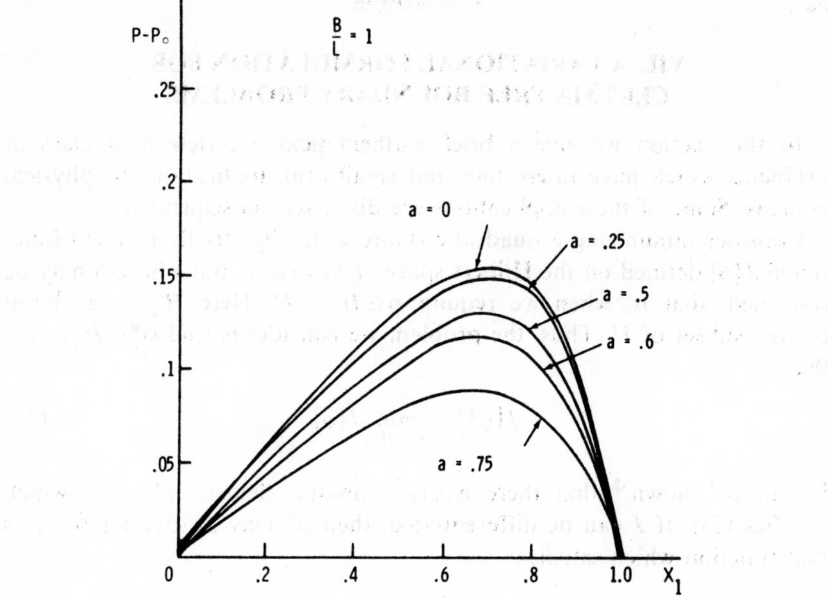

P-P₀
B/L = 1
.25
.2
.15
.1
.05
0
a = 0
a = .25
a = .5
a = .6
a = .75
.2
.4
.6
.8
1.0
X1

Fig. 8.

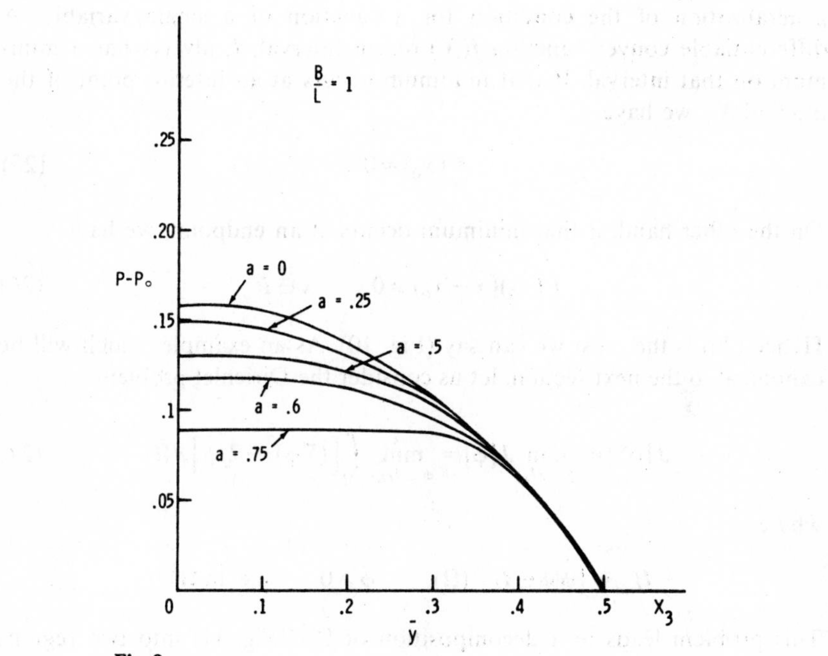

B/L = 1
.25
.20
P-P₀
.15
.1
.05
0
a = 0
a = .25
a = .5
a = .6
a = .75
.1
.2
.3
.4
.5
X3
y

Fig. 9.

VII. A VARIATIONAL FORMULATION FOR CERTAIN FREE BOUNDARY PROBLEMS

In this section we give a brief mathematical overview of a class of problems which have interesting and significant applications to physical systems. Some of these applications are discussed subsequently.

Consider minimizing a quadratic (more generally strictly convex) functional $J[\phi]$ defined on the Hilbert space H as before, but where ϕ may be restricted, that is, when we require $\phi \in H_{ad} \subseteq H$. Here H_{ad} is a closed convex subset of H. Thus, the problem we consider is find $\phi^* \in H_{ad}$ such that

$$J[\phi^*] = \min_{\phi \in H_{ad}} J[\phi] \tag{23}$$

It can be shown[32] that there exists a unique element $\phi^* \in H_{ad}$ which satisfies (23). If J can be differentiated, then ϕ^* may be characterized as that function which satisfies

$$J'[\phi^*](\phi - \phi^*) \geqslant 0 \tag{24}$$

for all $\phi \in H_{ad}$. This condition, termed a variational inequality, is the direct generalization of the condition for a function of a single variable. A differentiable convex function $f(X)$ on an interval, I, always has a minimum on that interval. If that minimum occurs at an interior point of the interval X_0, we have

$$f'(x_0) = 0 \tag{25}$$

On the other hand, if that minimum occurs at an endpoint, we have

$$f'(x_0)(x - x_0) \geqslant 0, \qquad x \in I \tag{26}$$

Hence (26) is the most we can say (Fig. 10). As an example, which will be canonical to the next section, let us consider the Dirichlet problem

$$J[\phi^*] = \min_{\phi \in H_{ad}} J[\phi] = \min_{\phi \in H_{ad}} \int_\Omega \left[(\nabla\phi)^2 - 2f\phi \right] d\Omega \tag{27}$$

where

$$H_{ad} = \left\{ \phi \mid \phi \in H_0^{12}(\Omega), \qquad \phi \geqslant 0 \qquad \text{a.e. in } \Omega \right\}$$

This problem leads to a decomposition of Ω[32] (Fig. 11) into two regions

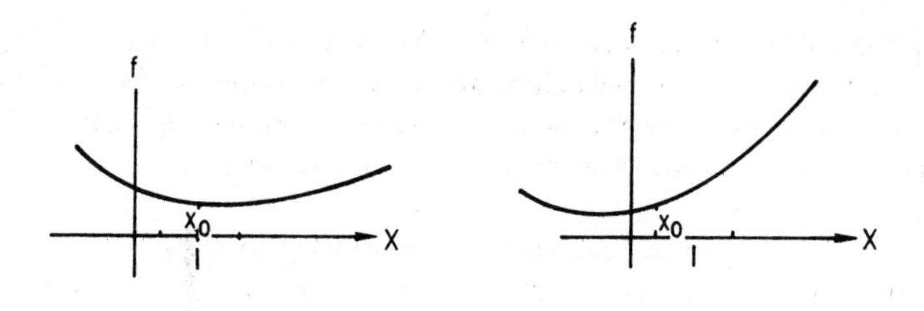

$$f'(X_0)\ (X - X_0) \geqslant 0, \quad X \in I$$

Fig. 10.

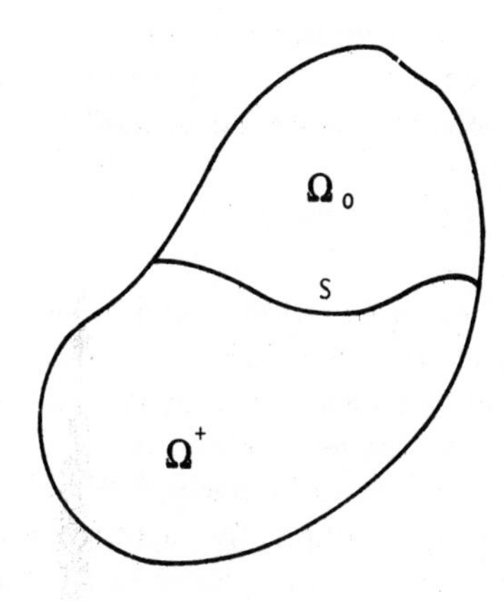

Fig. 11.

(one of which may be empty)

$$\begin{aligned} \Omega^+ &= \{x \mid x \in \Omega, \phi^*(x) > 0\} \\ \Omega_0 &= \{x \mid x \in \Omega, \phi^*(x) = 0\} \end{aligned} \tag{28}$$

In Ω_1 we have (in some sense)

$$\nabla^2 \phi^* = -f \tag{29}$$

Thus, both a solution and a free boundary, S, separating Ω^+ and Ω_0 are determined by this variational form. We further can prove that

$$\left. \frac{\partial \phi^*}{\partial n} \right|_{S(\text{from } \Omega^+)} = 0 \tag{30}$$

A question which is of great mathematical interest is the regularity of u in

Ω^+ and the smoothness of S (see for example Refs. 33 and 34). Boundary value problems such as these determined by the above variational formulation are known as unilateral boundary value problems. We can understand "unilateral" in the sense that only one-sided variations for ϕ are admissible on Ω_0.

The finite dimensional problem corresponding to these problems[25] is by no means trivial. The finite dimensional version of (27), for example, is to minimize the quadratic form J

$$J(\alpha^*) = \min_{\substack{\phi^n(x) \geqslant 0 \\ x \in \Omega}} J(\alpha) \tag{31}$$

For linear elements if $\phi_i \geqslant 0, 1 \leqslant i \leqslant n$ then $\phi^n(x) \geqslant 0, x \in \Omega$, that is, a certain "convexity" property is present. In this case then (31) becomes a standard quadratic programming problem[35] (minimize a quadratic form subject to linear constraints):

$$J(\alpha^*) = \min_{\alpha \geqslant 0} J(\alpha) \tag{32}$$

Conversely, if our approximating functions are not piecewise linear, the constraint α must satisfy becomes extremely nonlinear, and it is doubtful that this problem should be attacked as a mathematical programming problem. Indeed, recently, we have made some progress in devising algorithms to handle these problems effectively.[36] Finally we remark that the entire field of approximating functions by piecewise polynomials which satisfy certain constraints is relatively unexplored mathematical territory as was recently pointed out.[37] In the next section we discuss some particular physical problems leading to unilateral boundary value problems and some results which we have obtained.

VIII. SOME PARTICULAR FREE BOUNDARY PROBLEMS

Our first example is drawn from elasticity theory. Consider a membrane stretched over a finite region of the plane and loaded by a force $f(x,y)$. The equilibrium displacement u^+ of that membrane is such that the potential

$$J[u] = \int_\Omega \left(\frac{1}{2}(\nabla u)^2 - fu \right) d\Omega \tag{33}$$

is minimized.* Suppose that the membrane is constrained, that is, we

*Of course, in deriving[35] we have assumed that u is "small."

require $u(x,y) \geqslant g(x,y)$. Clearly this fits into the category of problems discussed in the last section. Of course, the equilibrium displacement of the membrane may be such that the constraint is inactive, that is, $u^+(x,y) > g$. On the other hand, a "contact patch" may be formed where u will be on the constrainting surface in parts (or all) of Ω as shown schematically in Fig. 12. Our variational formulation and approximation technique here gives an effective means of determining both the solution and the contact patch. Figures 13 and 14 show this for Ω a square with different constraints. We remark the method of Christopherson[38,39] was used to solve the resulting quadratic programming problem.

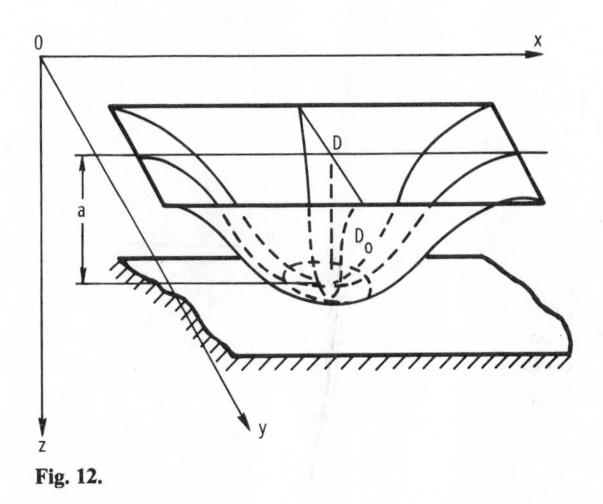

Fig. 12.

Another application which is of considerable interest in slow viscous flows, particularly, hydrodynamic lubrication and sealing, is the class of free boundary problems stemming from the minimization of a quadratic functional which may be derived from the Helmholz theorem.[7] This leads to the minimization of the quadratic functional:

$$J[P] = \int_\Omega \left[\frac{1}{2} H^3 (\nabla P)^2 - H,_x P \right] d\Omega \tag{34}$$

subject to the constraint that the pressure remain above the cavitation pressure of the fluid. Typical results are shown in Fig. 15. This topic is

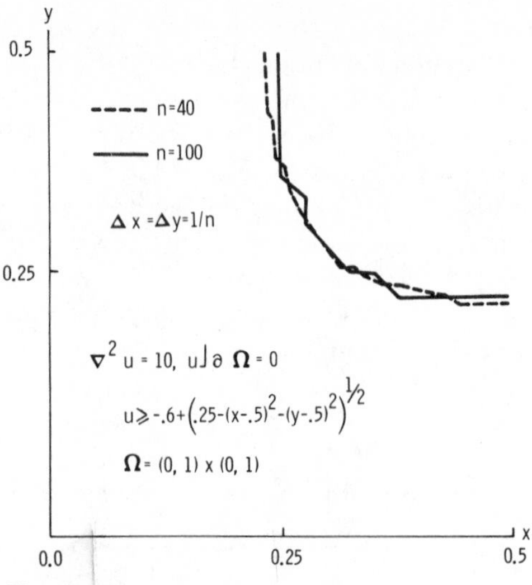

Fig. 13. Free boundary location.

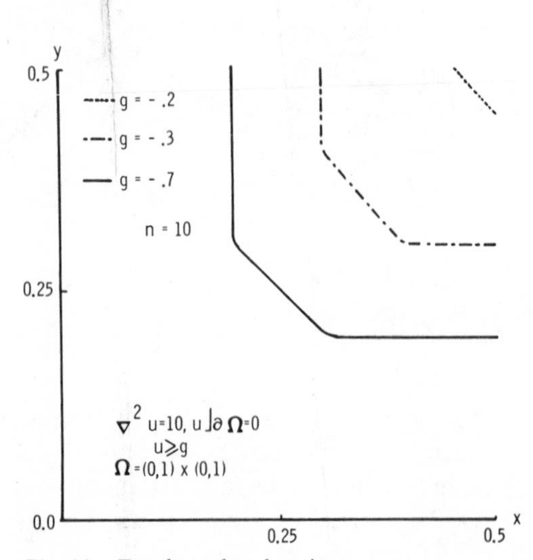

Fig. 14. Free boundary location.

discussed in Refs. 36 and 40. We remark, however, that the potential cavitation flows as discussed in Refs. 41 and 42 may also fit into this formulation.

Another application is the elastic–plastic torsion problem which is discussed in Refs 43 and 44. There it is shown that the elastic-plastic boundary is determined by minimizing a functional such as (35) subject to

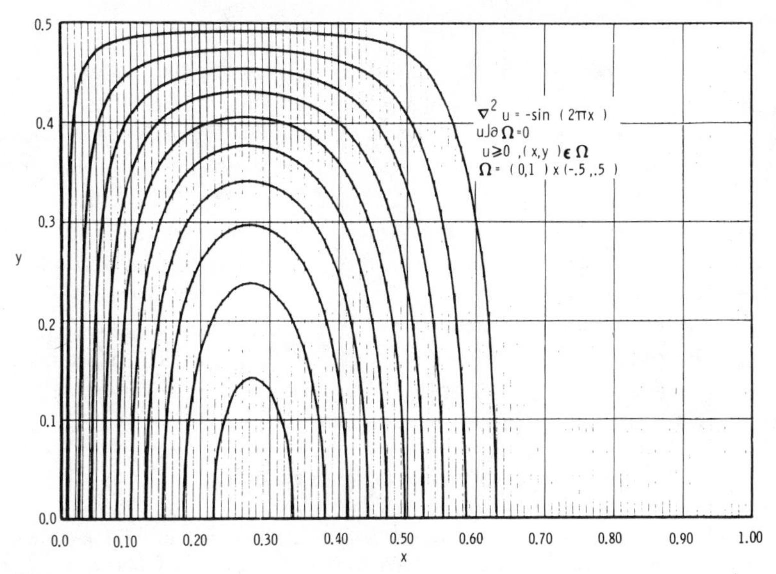

Fig. 15.

an inequality constraint:

$$u(x) \leqslant k_1 d(x, \partial\Omega) \qquad (\text{or } |\nabla u| \leqslant k_2) \tag{35}$$

where d denotes the distance function.

IX. APPLICATIONS TO STABILITY THEORY

We are currently, in the process of applying finite element methods to time-independent and time-dependent hydrodynamic, hydromagnetic, and chemical stability problems. Here, again, the local potential formulation may be used as in Ref. 45. Intuitively, the advantage of using these methods in some of these problems is varied. Firstly, complex basic solutions which are to be perturbed present no difficulties. Secondly, a large concentration of elements may be placed near critical points thus allowing for more information in those regions where it is necessary—a feature for which a global basis does not allow. The latter idea has been used recently[46] with finite difference methods to obtain numerical solutions to the Orr-Sommerfeld equation for plane Poiseuille flow at extremely high Reynolds numbers ($R \sim 10^6$). Hopefully, we will have results in the near future.

X. CONCLUSIONS

In this paper we have briefly reviewed some of the basic ideas underlying the variational formulation of certain physical systems. Emphasis was placed on the utility of these methods in obtaining approximate solutions to such problems. The local potential concept was tied to more modern choices of approximating spaces and, in particular, the finite element method. Examples showing the powerful "marriage" of these ideas were presented. The further application to stability problems was outlined.

The variational formulation of certain free-boundary problems was discussed along with direct techniques for obtaining approximate solutions. Though the potentials for those formulations were classical, it seems clear that extensions to the local potential formulation can be made. These topics will be studied in the near future.

Acknowledgments

The author wishes to thank Professor I. Prigogine for his stimulating conversation regarding variational formulations and his most generous invitation to attend this symposium. The author also wishes to acknowledge Dr. K. P. Oh and Dr. Dah-Chen Sun of the General Motors Research Laboratory for their assistance in the preparation of this paper.

References

1. C. B. Morrey, *Multiple Integrals in the Calculus of Variations*, Springer, New York, 1966.
2. A. Friedman, *Partial Differential Equations*, Holt, Rinehart, and Winston, New York, 1969.
3. S. Agmon, "The L_p approach to the Dirichlet Problem, I: Regularity Theoremes," *Ann. Scula Norm. Sup. Pisa*, **13**, 405, (1959).
4. S. G. Mikhlin, *Variational Methods in Mathematical Physics*, Macmillan, New York, 1964.
5. P. Glansdorff and I. Prigogine, *Physica*, **20**, 773, (1954).
6. P. Glansdorff and I. Prigogine, "On a General Evolution Criterion in Macroscopic Physics," *Physica*, **30**, 351 (1964).
7. P. Glansdorff and I. Prigogine, *Thermodynamic Theory of Structure, Stability and Fluctuations*, Wiley-Interscience, New York, 1971.
8. R. S. Schechter, "Variational Principles for Continuum Systems," in *Nonequilibrium Thermodynamics, Variational Techniques and Stability*, X. X. Donnelly, X. X. Herman, and I. Prigogine, Eds., University of Chicago Press, Chicago, 1966.
9. D. F. Hays and Harriet N. Curd, "Heat Conduction in Solids: Temperature Dependent Thermal Conductivity," *Int. J. Heat Mass Transfer*, **11**, 285, (1968).
10. H. W. Butler and R. L. Rackley, "Application of a Variational Formulation to Non-Equilibrium Fluid Flow," *Int. J. Heat Mass Transfer*, **10**, 1255 (1967).
11. S. M. Rohde, "Minimality Properties of the Local Potential," *Physica*, **61**, 53, (1972).
12. J. H. Bramble and A. H. Schatz, "Rayleigh-Ritz-Galerkin Methods for Dirichlet's Problem Using Subspaces without Boundary Conditions," *Commun, Pure Appl. Math.*, **23**, 653 (1970).
13. P. Glansdorff, "Application of the Local Potential to the Convergence of Vraitaional Techniques," *Physica*, **32**, 000 (1966).

14. R. Courant, "Variational Methods for the Solution of Problems of Equilibrium and Vibrations," *Bull. Am. Math. Soc.*, **49**, 1 (1943).
15. J. L. Synge, *The Hypercircle in Mathematical Physics*, Cambridge University Press, New York, 1957.
16. O. C. Zienkiewicz, *The Finite Element Method in Structural and Continuum Mechanics*, McGraw-Hill, New York, 1967.
17. C. A. Felippa and R. W. Clough, *The Finite Element Method in Solid Mechanics*, SIAM-AMS Proceedings V-II, 1970.
18. E. C. Lemmon and H. S. Heaton, "Accuracy, Stability, and Oscillation Characteristics of Finite Element Method for Solving Heat Conduction Equation," ASME Paper No. 69-WA/HT-35, ASME Winter Annual Meeting, Los Angeles, November 16–20, 1969.
19. A. F. Emery and W. W. Carson, "An Evaluation of the Use of the Finite-Element Method in the Computation of Temperature," *Trans. ASME*, May, 1971.
20. J. F. Booker and K. H. Huebner, "Application of Finite-Element Methods to Lubrication: An Engineering Approach," *Trans. ASME, Ser. F J. Lubrication Technol.*, Paper 72-Lub-N, 1972.
21. Ralph Ta-shun Cheng, "Numerical Solution of the Navier-Stokes Equations by the Finite Element Method," *Phys. Fluids*, **15**, No. 12, (1972).
22. S. C. Eisenstat and M. H. Schultz, "Computational Aspects of the Finite Element Methods," in *The Mathematical Foundations of the Finite Element Method with Applications to Partial Differential Equations*, Academic Press, New York, 1972.
23. R. S. Varga, "Hermite Interpolation-Type Ritz Methods for Two-Point Boundary Value Problems" in *Numerical Solution Partial Differential Equations*, New York, Academic Press, 1966.
24. J. L. Walsh, J. H. Ahlberg, and E. N. Nelson, "Best Approximation Properties of the Spline Fit," *J. Math. Mech.*, **11**, 225, (1962).
25. G. R. Birkhoff and C. R. DeBoor, *J. Math Mech.*, 5, 13, 1964.
26. S. M. Rohde and K. P. Oh, "A Unified Treatment of Thick and Thin Film Elastohydrodynamic Problems Using Higher Order Element Methods," to appear in *Proc. Royal Soc. London*.
27. D. C. Sun, "Analysis of Finite Porous Journal Bearings," to be submitted.
28. D. F. Hays, "An Extended Variational Formulation for Isothermal Gasdynamic Lubrication," *Trans. ASME, J. Lubrication Technol.*, Series F.
29. S. M. Rohde and K. P. Oh, "The Application of Higher Order Element Methods to Compressible Porous Bearing Problems," to appear in *Int. J. Num. Methods in Eng.*
30. N. Tipei and S. M Rohde, "Lubricant Films with Directional Variable Viscosity," *Trans. ASME*, 1974.
31. G. I. Taylor, "Rheology for Mathematicians," in *Proc. Intern. Congr. Rheology, 2nd*, (1953).
32. J. L. Lions, *Optimal Control of Systems Governed by Partial Differential Equations*, Springer-Verlag, New York, 1971.
33. H. Lewy and G. Stampaechia, "On the Regularity of the Solution of a Variational Inequality," *Commun. Pure Appl. Mathematics*, **XXII**, 153, (1969).
34. G. Stampaechia, "Forms Bilinearies Coercitives sur les Ensembles Convexes," *C. R. Acad. Sci. Paris*, **258**, 4413 (1964).
35. G. Hadley, *Nonlinear and Dynamic Programing*, Addison-Wesley, Reading, Mass., 1964.
36. S. M. Rohde and G. T. McAllister, "A Variational Formulation for a Class of Free Boundary Problems Arising in the Study of Certain Slow Viscous Flows," to appear in *Int. J. Eng. Science*.
37. J. T. Lewis, "Approximation with Convex Constraints," *SIAM Rev.*, **15**, No. 1, (1973).

38. D. G. Christopherson, *Proc. Inst. Mech. Eng.*, **146**, 135 (1941).
39. C. W. Cryer, "The Solution of a Quadratic Programming Problem Using Systematic Over-relaxation," *SIAM J. Control*, **9**, No. 3, (1971).
40. Guy Bayda, "Inequations Variationelles Elliptiques Avec Conditions aux Limites Periodiques Application a la Resolution de L'Equation de Reynolds," Thesis, L'Universite Claude Bernard de Lyon, 1972.
41. P. R. Garabedian and D. C. Spencer, " Extremal Methods in Cavitational Flow," *J. Rational Mech. Anal.* **1**, 359 (1952).
42. P. R. Garabedian, H. Lewy, and M. Schiffer, "Axially Symmetric Cavitational Flow," *Ann. Math.* **56**, 560 (1952).
43. T. W. Ting, "Elastic-Plastic Torsion of a Square Bar," *Trans. ASME*, **123**, 369 (1966).
44. H. Brezis, "Multiplicateur de Lagrange in Torsion Elasto-Plastique," *Arch. Rational Mechanics Anal.*, **29**, XII, (1972).
45. J. K. Platten, *Int. J. Eng. Sci.*, **9**, 37 (1971).
46. T. H. Hughes, "The Numerical Solution of the Orr-Sommerfeld Equation at Large Reynolds Number," Argonne National Laboratory Report ANL-7855, 1971.
47. F. M. Leslie, "Some Constitutive Equations for Anisotropic Fluids," *Quart. J. Mech. Appl. Math.*, **XIX**, Pt. 3, (1966).

GLANSDORFF-PRIGOGINE CRITERION AND STATISTICAL THEORY

F. SCHLÖGL

RWTH Aachen, Germany

I should like to make some remarks concerning the basis of the Glansdorff-Prigogine criterion in the framework of statistical physics. Nonlinear thermodynamics is the continuation of nonequilibrium thermodynamics from the linear "Onsager region" to the larger field of nonequilibrium states which on the one hand can still be described by relatively few macroscopic variables, yet on the other hand are so far away from equilibrium states that the dynamical equations are in general nonlinear with respect to the deviations from any equilibrium. As Professor Glansdorff pointed out in his talk, very surprising phenomena occur in this region, namely, the building up of dissipative structures. Thus nonlinear thermodynamics qualitatively is not only a continuation of linear nonequilibrium theory, but becomes a new exciting field of physics, chemistry, and biology.

Of course we are very much interested in finding general thermodynamic principles in this new field. Here the Glansdorff-Prigogine criterion is of outstanding importance. So far as we known today, it is the only new principle of such generality in this new field. The class of the steady states in nonlinear thermodynamics plays a prominent role as does the class of equilibria in other areas of thermodynamics. The latter of course form a subclass of the former. Not all steady states, however, are stable. Here the criterion comes into play. There can exist more than one stable steady state for the same boundary conditions. This is essentially connected with the building up of new structures. The fundamentality of the criterion can also be seen by the fact that the principle of minimal entropy production in linear thermodynamics and the second law of thermostatics can be considered as special statements of the criterion.

In the author's opinion the criterion is a fundamental new principle in macroscopic theory. That means that it cannot be deduced from other macroscopic principles in a rigorous manner. It can only be made plausible by accepting such assumptions as the existence of local equilibrium or others of similar quality. Nonlinear thermodynamics is essentially concerned with open systems and the approximate description of their states by restricted (frozen) equilibria as in thermostatics is not justified far from equilibrium.

In statistical theory it is possible to prove the criterion in a very general way. This proof shows once again the fundamentality and independence of the criterion as a very general principle. Essentially it makes use of a statistical measure $K(p,p')$ of two probability distributions p,p' over the same set of microstates i of a system.

$$K(p,p') = \sum_i p_i(\ln p_i - \ln p_i') \tag{1}$$

is defined in information theory as "Kullback information" or "information gain" and can be introduced as the mean bit-number of a communication which gives the change p' to p. It should be stressed that the proof uses only mathematical features of the measure and that it is independent of the semantic interpretation of K in thermodynamics. K has particular importance in thermodynamics of open systems because it can enclose the influence of the environment on the system by an adequate p'. So it can be shown that in statistical theory fundamental inequalities of thermostatics and thermodynamics of open systems follow from properties of the quantity K. In certain situations, for instance, K is the entropy which is produced in the interior of a system by the change from a state p to a state p'.

To given stochastic dynamics of the probability distribution $p(t)$ we get a stability criterion for a steady-state p' by using $K(p,p')$ as the Liapounoff function. This stochastic criterion is the basis of the macroscopic Glansdorff-Prigogine criterion.

It gives the latter if we assume that the distribution p is uniquely attached to the macroscopic variables which are mean values of microstate observables. These macroscopic variables in general are not only thermostatic variables but, moreover, typical nonequilibrium variables, as rate changes and vector flows.

This deduction of the Glansdorff-Prigogine criterion provides a more general form which is not restricted to infinitesimal deviations from the steady state. It can be extended to a sufficient, but not necessary, stability condition for final regions round the steady state.

Glansdorff and Prigogine have given the criterion in a general form and in a local form for fluid media as well. This more detailed form was deduced in the statistical theory by H. K. Janssen by the use of Mori distributions for the hydrodynamic states. To understand these distributions, the following explanation is necessary. If an equilibrium state is given by static macroscopic variables M^ν which are mean values of observables M_i^ν in the microstates i, the adequate probability distribution is the generalized canonical distribution

$$p_i = \exp\left(\phi - \sum_\nu \lambda_\nu M_i^\nu\right) \tag{2}$$

which belongs to maximal entropy to the given mean values. In a nonequilibrium state the $M^\nu(t)$ are functions of time. The distribution $p^a(t)$ which at any time t is constructed in this way corresponding to the values $M^\nu(t)$ may be called the "accompaning" canonical distribution. To describe the nonequilibrium states by such $p^a(t)$ is an approximation which, with respect to time only, is similar to the local equilibrium approximation, which is made with respect to time and space. The Mori distribution is justified for systems which show distinctly separate behavior in the short and long time scale. The short time behavior of an open system of this kind is not essentially different from an isolated system and can be described by an Liouville operator L. Therefore the Mori distribution $p(t)$ is constructed from the accompaning $p^a(t)$ by

$$p(t) = e^{-iL\tau} p^a(t-\tau) \tag{3}$$

where τ is long on the short time scale and short on the long time scale. The detailed construction shows that $p(t)$ is attached not only to the static variables M^ν, but to their time derivatives as well. If M^ν are functions of space in a hydrodynamic system, the stochastic stability criterion gives the local form of the macroscopic criterion for fluid media. This deduction does not use the approximation by a local equilibrium.

It should be mentioned that the statistical proof of the macroscopic criterion can be given in the same generality in quantum theory, where the probability distributions are replaced by statistical operators.

References

1. F. Schlögl, *Z. Phys.*, **243**, 303 (1971); *Z. Phys.*, **248**, 446 (1971).
2. H. K. Janssen, *Z. Phys.*, **253**, 176 (1972).
3. H. Mori, *Phys. Rev.*, **115**, 298 (1959).
4. F. Schlögl, *Z. Phys.*, **249**, 1 (1971).

NUMERICAL MODELS FOR CONVECTION

D. R. MOORE

Department of Applied Mathematics and Theoretical Physics
University of Cambridge, England

CONTENTS

Abstract

A computer model of two-dimensional convection in a Boussinesq fluid between slippery plates has been constructed. The dependence of the heat transport upon the Prandtl number has been examined. Three nonlinear instabilities in these flows have been found.

I. INTRODUCTION

Problems concerning unstable highly nonlinear fluid mechanical systems may be attacked by numerical modeling. Energy transport in a convecting fluid is one such problem which is of great interest to geophysicists and astrophysicists.

Linear theory may indicate when such systems become unstable, but cannot give quantitative results describing highly unstable systems. It is also unable to predict the existence of further instabilities which are nonlinear in origin. Detailed models, including many physical processes, are beyond the range of present-day computing techniques for all but the most simple systems. Idealized models, however, can be used to examine fundamental questions concerning the nature of convective flow. These models are sufficiently limited in their scope to be amenable to present machines and methods, yet yield useful insight.

Subsequent sections of this paper describe the formulation of such models, the methods used to solve them, and review results of simple investigations. These simple models are found to yield many fascinating nonlinear phenomena that need to be understood before embarking upon more elaborate descriptions of physical systems.

II. FORMULATION

In a two-dimensional Boussinesq fluid, the Navier-Stokes equation may be treated in a particularly simple form by considering the single component, ω, of the curl of the velocity field, **u**. The resulting explicit equation for the vorticity,

$$\frac{\partial \omega}{\partial t} = -\nabla \cdot (\omega \mathbf{u}) - g\alpha \frac{\partial F(T)}{\partial x} + \nu \nabla^2 \omega \tag{II.1}$$

the heat flow equation,

$$\frac{\partial T}{\partial t} = -\nabla \cdot (T\mathbf{u}) + \kappa \nabla^2 T \tag{II.2}$$

the equation of state,

$$\rho(T) = \rho_0[1 - \alpha F(T)] \tag{II.3}$$

and the incompressibility condition

$$\nabla \cdot \mathbf{u} = 0 \tag{II.4}$$

(where T is the temperature, ρ is the density, ν and κ are the viscous and thermal diffusivities, and α is the coefficient of thermal expansion) completely describe the time evolution of the flow from some initial configuration.

Equation (II.4) implies that the velocity field **u**, may be written

$$\mathbf{u} = (u, 0, w) = \left(-\frac{\partial \Psi}{\partial z}, 0, \frac{\partial \Psi}{\partial x} \right) \tag{II.5}$$

where Ψ is a scalar stream function. Ψ and ω are related by Poisson's equation

$$\nabla^2 \Psi = -\omega \tag{II.6}$$

The flow is confined to a rectangular region;

$$0 \leqslant x \leqslant L, 0 \leqslant z \leqslant d \tag{II.7}$$

in the x,z plane, where z is in the vertical direction and ω, T and **u** are constrained to behave as follows on the boundaries;

	$x=0,L$	$z=0,d$	
			(II.8)
ω	$=0$	$=0$	(II.8a, b)
T	$\frac{\partial T}{\partial x}=0$	$=T_b, T_t$	(II.8c, d)
u	$=0$	$\frac{\partial u}{\partial z}=0$	(II.8e, f)
w	$\frac{\partial w}{\partial x}=0$	$=0$	(II.8g, h)

Two numbers appear in the dimensionless form of the equations. These may be chosen to be: the Prandtl number,

$$p=\frac{\nu}{\kappa} \tag{II.9}$$

measuring the relative importance of the viscosity and conductivity in the fluid; and the Rayleigh number,

$$R=\frac{g\alpha F(T)d^3}{\kappa\nu} \tag{II.10}$$

measuring the overall degree of instability of the system. A third parameter, the dimensionless width,

$$\lambda=\frac{L}{d} \tag{II.11}$$

completes the description of the convecting region. When R, p, and λ are chosen, (II.1) and (II.2) allow initial distributions of ω and T to evolve in time. The efficiency of this flow in transporting heat energy can be measured by comparing the total heat flow through the layer with that which would be carried by conduction in the absence of motion. This ratio is the Nusselt number, N, and if a steady final state is found such that $\partial\omega/\partial t=\partial T/\partial t=0$, N is constant throughout the layer. For a more detailed derivation of these equations, see Moore and Weiss.[1,2]

III. NUMERICAL METHOD

T and ω are represented by numbers at points on a rectangular staggered mesh covering the region $0\leqslant x\leqslant L$, $0\leqslant z\leqslant d$. The differential equations stated in Section II are approximated by space- and time-centered second-order difference equations. These are used to predict the values of ω and T at the meshpoints for a time $t+\delta t$ in the future from those defined on the mesh at time t, subject to the boundary conditions (II.8). The stream function is then constructed from the new values of ω by solving an implicit finite difference approximation to Poisson's equation (II.6). This process is repeated over and over, evolving ω and T from some initial configuration at time $t=0$ to some final state. Exact details are given in Moore, Peckover, and Weiss.[11]

These methods were used on grids with 12, 24, or 48 intervals in the region $0 \leqslant x \leqslant L$ and up to 200 intervals for $0 \leqslant x \leqslant d$. Typically the number of vertical intervals was eight times the Nusselt number. This ensured adequate resolution of the two horizontal boundary layers. Usually one convective roll was fitted onto the grid. Some calculations took up to 40,000 timesteps to converge to a steady state, the maximum permissible time step being bounded by the lesser of the limits set by the Courant-Friedrich-Levy stability criteria and the accuracy criteria for diffusion.

IV. RESULTS

In this section various instabilities found in simple convection models are reviewed. We start with the simplest model, that considered by Lord Rayleigh.[3] For this model

$$F(T)=T-T_0 \tag{IV.1}$$

and the Rayleigh number is

$$R=\frac{g\alpha\Delta T d^3}{\kappa\nu} \quad \text{where } \Delta T=T_b-T_t \tag{IV.2}$$

The linear instability of this model is well understood and has been extensively treated in the literature (c.f. Ref. 4). The system becomes unstable to infinitesimal perturbations of width $\lambda=\sqrt{2}$ when $R>\frac{27}{4}\pi^4$. Such disturbances grow monotonically in amplitude.

Our research has been concerned with fully developed two-dimensional convection when $R \gg R_c$. Figure 1 presents some results for this simple model as values of the Nusselt number plotted against the Prandtl number for various values of the Rayleigh number. The Nusselt number is seen to be very insensitive to the value of the Prandtl number. Like Morgan,[5] we find three regimes of behavior for large Rayleigh numbers: a low, an intermediate, and a high Prandtl number regime. We chose, however, different boundary conditions for the velocity field and consider implicitly in our model up to 48 horizontal convective roll modes, so our results are not directly comparable with his.

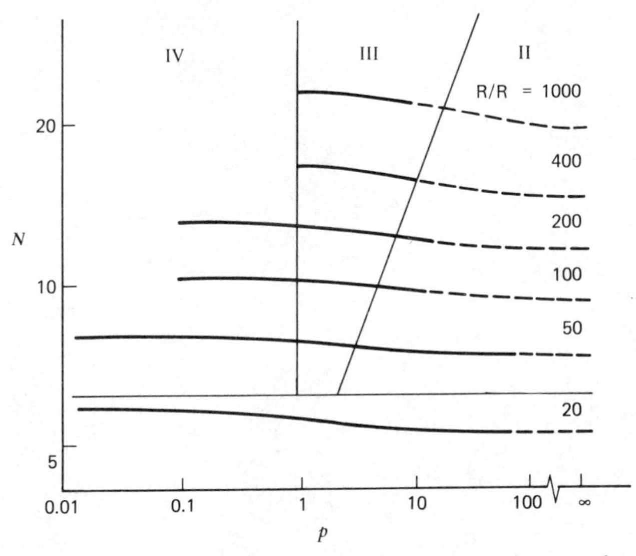

Fig. 1. Nusselt number, N, as a function of Prandtl number, p, for several values of the Rayleigh number, R/R_c. The straight lines divide the plane into the regions: *I*: low Rayleigh number convection; *II*: viscous convection; *III*: mixed viscous and advective convection; and *IV*: advective convection.

Examination of the streamlines and isotherms for the case $R/R_c = 400$ for values of p in each of the three regimes ($p = 1, 6, 8, \infty$) in Figure 2 shows why the Nusselt number is so insensitive to such tremendous variation in Prandtl number. Both Ψ and T variables are indistinguishable in form. Only the vorticity fields reveal the differences in the physical processes taking place. In the viscous regime the vorticity is dissipated where it is generated, that is, where the horizontal temperature gradients are largest. As viscosity gets weaker, the vorticity is carried by the flow out of these regions and into the boundary layers at the top and bottom surfaces. Decreasing the viscosity still further permits the vorticity to be swept several times around the cell before being dissipated. This also gives it time to diffuse into the center where it forms a constant pool. Thus the appearance is that of a plateau with raised edges, implying that the fluid in the center rotates as a solid body. The boundary between the viscous and intermediate regimes appears to be given, crudely, by the expression

$$\frac{R}{R_c} \approx p^{3/2} \Rightarrow N \approx p^{\frac{1}{2}} \qquad \text{for} \quad N > 6 \qquad \text{(IV.3)}$$

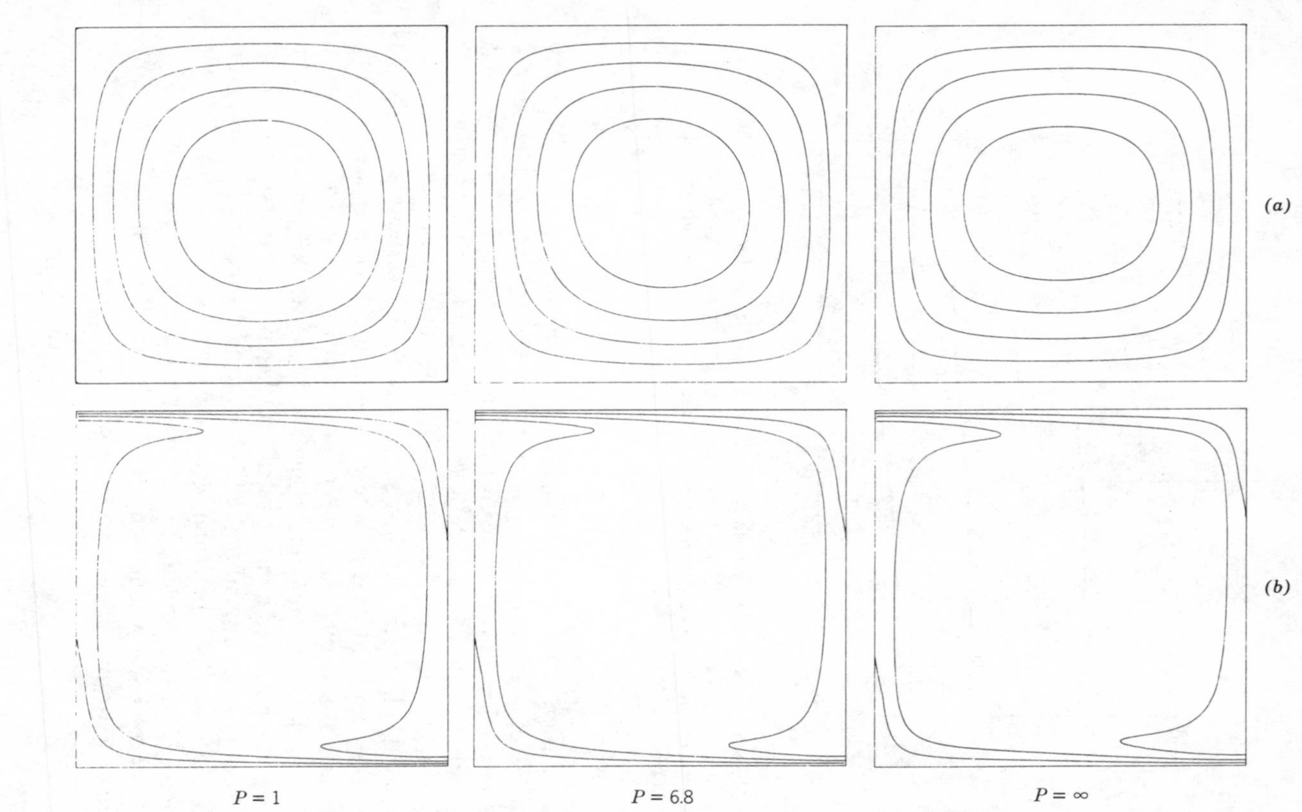

Fig. 2. (*a*) Streamlines; (*b*) isotherms; and (*c*) vorticity profiles for the advective, mixed and viscous regimes; $p = 1$, $\underline{6}.8$, and ∞ for $R/R_c = 400$ and $\lambda = 1$.

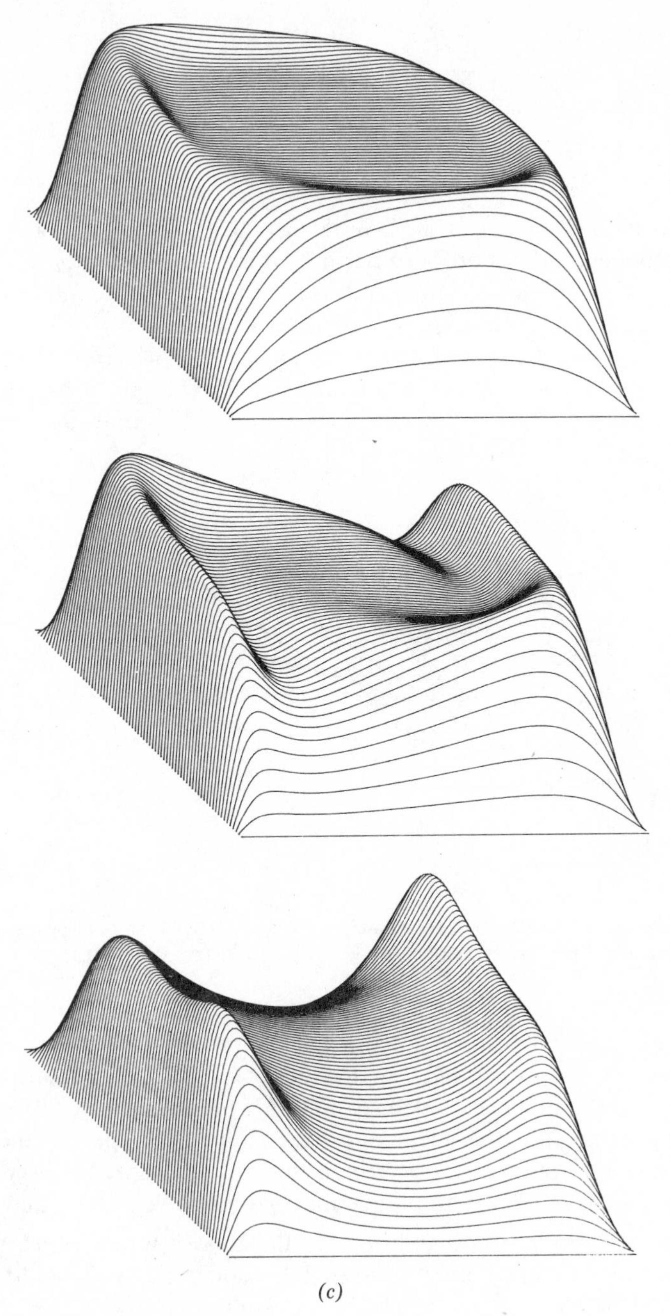

(c)

Fig. 2. (*Continued*)

In the viscous regime

$$N \sim R^{0.33} \tag{IV.4}$$

and in the advective regime

$$N \sim R^{0.37} \tag{IV.5}$$

Schematically, we can depict the behavior as shown in Fig. 3. For very low Prandtl numbers, advection becomes important as soon as the isotherms are distortled and the $R^{0.37}$ dependence of the Nusselt number is immediate. For larger values of p, the efficiency of advection is impaired, and the viscous type of behavior persists for higher R. Ultimately, for $p=\infty$, the viscous behavior is found for all values of R. The transition point, R_t, from the $R^{0.33}$ to the $R^{0.37}$ behavior moves up the $R^{0.33}$ line as p increases, as implied by the empirical equation (IV.3).

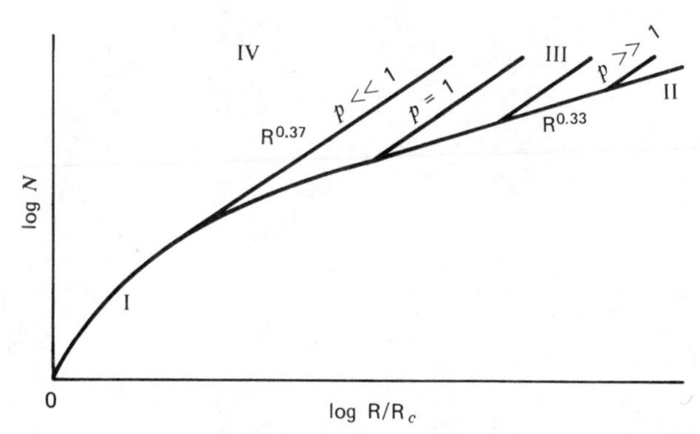

Fig. 3. Schematic representation of the Nusselt number as a function of the Rayleigh number for various values of the Prandtl number for the three regimes (not to scale!).

Nonlinear instabilities, not predictable by linear theory, can appear. Oscillations of the values of T and ω occur in the intermediate regime when R is large. For cells narrower than a critical width $\lambda_c(R,p)$, these oscillations decay in amplitude, and a steady state is reached. If the width of the convecting region is greater than this critical value, the oscillations persist and grow to a finite amplitude that is a function of R, p, and λ. For the case $p=6.8$, $R/R_c=200$, and $\lambda=\sqrt{2}$, the Nusselt number measured at the boundaries fluctuated by 20%, while the velocity fields varied by 5%. It is unclear whether such an instability indicates a preference for a three-dimensional structure for T and ω that might be steady, or if this is the onset of unsteady flow in all possible geometries (turbulent convection?), for this range of parameters and boundary conditions. Fuller details of this instability can be found in Moore and Weiss.[1]

When $F(T)$ depends quadratically upon the temperature, as it does for the density of water around the maximum density point at 4°C, a phenomenon develops that is of considerable astrophysical and geophysical interest. The convecting region is bounded on one side by a stable layer of the same fluid into which the motion penetrates. This penetrative convection has been considered in a detailed linear analysis by Veronis;[6] experimentally by Townsend,[7] and Myrup et al. Gross, and Hoo;[8] in a one-mode analysis by Musman;[9] and in a two-dimensional numerical model by Moore and Weiss.[1,2]

Consider $F(T)$ of the form

$$F(T) = -(T - T_0)^2 \tag{IV.6}$$

If the bottom temperature $T_b < T_0$, and the top temperature $T_t = T_b + \Lambda(T_b - T_0) > T_0$, the configuration depicted schematically in Fig. 4 results.

The Rayleigh number for the entire cell is determined by the total height. The true degree of instability of the convecting region is determined by its thickness, d/Λ. This situation gives rise to an instability of the 'snap through' variety described by Roberts.[10] The form of this instability

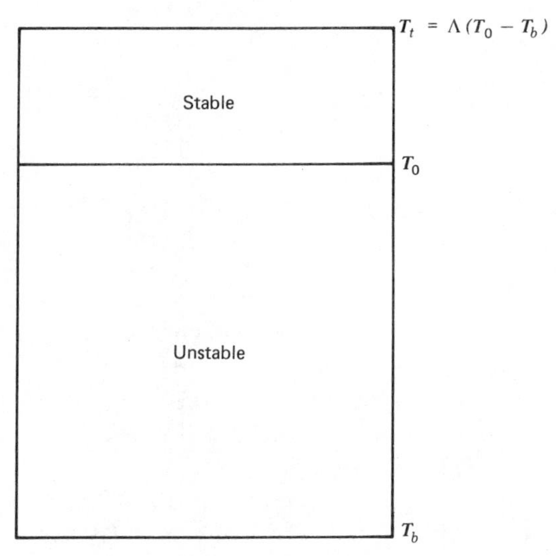

Fig. 4. Schematic depiction of an unstable region bounded above by a stable region. $T_b < T_0 < T_t$.

can be seen from the graph of the Nusselt number as a function of the Rayleigh number for this system (see Fig. 5). If a nonconvecting equilibrium state is taken and the Rayleigh number is increased beyond the point $R = R_c$, the onset of motion distorts the mean temperature profile so that the unstable region can grow in size. As the degree of instability in this region depends on the cube of the thickness, convection increases dramatically. An equilibrium state is achieved only for some large value of N such that the conducted heat flux through the thinner stable layer balances the conducted and convected heat flux carried through the thicker unstable layer. If such a convecting state is taken as an initial condition and the Rayleigh number is reduced, the distorted temperature profile permits convection to continue even when the undistorted unstable layer does not convect. There is a minimum value of R for which such a state can exist, and the corresponding Nusselt number is substantially greater than unity. If R is decreased below this point, the system abruptly decays into the conducting state $N = 1$.

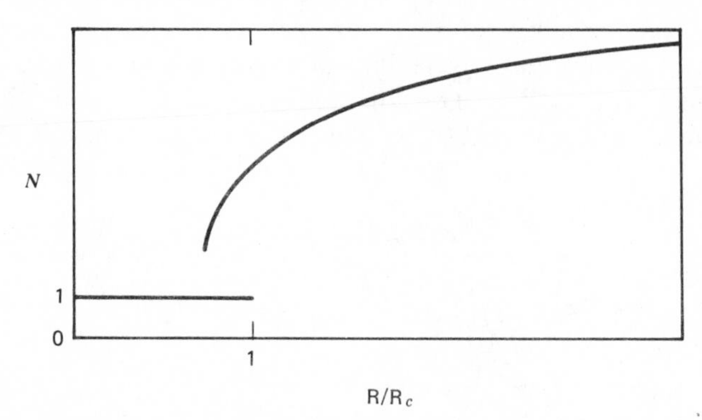

Fig. 5. Nusselt number as a function of the Rayleight number for an unstable penetrative system.

Another nonlinear instability has been encountered in this model. If R is increased the stable layer becomes thinner and more stably stratified. The motion in the convecting region becomes more vigorous and the period of the gravity oscillations of the stable region becomes commensurate with the turnover time of the unstable layer. When this occurs, the two regions couple dynamically and unsteady motion results. The stable layer is seen to slosh to and fro while the magnitudes of ω and T varies. Removing the stable layer stops the oscillations, demonstrating that this instability is quite distinct in nature from that described for the Rayleigh-Bénard convection. There is experimental indication of this type of instability.[7,8] This model is described in detail by Moore and Weiss.[2]

These highly idealized models are far removed from most real systems. Yet they contain some of the elements of the basic physics governing their behavior. If the surprising and unexpected nonlinear phenomena that appear can be understood at the physical level, these processes may be identified at work in real systems, simplifying the analysis of their behavior and their description.

Acknowledgments

I wish to thank Dr. N. O. Weiss for his collaboration in this research.

References

1. D. R. Moore, and N. O. Weiss, *J. Fluid Mech.*, **58**, 289 (1973).
2. D. R. Moore, and N. O. Weiss, *J. Fluid Mech.*, **61**, 553 (1973).
3. Lord Rayleigh, *Phil. Mag.* **32**, 529 (1916).
4. S. Chandrasekhar, *Hydrodynamic and Hydromagnetic Stability*, Clarendon Press, Oxford, 1961.
5. J. W. Morgan, 1974 (paper presented at the Brussels symposium).
6. G. Veronis, *Astrophys. J.*, **137**, 641 (1963).
7. A. A. Townsend, *Quart. J. Royal Meteorol. Soc.*, **90**, 248 (1964).
8. L. Myrup, D. Gross, L. S. Hoo, and W. Goddard, *Weather*, **25**, 150 (1970).
9. S. Musman, *J. Fluid Mech.*, **31**, 343 (1968).
10. P. Roberts, 1974 (paper presented at the Brussels symposium).
11. D. R. Moore, R. S. Peckover, and N. O. Weiss, *Computer Phys. Commun.*, **6**, 198 (1974).

THE EFFECT OF PRANDTL NUMBER ON FINITE AMPLITUDE BÉNARD CONVECTION

J. C. MORGAN

Department of Mathematics and Physics
Hull College of Technology
Kingson-upon-Hull, England

CONTENTS

Abstract

The equations obtained by Roberts[1] for Bénard convection in hexagonal cells are solved asymptotically for large Rayleigh number, R, when both boundaries are free. Solutions are found in the form of a mainstream, in which the effects of diffusion are small, together with thermal layers at the horizontal surfaces. The convection equations are dependent upon the Prandtl number, p, of the working fluid. When $NRa^2 \ll O(p^2/C^2)$ (N the Nusselt number), the viscous terms dominate the mainstream. This is only true for large Prandtl numbers. The solutions are similar to those found by Roberts[1] for rolls. When $NRa^2 \gg O(p^2/C^2)$, the advective terms dominate the mainstream. This is possible for p small, moderate, or large. Explicit expressions for the Nusselt number, N, as $R \to \infty$ are derived.

I. INTRODUCTION

Using the evolutionary criterion of Glansdorff and Prigogine,[2] Roberts[1] derived the equations for stationary Bénard convection. The equations for rolls are independent of the Prandtl number, p, and in the above mentioned paper the equations are solved asymptotically as $R \to \infty$. The equations for hexagonal cells, however, are dependent upon the Prandtl number. These equations are solved for two free boundaries. As $R \to \infty$, the horizontal mean temperature DT_0 is small in the mainstream and thermal layers form at the horizontal boundaries.

For large p the influence of the advective terms on the mainstream is small when $NRa^2 \ll p^2/C^2$, where N is the Nusselt number, a the horizontal wave number, and $C = 1/\sqrt{6}$. The thermal layers are similar to those found by Roberts for rolls, and the Nusselt number is given by $N \sim O[(Ra^2)^{1/3}]$.

When $NRa^2 \gg O(p^2/C^2)$, the main stream is dominated by the advective terms. This is possible for p small, moderate, or large. The Nusselt number is given by $N \sim O\{[(pRa^2/C)\ln(pRa^2/C)]^{1/5}\}$. For a larger Rayleigh number, if p is large or moderate (or a smaller Rayleigh number if p is small), we have $N \sim O\{[(pRa^2/C)\ln(Ra^2C^2/p^{3/2})]^{1/5}\}$. For $pR \sim O(1)$ the last expression shows that N decreases as p increases.

Moore and Weiss,[3] in their numerical solution for two free boundaries, found $N \sim O(R^{1/3})$ (independent of p) when $p \gg 1$ and $R < O(p^{3/2})$. For $R > O(p^{3/2})$, with p large or $p < 1$, they found $N \sim O(R^{0.365})$. Also Gough, Spiegel, and Toomre[4] found, for three dimensional convection with free boundaries, $N \sim O(R^{1/3})$ for p large but $N \sim O[(R \ln R)^{1/5}]$ when p is of order 1.

II. THE CONVECTION EQUATIONS

The equations for Bénard convection under the Boussinesq approximation and in the dimensionless form as used by Roberts[1] are

$$\operatorname{div} \mathbf{u} = 0 \tag{II.1}$$

$$\frac{\partial \mathbf{u}}{\partial t} + \frac{(\mathbf{u} \cdot \operatorname{grad} \mathbf{u})}{p} = -\operatorname{grad} \overline{w} + \nabla^2 \mathbf{u} + RT\boldsymbol{l}_z \tag{II.2}$$

$$p\frac{\partial T}{\partial t} + \mathbf{u} \cdot \operatorname{grad} T = \nabla^2 T \tag{II.3}$$

with $w = du/dz = dv/dz = 0$ at a free boundary; $\mathbf{u} = (u, v, w)$ is the fluid velocity, T the temperature, $\overline{w}$ the modified pressure, and $\boldsymbol{l}_z$ the unit vector in the direction of z. The parameters R and p are the Rayleigh and Prandtl numbers. Use is made of the evolutionary criterion of Glansdorff and Prigogine[2] to obtain the equations applicable to a stationary state. To achieve this Roberts[1] expressed the temperature T and the velocities as follows:

$$T = T_0(z) + F(z)f(x,y) \tag{II.4}$$

and

$$\mathbf{u} = \left(\frac{DW[\partial f/\partial x]}{a^2}, \frac{DW[\partial f/\partial y]}{a^2}, Wf \right) \tag{II.5}$$

where $D \equiv d/dz$, and $f(x,y)$ defines the cell planform whose horizontal wave number is a.

The equations obtained are

$$(D^2-a^2)^2 W = Ra^2F + \frac{C}{p}[WD(D^2-a^2)W + 2DW(D^2-a^2)W] \quad \text{(II.6)}$$

$$(D^2-a^2)F = WDT_0 + C[FDW + 2WDF] \quad \text{(II.7)}$$

$$DT_0 = -N + FW \quad \text{(II.8)}$$

and

$$N = 1 + \int_0^1 FW\,dz \quad \text{(II.9)}$$

where N is the Nusselt number. The thermal conditions require that $T_0 = 0$ at $z = 0$ and $T_0 = -1$ at $z = 1$. At a free boundary $W = D^2W = F = 0$. In (II.6) and (II.7) C is a constant, which is equal to zero for rolls and $1/\sqrt{6}$ for hexagonal cells.[1] Eliminating F from (II.6) and (II.7), we obtain

$$(D^2-a^2)^3 W - \frac{C}{p}(D^2-a^2)L + NRa^2W = W^2(D^2-a^2)^2 W - \frac{C}{p}W^2L$$

$$+ CDW(D^2-a^2)^2 W \quad - \frac{C^2}{p}DWL + 2CWD(D^2-a^2)^2 W - \frac{2C^2}{p}WDL$$

$$\text{(II.10)}$$

where the advection terms $L \equiv WD(D^2-a^2) + 2DW(D^2-a^2)W$ represent the convection of the vorticity $(D^2-a^2)W$. Supposing that W is large when R is large, we see that the dominant terms of (II.10) are

$$W(D^2-a^2)^2 W = NRa^2 + \frac{C}{p}WL \quad \text{(II.11)}$$

This is the mainstream equation. It is equivalent to $DT_0 \equiv 0$ and can be obtained from (II.6) and (II.8). Thermal boundary layers will develop at the horizontal boundaries at $z = 0$ and $z = 1$. From (II.6) and (II.9) we obtain

$$\int_0^1 \left\{ W\left[(D^2-a^2)^2 W - \frac{CL}{p} \right] - NRa^2 \right\} dz = -Ra^2 \quad \text{(II.12)}$$

For inviscid or infinite conducting fluids, ($p=0$) with $pRa^2 \sim O(1)$ and $W \sim O(1)$ (II.6) and (II.9) give $N=1$ since $\int_0^1 WL\,dz=0$. Thus for p small with pRa^2 and $W \sim O(1)$ there will be no thermal layers.

III. THE ASYMPTOTIC SOLUTION AS $Ra^2 \to \infty$ FOR TWO FREE BOUNDARIES $[a \sim O(1)]$

A. The Mainstream Dominated by the Viscous Term

If the viscous term dominates, the mainstream equation becomes

$$(D^2-a^2)^2 W \sim \frac{NRa^2}{W} \tag{III.1}$$

A solution at $z=0$ (also at $z=1$) for a free boundary is

$$W \sim (NRa^2)^{1/2}\left[cz - \frac{z^3}{6c}\ln z^{-1} + \cdots\right] \tag{III.2}$$

where c is a constant that can be determined from the main stream. From (III.2) the advection term is negligible in the mainstream if $NRa^2 \ll p^2/C^2$. Thus this solution is valid only for large p. In (III.2) let $z=\epsilon\eta$, where ϵ is the thickness of the thermal layer. W is then

$$W \sim (N\,Ra^2)^{1/2}\epsilon\left[c\eta - \frac{\eta^3}{6c}\epsilon^2 \ln\epsilon^{-1} + \frac{\eta^3 \ln\eta}{6c}\epsilon^2 + \cdots\right] \tag{III.3}$$

Therefore in the thermal layer let W be

$$W \sim A^{1/2}\epsilon\left[f - h\epsilon^2 \ln\epsilon^{-1} + \frac{g}{c}\epsilon^2 + \cdots\right] \tag{III.4}$$

where $A = NRa^2$. Ignoring terms in $\epsilon^2 a^2$, (II.10) is now

$$D^6 f \mp \frac{C}{p}A^{1/2}\epsilon^2 D^2 L + \epsilon^6 NRa^2 f = \epsilon^4 A f^2 D^4 f \mp \frac{C}{p}\epsilon^6 A^{3/2} f^2 L$$

$$\pm C\epsilon^2 A^{1/2} Df D^4 f - \frac{C^2}{p}\epsilon^4 A Df L \pm 2C\epsilon^2 A^{1/2} f D^5 f - 2\frac{C^2}{p}\epsilon^4 A^{1/2} f DL \tag{III.5}$$

where the lower sign applies to the thermal layer at $z=1$ and $D \equiv d/d\eta$.

From (III.5) let $\epsilon^4 A = 1$, i.e.,

$$\epsilon^4 NRa^2 = 1 \tag{III.6}$$

We now have

$$D^6 f \mp \frac{C}{p} D^2 L + \epsilon^2 f = f^2\left[D^4 f \mp \frac{C}{p} L\right] \pm CDf\left[D^4 f \mp \frac{C}{p} L\right] \pm 2CfD\left[D^4 f \mp \frac{C}{p} L\right] \tag{III.7}$$

If terms in (C/p) are ignored (p being large), (III.7) becomes

$$D^6 f + \epsilon^2 f = f^2 D^4 f \pm CDfD^4 f \pm 2CfD^5 f \tag{III.8}$$

The equation for f is

$$D^6 f = f^2 D^4 f \pm CDfD^4 f \pm 2CfD^5 f \tag{III.9}$$

with $f = D^2 f = D^4 f = 0$ at $\eta = 0$ and $f \to c\eta$ as $\eta \to \infty$. Since (III.9) has two singular solutions, $f = A/(\eta - \eta_0)$ where $A^2 - 11CA - 30 = 0$, the unique solution is $f = c\eta$. The equation for h is

$$D^6 h = c^2 \eta^2 D^4 h \pm cCD^4 h \pm 2cC\eta D^5 h \tag{III.10}$$

with $h = D^2 h = D^4 h = 0$ at $\eta = 0$ and $h \to \eta^3/6c$ as $\eta \to \infty$. The unique solution is $h = \eta^3/6c$. Since terms in $\epsilon^2 a^2$ and $\epsilon^4 a^4$ in (II.10) vanish when $f = c\eta$, the equation for g is

$$\frac{1}{c} D^6 g + c\eta = c\eta^2 D^4 g \pm CD^4 g \pm 2C\eta D^5 g \tag{III.11}$$

with $g = D^2 g = D^4 g = 0$ at $\eta = 0$, and $g \to \eta^3 \ln \eta / 6$. Assuming $c = 1$ and letting $\eta = y/\sqrt{6}$ and $D^4 g = U$, (III.11) reduces to

$$D^2 U \mp 2yDU \mp (1 \pm 6y^2) U = \frac{-y}{6\sqrt{6}} \tag{III.12}$$

with $U = 0$ at $y = 0$ and $U \to 0$ as $y \to \infty$. Now letting $U(y) = V(y)\exp(\pm y^2/2)$, (III.12) reduces to

$$D^2 V \mp 7y^2 V = -y \exp\left(\frac{\mp y^2}{2}\right) \tag{III.13}$$

with $V=0$ at $y=0$ and $V \to 0$ as $y \to \infty$. This equation can be solved in terms of $y^{1/2}I_{1/4}(\sqrt{7}\,y^2/2)$ and $y^{1/2}K_{1/4}(\sqrt{7}\,y^2/2)$ [or $y^{1/2}J_{1/4}(\sqrt{7}\,y^2/2)$ and $y^{1/2}Y_{1/4}(\sqrt{7}\,y^2/2)$] where $I_{1/4}$, $K_{1/4}$, $J_{1/4}$, and $Y_{1/4}$ are Bessel functions of order 1/4.[5] Substituting the solution for W into (II.12) and taking into account that the integrand is zero in the mainstream, we obtain

$$\epsilon^3 Ra^2 = \int_0^\infty (1-\eta D^4 g)\,d\eta + \int_0^\infty (1-\eta D^4 \mathbf{g})\,d\eta \qquad \text{(III.14)}$$

where $\mathbf{g}$ is the solution at $z=1$. Thus we have

$$\epsilon \sim O\left[(Ra^2)^{-1/3}\right] \qquad \text{(III.15)}$$

and, using (III.6),

$$N \sim O\left[(Ra^2)^{1/3}\right] \qquad \text{(III.16)}$$

This solution is similar to that given by Roberts for rolls. Since $NRa^2 \ll p^2/C^2$, from (III.16) $Ra^2 \ll (p/C)^{3/2}$.

B. The Mainstream Dominated by the Advective Terms

If the advective terms dominate the mainstream, then

$$W[WD(D^2-a^2)W + 2DW(D^2-a^2)W] \sim -\frac{NRa^2p}{C} \qquad \text{(III.17)}$$

The solution at $z=0$ is

$$W \sim \left(\frac{3NRa^2p}{C}\right)^{1/3} z(\ln z^{-1})^{1/3} \qquad \text{(III.18)}$$

while the solution at $z=1$ is

$$W \sim -\left(\frac{3NRa^2p}{C}\right)^{1/3} Z(\ln Z^{-1})^{1/3} \qquad \text{(III.19)}$$

where $Z=1-z$. From (III.18) and (III.19) the advective terms are dominant in the mainstream if $NRa^2 \gg 9p^2/C^2$. This is possible for small, moderate, or large Prandtl numbers. In the thermal layers let $z=\epsilon\eta$ so that

W becomes

$$W = \pm A^{1/3}\epsilon\left(\eta - \frac{\eta \ln \eta}{3 \ln \epsilon^{-1}} + \cdots\right) \quad \text{(III.20)}$$

where $A = (3NRa^2p \ln \epsilon^{-1}/C)$. Thus in the thermal layer let W be

$$W = A^{1/3}\epsilon\left[f - \frac{g}{\ln \epsilon^{-1}} + \cdots\right] \quad \text{(III.21)}$$

Substituting into (II.10) and neglecting terms in $\epsilon^2 a^2$ results in

$$D^6 f - \frac{\epsilon^2 A^{1/3} C}{p} D^2 L + \epsilon^6 NRa^2 f = \epsilon^4 A^{2/3} f^2\left(D^4 f - \frac{\epsilon^2 A^{1/3} C}{p} L\right)$$
$$+ C\epsilon^2 A^{1/3} Df\left(D^4 f - \frac{\epsilon^2 A^{1/3} C}{p} L\right)$$
$$+ 2C\epsilon^2 A^{1/3} fD\left(D^4 f - \frac{\epsilon^2 A^{1/3} C}{p} L\right) \quad \text{(III.22)}$$

This equation applies to the thermal layers at $z=0$ and $z=1$. There are now two cases to consider. Either

$$(a) \quad \epsilon^2 A^{1/3} = 1, \qquad \text{i.e.,} \qquad 3\epsilon^6 NRa^2 \ln \epsilon^{-1} = C/p \quad \text{(III.23)}$$

or

$$(b) \quad \frac{\epsilon^2 A^{1/3} C}{p} = 1, \qquad \text{i.e.,} \qquad 3\epsilon^6 NRa^2 \ln \epsilon^{-1} = \left(\frac{p}{C}\right)^2 \quad \text{(III.24)}$$

1. *Case* (a)

For this case to exist we must have $NRa^2p/C \gg 1$. Equation (III.22) now becomes

$$D^2\left(D^4 f - \frac{C}{p} L\right) + \frac{Cf}{3p \ln \epsilon^{-1}} = f^2\left(D^4 f - \frac{C}{p} L\right) + Df\left(D^4 f - \frac{C}{p} L\right)$$
$$+ 2fD\left(D^4 f - \frac{C}{p} L\right) \quad \text{(III.25)}$$

where $L = fD^3f + 2DfD^2f$. The equation for f is

$$D^2\left(D^4f - \frac{C}{p}L\right) = f^2\left(D^4f - \frac{C}{p}L\right) + Df\left(D^4f - \frac{C}{p}L\right) + 2fD\left(D^4f - \frac{C}{p}L\right) \tag{III.26}$$

with $f = D^2f = D^4f = 0$ at $\eta = 0$ and $f \to \eta$ as $\eta \to \infty$.

Equation (III.26) is satisfied by the solution of

$$D^4f - \frac{C}{p}(fD^3f + 2DfD^2f) = 0 \tag{III.27}$$

with $f = D^2f = D^4f = 0$ at $\eta = 0$ and $f \to \eta$ as $\eta \to \infty$. Since (III.27) has a singular solution $f = -(2.4p/C)/(\eta - \eta_0)$, the unique solution is $f = \eta$. The equation for g is then

$$D^2U - 2\eta DU - (1 + \eta^2)U = \frac{C\eta}{3p} \tag{III.28}$$

with $g = D^2g = D^4g = 0$ at $\eta = 0$, and $g \to (\eta/3)\ln\eta$ as $\eta \to \infty$ and where $U = D^4g - (C/p)(\eta D^3g + 2D^2g)$. $U = 0$ at $\eta = 0$ and $U \to 0$ as $\eta \to \infty$. This equation can be solved in terms of Bessel functions as in Section (III.A. Substituting (III.21) into (II.12) gives

$$\epsilon^5 Ra^2 \ln\epsilon^{-1} = 2\int_0^\infty \left(\eta U + \frac{C}{3p}\right) d\eta \tag{III.29}$$

Thus we have

$$\frac{\epsilon^5 p Ra^2 \ln\epsilon^{-1}}{C} \sim O(1) \tag{III.30}$$

i.e.,

$$\epsilon \sim O\left\{\left[\left(\frac{pRa^2}{C}\right)\ln\left(\frac{pRa^2}{C}\right)\right]^{-1/5}\right\} \tag{III.31}$$

and from (III.23)

$$N \sim O\left\{\left[\left(\frac{pRa^2}{C}\right)\ln\left(\frac{pRa^2}{C}\right)\right]^{1/5}\right\} \tag{III.32}$$

Since $NRa^2 \gg O(p^2/C^2)$, we must have $Ra^2(\ln pRa^2/C)^{1/6} \gg O[(p/C)^{3/2}]$.

If p is small and $pRa^2 \sim O(1)$, from (III.31) there is no boundary layer as was mentioned in Section II.

2. *Case* (b)

From (III.24) we must now have $NRa^2 \gg p^2/C^2$. Equation (III.22) reduces to

$$D^2(D^4f - L) + \frac{p^2 f}{3C^2 \ln \epsilon^{-1}} = \frac{p^2}{C^2} f^2 (D^4 f - L) + pDf(D^4 f - L) + 2pfD(D^4 f - L) \quad \text{(III.33)}$$

The equation for f is satisfied if we let

$$D^4 f - (fD^3 f + 2DfD^2 f) = 0 \quad \text{(III.34)}$$

with $f = D^2 f = D^4 f = 0$ at $\eta = 0$ and $f \to \eta$ as $\eta \to \infty$. The unique solution is $f = \eta$. The equation for g is then

$$D^2 U - 2p\eta DU - \left[p + \frac{p^2}{C^2}\eta^2 \right] U = \frac{p^2}{3C^2}\eta \quad \text{(III.35)}$$

with $g = D^2 g = D^4 g = 0$ at $\eta = 0$, and $g \to \eta \ln \eta / 3$ as $\eta \to \infty$ and where $U = D^4 g - (\eta D^3 g + 2D^2 g)$. $U = 0$ at $\eta = 0$ and $U \to 0$ as $\eta \to \infty$. Letting $\eta = \zeta / p^{1/2}$, (III.35) becomes

$$D^2 U - 2\zeta DU - (1 + 6\zeta^2) U = 2p^{1/2}\zeta \quad \text{(III.36)}$$

with $U = 0$ at $\zeta = 0$ and $U \to 0$ as $\zeta \to \infty$. This equation can again be solved in terms of Bessel functions. The solution is

$$\int_0^\infty \left[\frac{1}{\left(3\sqrt{2}\,(7)^{1/4} x^{1/2}\right)} - \sqrt{2}\,(7)^{-5/4} x^{1/4} \exp\left(\frac{x}{\sqrt{7}}\right) \times \left\{ K_{1/4}(x) \int_0^x x^{1/4} \exp\left(\frac{-x}{\sqrt{7}}\right) I_{1/4}(x)\, dx + I_{1/4}(x) \int_x^\infty x^{1/4} \exp\left(\frac{-x}{\sqrt{7}}\right) K_{1/4}(x)\, dx \right\} \right] dx = U/p^{1/2}$$

where $x=\sqrt{7}\,\zeta^2/2$. Substituting into (2.12) gives

$$\frac{\epsilon^5 C^2 Ra^2 \ln\epsilon^{-1}}{p} = 2\int_0^\infty \left(\zeta U + \frac{p^{1/2}}{3}\right) d\zeta \tag{III.37}$$

Thus

$$\epsilon^5 Ra^2 \ln\epsilon^{-1} \sim O\left(\frac{p^{3/2}}{C^2}\right) \tag{III.38}$$

i.e.,

$$\epsilon \sim 1.543\left[\left(\frac{Ra^2}{p^{3/2}}\right)\ln\left(\frac{Ra^2}{p^{3/2}}\right)\right]^{-1/5} \tag{III.39}$$

and from (3.24)

$$N \sim 2.222\left[(pRa^2)\ln\left(\frac{Ra^2}{p^{3/2}}\right)\right]^{1/5} \tag{III.40}$$

For p small and $pRa^2 \sim O(1)$ the thermal layers exist in this case and N decreases as p increases. When the thermal thickness ϵ is the same in both cases (a) and (b), from (III.23) and (III.24) we have $(NRa^2)_b = (p/C)^3 (NRa^2)_a$. Thus for p moderate or large, case (b) occurs at a larger Rayleigh number than (a), while the reverse is true for small p. Also for p large, from (III.39) $Ra^2 > p^{3/2}$.

References

1. P. H. Roberts, "On Non-linear Bénard Convection," in *Non-equilibrium Thermodynamics, Variational Techniques, and Stability*, R. J. Donnelly, R. Herman, and I. Prigogine, Eds., University of Chicago Press, Chicago, 1966.
2. P. Glansdorff and I. Prigogine, *Physica*, **30**, 351 (1964).
3. D. R. Moore and N. O. Weiss, "Two-Dimensional Rayleigh-Bénard Convection," *J. Fluid Mech.*, **58**, 289 (1973).
4. D. O. Gough, E. A. Spiegel, and J. Toomre, *Model Equations for Turbulent Convection*, in press, 1973.
5. G. M. Murphy, *Ordinary Differential Equations and Their Solutions*, Van Nostrand Reinhold, Princeton, 1960.

LIGHT SCATTERING FROM NONEQUILIBRIUM FLUID SYSTEMS

JEAN-PIERRE BOON

Faculté des Sciences,
Université Libre de Bruxelles,
Bruxelles, Belgium

CONTENTS

Abstract

Instability phenomena have been studied in a wide range of situations and one of the prime objects of hydrodynamic stability theory has been the analysis of the metastable marginal states characterizing transitions from the stability regime to an instability domain. It is only recently, however, that attention has been focused on the processes of initiation of nonequilibrium instabilities. Because instabilities are triggered by thermal fluctuations, light scattering which probes such fluctuations appears as an appropriate tool to investigate pretransitional phenomena as well as the transition region. Two cases of fluid instabilities, namely turbulence and convective instability, are discussed in the present paper along these lines.

I. INTRODUCTION

Consider a fluid system subject to an increasing external field which for a critical value drives the system into a state of instability. The transition from the stability regime to the instability region occurs when the external force reaches a critical value at which internal dissipation processes are no longer sufficient to sustain the potentially unstable state created by the driving force. At the critical point, one (or several) fluctuation(s), instead of decaying through internal dissipation, is (are) amplified, leading to a macroscopic readjustment of the whole system. At this point the fluid passes from its initial steady state to the instability regime, the transition being characterized by the existence of a state of marginal stability. The two examples of fluid instabilities discussed herein are (1) turbulence and

(2) convective instability (known in classical physics as the Bénard problem).

The field of light scattering from nonequilibrium fluid systems is presently still at a very early stage, but may prove to be quite promising in the future. Indeed since instabilities are triggered by thermal fluctuations, light-scattering spectroscopy, which probes these fluctuations, should appear as an appropriate method to investigate the onset of the instability regime. Furthermore an interesting analogy arises from the observation that such transitions seem to bear striking resemblance to structural phase transitions, which are triggered by "soft modes."

Theoretically, a simple analytical treatment can be given to study the onset of the convective instability, whereas the spectral distribution of the light scattered from a turbulent fluid is far less simple. From the experimental viewpoint, one faces almost the opposite situation: although scarce, experimental results are available for the case of turbulence, whereas the first experimental study of the Bénard instability by light scattering has been performed only very recently, and the investigation of the *onset* of the Bénard instability is still in progress at present.

Nevertheless, one should not underestimate the approach to fluid instabilities from the study of their spectral distribution characteristics as this domain may appear as a very interesting one for several reasons: Light scattering,* contrary to classical techniques, provides means of investigation of instabilities without perturbing the fluid whose properties are to be measured; the modifications of the spectral distribution during the onset of the instability and at the transition should reveal valuable information on the dynamics governing these phenomena; interesting analogies with phase transitions could shed new light on the mechanism of instability phenomena in fluid systems.

II. TURBULENCE

From experimental observation, it is well known that at sufficiently low velocities, a fluid exhibits smooth streamline flow, which characterizes the *laminar regime*: the fluid is steady and remains in the *stability region* as long as small changes of the external conditions induce no qualitative modification in the flow pattern. On the other hand, at sufficiently high velocities, the flow is chaotic and/or irregular, with rapid changes in time even under constant external conditions: here the fluid exhibits *turbulence* and is in the *instability regime*. The regime transition, from laminar to

*For a review of laser light-scattering spectroscopy in fluid systems the reader is referred to the recent chapter by Fleury and Boon[1] in this series.

turbulent, occurs when the dimensionless quantity $R_E = ul/\nu$, called the Reynolds number (u = flow velocity; l = diameter of the tube; ν = kinematic viscosity) reaches a critical value R_E^C, depending on the geometry of the system and on the flow configuration. Because of its complexity, turbulence constitutes "a complicated phenomenon which has remained one of the last mysteries of classical physics."[2]

Two different cases should be distinguished and are reviewed. Either, light scattering is used as a probe of the thermal fluctuations in the fluid which are modified by the presence of an external force, or one analyzes the light scattered from scattering centers (e.g., small polystyrene spheres) suspended in the moving fluid to enhance the scattered signal. Only the latter technique has been used in the laboratory to date for the study of liquid flows. The first problem investigated by laser light scattering in this way was localized laminar flow.[3] The light scattered by a particle traveling with the fluid velocity, **u**, is Doppler shifted as

$$\omega = \mathbf{k}\cdot\mathbf{u} = ku\cos\alpha$$

$$\cong 2k_0 u \cos\alpha \sin\theta/2 \tag{1}$$

where θ is the scattering angle and α is the angle between the wave vector (**k**) and the direction of the flow (**u**). From (1) the velocity profile can be determined by measuring the frequency shift at different radial positions from the center of the sample tube. In addition, the method provides a technique for measuring very slow fluid velocities ($\leqslant 5\times10^{-2}$ cm sec^{-1}).

A similar technique has been used to investigate the turbulent regime.[4] In this regime, not only does the frequency shift increase with increasing flow velocity (i.e., for larger Reynolds number) but also the spectral linewidth exhibits interesting features. In the laminar regime the linewidth (in addition to the width due to Browian diffusion) is mainly due to the finite angular width, $\Delta\theta$, of the light beam, that is, (considering the case where **k** is parallel to **u**; $\alpha = 0$)

$$\Delta\omega = k_0 u \cos\theta\,\Delta\theta \tag{2}$$

which is relatively small as long as the fluid remains in the steady state ($R_E < R_E^C$). Now considering the ratio $\Delta\omega/\bar{\omega}$, where $\bar{\omega}$, taken at the smoothed maximum of the spectral line, corresponds to the axial velocity, one observes a dramatic broadening of the shifted line at the regime transition (see Fig. 1). More recent experiments[5,6] using statistical spectroscopy as well as intensity spectroscopy have confirmed the nearly Gaussian character of the velocity distribution and provide information on

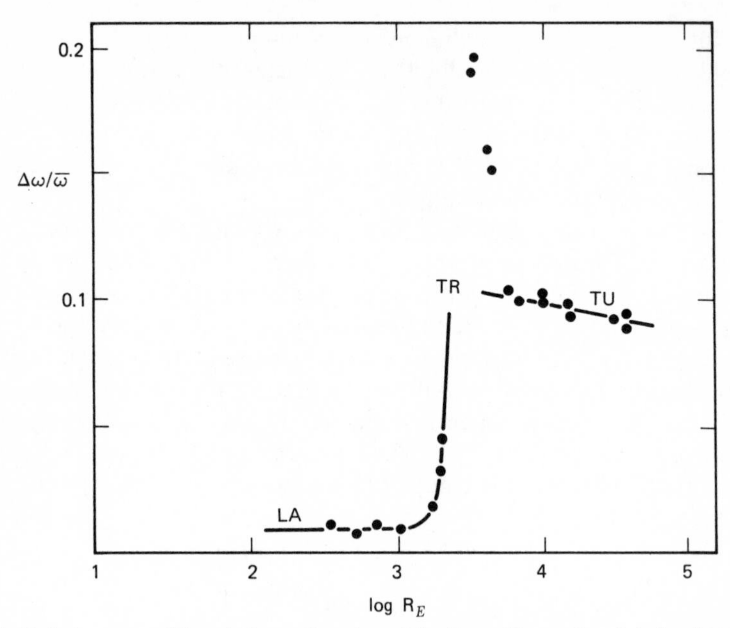

Fig. 1. Relative spectral linewidth as a function of the Reynolds number. LA: laminar regime ($R_E < R_E^c$); TU: turbulent regime ($R_E > R_E^c$); RR: transition region of fluctuating flow. (After Goldstein and Hagen, 1967.)

the spatial coherence of turbulent systems.

Theoretical investigations of light scattering by particles suspended in a turbulent fluid have been attempted. For instance Bertolotti et al.[7] have calculated the spectral intensity under the hypothesis that the particles follow the motion of the fluid exactly. A decomposition of the fluid velocity into its mean and fluctuating parts, $\mathbf{u}_0$ and $\delta\mathbf{u}(\mathbf{r},t)$, respectively, yields the formal result

$$I(\mathbf{k},\omega) = A(\mathbf{k}) \int_0^\infty dt \exp[-it(\omega-\omega_0+\mathbf{k}\cdot\mathbf{u}_0)]$$

$$\times \sum_{i,j} \left\langle \exp\left\{ -i\mathbf{k}\cdot\left[\mathbf{r}_j(0) - \mathbf{r}_i(0) - \int_0^t d\tau\, \delta\mathbf{u}(\mathbf{r}_j,\tau) \right]\right\}\right\rangle \tag{3}$$

where $A(\mathbf{k})$ is the amplitude factor and the brackets denote an ensemble average over the turbulent state. Equation (3) indicates indeed that the Doppler shift associated with the drift motion is $\mathbf{k}\cdot\mathbf{u}_0$. However no theoretical treatment, to date, has led to an explicit calculation of the spectral linewidth to explain its sudden increase at the critical Reynolds number.

A formal study of turbulence by spectral fine structure of scattered light has been performed by Frisch[8] considering the fluctuations in the fluid itself without injected scattering centers. The extra increment of relative scattered light intensity, $\Delta I(\mathbf{k},\omega)$, from a sample of turbulently moving fluid, over that exhibited in the absence of this motion is simply obtained from the correlation function of the dielectric fluctuations resulting from turbulence, since it may be assumed that there are no cross terms between the thermal fluctuations and the turbulent fluctuations. With the assumption that the fluid is incompressible, and disregarding self-heating in flow and heat flow by conduction (i.e., neglecting $\partial\varepsilon/\partial T$), only the contribution from pressure fluctuations need be retained to yield

$$\Delta I(\mathbf{k},\omega)\propto\left(\frac{\partial\epsilon}{\partial p}\right)_S^2\langle\delta p(-\mathbf{k})\,\delta p(\mathbf{k},\omega)\rangle \tag{4}$$

where δp is the turbulent pressure fluctuation and the brackets denote some "appropriate" ensemble average. Using the Navier-Stokes equation for a turbulent fluid approximating isotropic turbulence, (4) may be rewritten as

$$\Delta I(\mathbf{k},\omega)\propto\left(\frac{\partial\epsilon}{\partial p}\right)_S^2\frac{k_ik_jk_lk_m}{k^4}\langle\delta\pi_{ij}(-\mathbf{k})\delta\pi_{lm}(\mathbf{k},\omega)\rangle \tag{5}$$

where $\delta\pi$ is the fluctuation in the Reynolds stress tensor. Frisch also predicts that the relative width $\Delta\omega/\overline{\omega}$ is of order u_0c^{-1} (where c is the velocity of light) and he anticipates that "light scattering experiments could shed considerable light on the small scale of turbulence and its decay rate." Montroll[2] also gives a Fourier analysis of the Navier-Stokes equation for a turbulent fluid and he constructs a schematic mechanism for energy transfer from small to large wave numbers: the cascade process occurring in the turbulent regime starts with unstable large eddies ($\sim$ small k) developed by some driving process and decaying into smaller and smaller eddies ($\sim$ large k), terminating in this way the energy transfer cascade through internal dissipation processes. Montroll's theory connects with a recent tentative picture proposed by Martin[9] to analyze the vorticity correlations in a turbulent medium in analogy with the order parameter fluctuations at a phase transition.

A bibliographical survey of the subject can be found in a recent paper by Crosignani and Di Porto.[10]

III. THE BÉNARD PROBLEM

The phenomenon of convective instability occurs in a fluid layer heated from below (provided that α, the thermal expansion coefficient of the fluid, is positive) when the temperature gradient reaches a critical value. Indeed when a fluid layer is heated from below, the system undergoes a top-heavy arrangement which is potentially unstable. Therefore the fluid has a tendency to redistribute itself, which is not possible as long as internal dissipation is able to sustain the temperature gradient. As a consequence the temperature gradient must exceed a certain value before the instability can manifest itself. When the temperature gradient reaches this critical value (depending on the properties of the fluid and on the thickness of the layer) convection arises which permits the fluid to adjust. Convection however begins in a peculiar way, in that a stationary instability occurs, which results in an arrangement of convection cells or convection rolls, first observed by Bénard in 1900. When one of these spectacular patterns is observed, the fluid is in a marginal state (known as the Bénard instability) which is quite well described from the point of view of the macroscopic theory.[11] A number of experimental studies performed essentially by classical thermal conductivity methods have been reported on the stability criteria and on the determination of the parameters characterizing the transition between the stability region and the instability domain (for a bibliography see Ref. 12). However it is only very recently that the accurate determination of the convective velocity field in the Bénard instability was undertaken for the first time.[13] The local velocity of the convective fluid was measured with a laser velocimeter by means of the real fringes method. The separation between the parallel horizontal fringes produced by the intersection of two laser beams at the mid-height of the fluid layer is given by $\Delta l=\lambda_0[2n(1-\cos^2\frac{1}{2}\varphi)^{1/2}]^{-1}$, where λ_0 is the wavelength of the incident light; n is the refractive index of the fluid, and φ is the angle between the incident beams. A particle (suspended in the fluid and moving with vertical velocity v_z) passing through the fringes lattice scatters the light which is then modulated with a period $\Delta l/v_z$. Therefore the spectral analysis of the Doppler scattered intensity exhibits a peak whose position is given by $\Delta\nu=v_z/\Delta l=2nv_z(1-\cos^2\frac{1}{2}\varphi)^{1/2}\lambda_0^{-1}$. Hence measuring the frequency shift $\Delta\nu$ provides a direct measure of the convective velocity field. In addition with this method one can observe the spatial periodicity of the structure of the convection pattern by a continuous displacement of the relative position of the cell with respect to the intersection point of the incident beams. When the light-scattering experiment is repeated for different values of the temperature gradient, ΔT, above its critical value ΔT^c, the power law: $v_z \propto (\Delta T-\Delta T^c)^\gamma$ can be checked. The classical theory[14] predicts $\gamma=\frac{1}{2}$; the best fit to the experimental data[13] yields $\gamma=0.50\pm0.02$ in the range $1.08<\Delta T/\Delta T^c<5.2$ (Fig. 2).*

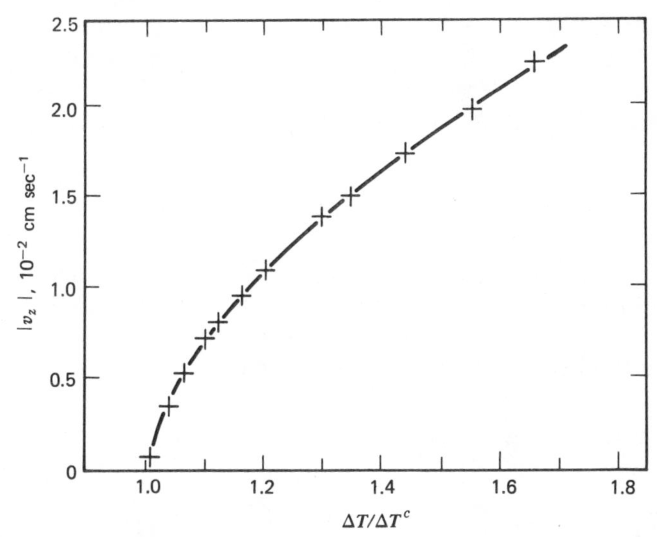

Fig. 2. The convective velocity as a function of the temperature gradient. (After Bergé and Dubois, 1974.)

We now turn to the second problem encountered in the study of the convective instability, one which concerns the domain $0<\Delta T/\Delta T^c\leqslant 1$, that is, the pretransitional phenomena region below the instability critical point. Here it appears that the normal mode analysis used in the classical macroscopic treatment is not appropriate to investigate the mechanism governing the onset of the instability. The question indeed arises as to the dynamics of the fluctuations in a fluid which experiences an increasing vertical linear temperature gradient, $\nabla T^s=-\beta\hat{\mathbf{n}}, \hat{\mathbf{n}}=(0,0,1)$. Here the superscript s refers to the steady state characterized by the equations:

$$T^s=T_0-\beta z \tag{6}$$

$$\nabla p^s=-g\rho^s\hat{\mathbf{n}} \tag{7}$$

Note added in proof*: The Taylor instability in a rotating fluid was recently studied also by means of laser light spectroscopy [J. P. Gollub and M. H. Freilich, *Phys. Rev. Lett.*, **33, 1465 (1974)].

where g is the gravitational constant, and the subscript 0 denotes the reference state ($z=0$). Boon and Deguent[15] have presented a spectral analysis of the problem and suggest the use of light scattering to probe the onset of the convective instability. A complete treatment of the spectral analysis of the problem has been given subsequently by Boon and Lekkerkerker.[16] In the present case, there is isotropy breaking due to the vectorial nature of the temperature gradient and of the gravitational force. As a consequence of the particular symmetry of the system, it is appropriate to choose the following set of variables to describe the velocity field:

$$\phi = \operatorname{div} v; \qquad \psi = (\operatorname{curl} v)_z; \qquad \xi = (\operatorname{curl} \operatorname{curl} v)_z$$

ϕ and ξ are scalars which therefore can couple to the thermodynamic variables, δs and δp, whereas ψ does not as it is a pseudoscalar and may therefore be discarded here. Then the starting set of hydrodynamic equations for a fluid subject to an adverse linear temperature gradient reads

$$\begin{aligned}
\partial_t \delta\rho(\mathbf{r},t) &= -\rho_0 \phi(\mathbf{r},t) \\
\partial_t \phi(\mathbf{r},t) &= -\rho_0^{-1}\nabla^2 \delta p(\mathbf{r},t) + (\nu+\nu')\nabla^2\phi(\mathbf{r},t) - \frac{g}{\rho_0}\partial_z \delta\rho(\mathbf{r},t) \\
\partial_t \xi(\mathbf{r},t) &= \nu\nabla^2\xi(\mathbf{r},t) + \frac{g}{\rho_0}(\partial_x^2 + \partial_y^2)\delta\rho(\mathbf{r},t) \\
\partial_t \delta s(\mathbf{r},t) &= \frac{\lambda}{\rho_0 T_0}\nabla^2 \delta T(\mathbf{r},t) + \beta C_p T_0^{-1} v_z
\end{aligned} \tag{8}$$

where $\lambda = \kappa(\rho_0 C_p)^{-1}$; $\nu = \eta\rho_0^{-1}$; $\nu' = (\eta/3 + \zeta)\rho_0^{-1}$; and $\nabla^2 v_z = \partial_z\phi - \xi$.

In addition, from first-order perturbation theory as applied to the equation of state of the system

$$\begin{aligned}
\delta\rho &= -\frac{\alpha T_0 \rho_0}{C_p}\delta s + \rho_0 X_s \delta p \\
\delta T &= \frac{T_0}{C_p}\delta s + \frac{\alpha T_0}{\rho_0 C_p}\delta p
\end{aligned} \tag{9}$$

Equations (9) are Laplace-Fourier transformed to yield the dispersion equation which is then solved by perturbation theory. Here the perturbation terms are those describing the effect of the gravitational force (i.e., the terms containing g) and those which arise because of the temperature gradient (i.e., the terms containing β). The use of a perturbation procedure is justified here as one seeks to obtain information about the *onset* of the instability, that is, about the behavior of the system as departing from the state of equilibrium. It is found that the hydrodynamic matrix exhibits four modes. The two propagating (Brillouin) modes are independent of the other modes (at least within the usual limit of validity of a calculation to order k^2) and—within the same limit—are not affected by the external force. We shall therefore ignore these modes in the remainder. The two nonpropagating modes read

$$s_{\pm} = -\frac{k^2}{2}\left\{(\nu+\lambda)\pm\left[(\nu-\lambda)^2+\frac{4\nu\lambda R_A}{R_A^C(k)}\right]^{1/2}\right\} \tag{10}$$

where R_A is the Rayleigh number (the dimensionless temperature gradient) and $R_A^C(k)$, its critical value (determined by the boundary conditions), with

$$\frac{R_A}{R_A^C(k)} = \frac{\alpha\beta g}{\nu\lambda}\,\frac{k^2-k_z^2}{k^6} \tag{11}$$

the subscript z indicating the vertical component of the wave vector. These modes are coupled in such a way that (as for most liquids $\nu > \lambda$) the thermal diffusivity mode, s_- behaves like a "soft mode" (i.e., exhibits narrowing) simultaneously with the modification (i.e., broadening) of the "vorticity mode" s_+ when approaching the instability critical point (i.e., when R_A increases from 0 towards R_A^C). The above set of equations is then easily solved for $\widetilde{\delta s}(k,s)$ (the Laplace-Fourier transform of δs) from which the appropriate correlation functions are constructed to yield the spectral distribution. Restriction being made to the central components (for the obvious reasons expressed above) one obtains for the normalized light-scattering spectrum

$$I(k,\omega) = \left(\frac{\partial\varepsilon}{\partial s}\right)_p^2 \langle|\delta s(k)|^2\rangle\left(A_-\frac{|s_-|}{\omega^2+s_-^2}+A_+\frac{|s_+|}{\omega^2+s_+^2}\right) \tag{12}$$

with

$$A_- = 1-A_+ = (\nu k^2-s_-)^2\left[(\nu k^2-s_-)^2+\nu\lambda k^4\frac{R_A}{R_A^C(k)}\right]^{-1} \tag{13}$$

Note that the vorticity component does not exist in the absence of the external force; indeed when R_A (or ΔT^s)=0, $A_+=0$ and $A_-=1$, and in (12), the second term vanishes, whereas the first term takes the usual value $\lambda k^2[\omega^2+(\lambda k^2)^2]^{-1}$ as for a fluid at equilibrium. It is thus predicted that the presence of an external force induces an additional central peak. The spectrum is illustrated in Fig. 3*a*. Figure 3*b* shows the behavior of the characteristic correlation times (i.e., the reciprocal linewidth of the central components) when the temperature gradient increases from zero to its critical value ΔT^C.

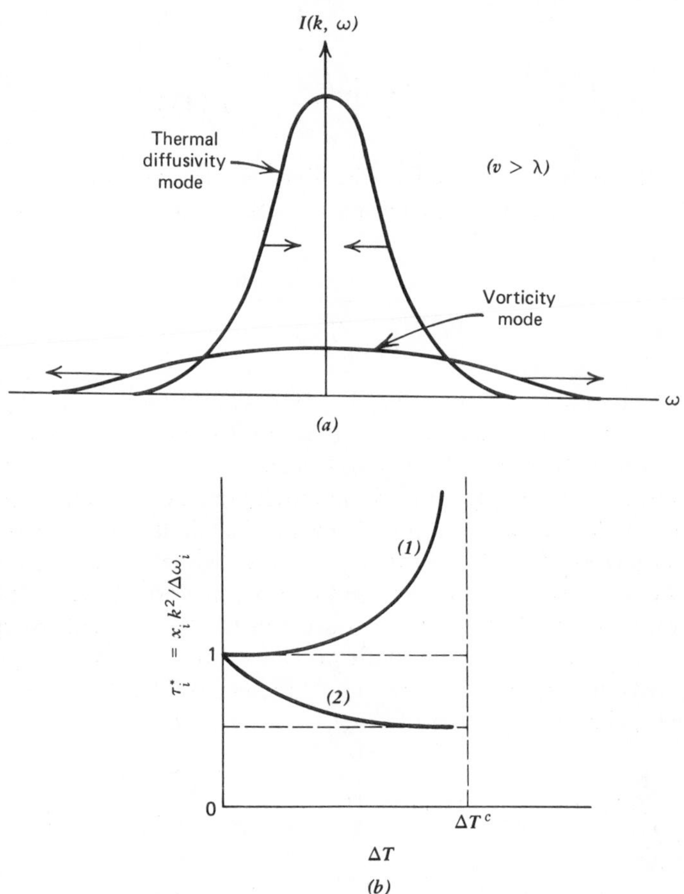

Fig. 3. Illustration of the light-scattering spectrum according to (13); the arrows indicate the modification of the peaks when R_A increases (*a*). Representation of the behavior of the corresponding correlation times as a function of the temperature gradient (*b*): (1) $i=1$, $x_1=\lambda$; (2) $i=2$, $x_2=\nu$.

As a particular example, let us consider the case of a usual liquid around 300°K (e.g., water or toluene) for which the thermal diffusivity is sufficiently small as compared to the shear viscosity so that to a first approximation only the mode s_- is to be retained (since $\lambda/\nu \ll 1$). Then the half-width of the spectrum is given by

$$\Delta\omega_- = \lambda k^2 \left[1 - \frac{R_A}{R_A^C(k)} \right] \tag{14}$$

This result is most interesting in that it predicts a dramatic narrowing of the spectral line when approaching the instability critical point ($\Delta\omega_- = \lambda k^2 \rightarrow 0$ when $R_A / R_A^C = 0 \rightarrow 1$), indicating in this way a "soft mode" behavior of the type encountered in structural phase transitions. Although the wavelength of the characteristic instability soft mode is large (of the order of the vertical dimension of the system), the presently available techniques used in light scattering (in particular statistical spectroscopy) should allow the probing of modes which are expected to be effected by the external gradient. The above theory has been extended by Boon,[17] to the case of binary systems: in the limit of dilute solutions the diffusion peak (concentration fluctuations mode) also exhibits narrowing when approaching the convective instability critical point, but the effect is much weaker than for the thermal diffusivity peak (entropy fluctuations mode). Unless the instability soft mode itself can be probed directly, the effect observed from the other proper modes (with larger k, hence easier to probe) is expected to be much smaller. Because the smallness of the effect is partly connected with the value of the thermal conductivity of the fluid, Boon suggests a possible way around this obstacle by considering a Brownian system (to which case the theory of dilute systems is applicable). In addition to the fact that the scattering from suspended particles is much more intense than the scattering from thermal fluctuations (see section on turbulence), utilizing a Brownian system would have the effect of replacing ν and λ by the diffusion constant, D, of the Brownian particles, which is smaller by several orders of magnitude. If for a Brownian system there exists a corresponding Rayleigh number, (11), where $\nu\lambda$ appearing in the denominator should be replaced by D^2, then the effect would be significantly larger even for large wave-number modes.

The problem of fluctuations near convection threshold has also been analyzed by Zaitsev and Shliomis[18] on the basis of the hydrodynamic fluctuation theory by the introduction of the concept of fluctuation velocities and temperatures excited by random forces. As a consequence the approach of these authors differs from the treatment of Boon and coworkers by the presence of injected random terms in the starting set of hydrodynamic equations. In this way the Russian workers are able to treat the problem for a system with arbitrary geometry and rigid boundaries at the cost of mathematical complexity, whereas the theory presented above has the advantage of analytical simplicity with the restriction of the hypothesis that (8) is Fourier transformable in the variable z. Although the analysis by Zaitsev and Shliomis does not lead to an explicit evaluation of the spectral distribution, most important is the conclusion that both approaches confirm the belief that a divergence of the characteristic fluctuations power spectrum should occur at the convection threshold. This prediction which follows from linear hydrodynamic theory has recently been confirmed by Kawasaki[19] who has shown that for a fluid layer subject to symmetric boundary conditions, nonlinearities have a negligible effect upon the dynamics of fluctuations even in the vicinity of the transition critical point.

IV. GENERAL COMMENTS

The bracket used in the analysis of nonequilibrium systems to define the correlation functions for the evaluation of the scattering function denotes some "appropriate" ensemble average. Although the procedure is strictly valid only for systems at thermodynamic equilibrium, the extension of the averaging procedure should be applicable, since local thermodynamic equilibrium is assumed. However, when deviations from equilibrium are no longer small, such procedure may have to be revised, as in the immediate vicinity of the convection threshold, where the experimental study would indeed appear to be very helpful.

Finally let us mention that an analysis similar to the treatment used by Boon and coworkers, has recently been presented by Deutch et al.[20] to investigate chemical oscillations and dissipative structures by studying the spectrum of light scattered from a system of coupled nonlinear chemical reactions for steady states far from equilibrium. When compared to the spectrum obtained from reacting systems at equilibrium, new spectral features are qualitatively predicted which consist mainly of splittings of the chemical lines and dispersive (non-Lorentzian) components.

It is also important to notice that the analogy between the phenomena considered herein and phase transition phenomena is essentially phenomenological to the extent that the soft mode picture provides a satisfactory description in both cases. The analogy should be understood as a translation from the molecular description language used for phase transitions to the hydrodynamic mode jargon used for instability phenomena. On the other hand, it is worthwhile to mention that at the level of microscopic analysis there is presently no evidence for an actual analogy between hydrodynamic or thermal instabilities and phase transitions for the very reason that the microscopic mechanism governing the evolution of a system towards an instability point remains at present a totally open question.

Acknowledgments

This research was supported by the Fonds National de la Recherche Scientifique (FNRS), Belgium.

References

1. P. A. Fleury and J. P. Boon, in *Advances in Chemical Physics*, **24**; I. Prigogine and S. A. Rice, Eds., Interscience, New York, 1973.
2. E. W. Montroll, in *Contemporary Physics*, **1**, p. 273 (1969).
3. Y. Yeh and H. Z. Cummins, *Appl. Phys. Lett.*, **4**, 176 (1964).
4. R. J. Goldstein and W. F. Hagen, *Phys. Fluids*, **10**, 1350 (1967).
5. E. R. Pike, D. A. Jackson, P. J. Bourke, and D. I. Page, *J. Phys.* (*E*), **1**, 727 (1968).
6. P. J. Bourke et al., *Phys. Lett.*, **28A**, 692 (1969).
7. M. Bertolotti, B. Crosignani, P. Di Porto, and D. Sette, *J. Phys.* (*A*), **2**, 126 (1969).
8. H. L. Frisch, *Phys. Rev. Lett.* **19**, 1278 (1967).
9. P. C. Martin, unpublished manuscript entitled "A Tentative Picture for Fully Developed Turbulence," 1972.
10. B. Crosignani and P. Di Porto, in *Photon Correlation and Light Beating Spectroscopy*, H. Z. Cummins and R. Pike, Eds., Plenum Press, New York, 1974.
11. S. Chandrasekhar, *Hydrodynamic and Hydromagnetic Stability*, Clarendon Press, Oxford, 1961, Chapt. 2.
12. M. G. Velarde, in *Hydrodynamics*, *Les Houches* 1973, R. Balian, Ed., Gordon and Breach, New York, 1975.
13. P. Bergé and M. Dubois, *Phys. Rev. Lett.*, **32**, 1041 (1974); M. Dubois and P. Bergé, preprint "Spatial and Critical Dependence of the Convective Velocity in the Rayleigh-Bénard Problem," 1975.
14. L. D. Landau (1944), in *Fluid Mechanics*, L. D. Landau and E. M. Lifshitz, Eds., Addison Wesley, New York, 1958, Chapt. 3.
15. J. P. Boon and P. Deguent, *Phys. Lett.*, **39A**, 315 (1972).
16. J. P. Boon and H. N. W. Lekkerkerker, in *Anharmonic Lattices, Structural Transitions, and Melting*, Universitetsforlaget, Oslo, 1973; H. N. W. Lekkerkerker and J. P. Boon, *Phys. Rev.* A, **10**, 1355 (1974).
17. J. P. Boon, *Phys. Chem. Liquids*, **3**, 157 (1972).
18. V. M. Zaitsev and M. Q. Shliomis, *Soviet Phys. JETP*, **32**, 866 (1971).
19. K. Kawasaki, *J. Phys.* (*A*), **6**, L4 (1973).
20. J. M. Deutch, S. Hudson, P. J. Ortoleva, and J. Ross, *J. Chem. Phys.*, **57**, 4327 (1973).

MAGNETIC FIELDS AND CONVECTION

N. O. WEISS

Department of Applied Mathematics and Theoretical Physics, University of Cambridge, England

Abstract

The nonlinear equations governing convection in the presence of a magnetic field have been integrated numerically for two-dimensional flow in a Boussinesq fluid confined between free boundaries. Small disturbances are described by the linearized equations, which allow overstable solutions. Some nonlinear solutions show finite amplitude oscillations even for monotonically growing linear modes.

CONTENTS

I. INTRODUCTION

It is well known that the principle of the exchange of stabilities is valid for Rayleigh-Bénard convection[1]: instability sets in via a stationary state of marginal stability. For more complicated systems, the marginal state may be oscillatory.[2] Overstable modes appear for thermosolutal convection,[3] in rotating systems and in the presence of a magnetic field. This last problem is particularly important in astrophysics.

The development of linear modes into finite amplitude convection can be studied in laboratory experiments or by computing solutions to the nonlinear equations. Since the parameters relevant for astrophysical plasmas cannot be modeled in the laboratory it is necessary to rely on numerical experiments. In this paper the results of a series of computations on convection in a magnetic field are related to linear theory. These calculations show that monotonically growing instabilities can develop into finite amplitude oscillations.

It is not surprising that linear theory may provide a misleading description of finite-amplitude convection; indeed, some overstable modes can grow into stationary convection.[3] The aim of these numerical experiments is to gain a qualitative understanding of the nonlinear problem which can be applied, for instance, to the structure of penumbral filaments in sunspots.[4]

II. THE MODEL PROBLEM

Let us restrict our attention to two-dimensional convection in a Boussinesq fluid contained between free boundaries, in the presence of an imposed vertical magnetic field. Then

$$\frac{\partial T}{\partial t} = -\nabla \cdot (T\mathbf{u}) + \kappa \nabla^2 T \tag{1}$$

$$\frac{\partial \mathbf{B}}{\partial t} = \nabla \times (\mathbf{u} \times \mathbf{B}) + \eta \nabla^2 \mathbf{B} \tag{2}$$

$$\frac{\partial \boldsymbol{\omega}}{\partial t} = \nabla \times (\mathbf{u} \times \boldsymbol{\omega}) + \nabla \times (\mathbf{j} \times \mathbf{B}) - \alpha \nabla T \times \mathbf{g} + \nu \nabla^2 \boldsymbol{\omega} \tag{3}$$

and

$$\nabla \cdot \mathbf{u} = 0, \qquad \nabla \cdot \mathbf{B} = 0 \tag{4}$$

where $\mathbf{u}$ is the velocity, T the temperature and $\mathbf{B}$ the magnetic field, while the vorticity $\boldsymbol{\omega} = \nabla \times \mathbf{u}$ and the current $\mathbf{j} = \mu^{-1} \nabla \times \mathbf{B}$; κ, η, and ν are thermal, magnetic, and viscous diffusivities, α is the coefficient of thermal expansion, and μ is the permeability of the fluid. We shall assume that the z-axis is vertical and that motion is confined to the xz-plane and independent of the y-coordinate.

These equations can be solved in the region $0 < x < L$, $0 < z < d$ subject to the following boundary conditions:

$$\begin{gathered} T = \Delta T (z=0), \qquad T = 0 (z=d), \qquad \frac{\partial T}{\partial x} = 0 (x = 0, L) \\ B_x = 0 \qquad (x = 0, L; z = 0, d) \\ u_z = \frac{\partial u_x}{\partial z} = 0 (z = 0, d), \qquad u_x = \frac{\partial u_z}{\partial x} = 0 (x = 0, L) \end{gathered} \tag{5}$$

The stress-free condition does not apply to laboratory experiments, though it simplifies the problem and is relevant in astrophysics. The corresponding condition on the magnetic field is that **B** should be vertical at the boundaries. Thus we consider an infinite plane layer of depth d, containing periodic convection rolls each of width L. This configuration can be described by the dimensionless width

$$\lambda = \frac{L}{d} \tag{6}$$

and by four dimensionless parameters, the Rayleigh number

$$R = \frac{g\alpha\Delta T d^3}{\kappa\nu} \tag{7}$$

the Chandrasekhar number

$$Q = \frac{B_0^2 d^2}{\mu\rho\nu\eta} \tag{8}$$

and the Prandtl numbers

$$p = \frac{\nu}{\kappa}, \qquad p_m = \frac{\nu}{\eta} \tag{9}$$

Here $\mathbf{B}_0$ is the initially uniform vertical field, and ρ the density of the fluid. Once convection takes place, its efficacy is measured by the dimensionless Nusselt number

$$N = \frac{Fd}{\kappa\Delta T} \tag{10}$$

where F is the total thermometric flux.

III. LINEAR THEORY

Equations (1) to (3) can be linearized for small disturbances. The normal modes vary as $\exp st$, where the growth rate s satisfies a cubic characteristic equation.[1] Of the three roots one is always real and negative, corresponding to a damped solution; the other two are complex conjugates. For $\kappa < \eta$ ($p > p_m$) instability sets in via a neutral state with $s = 0$ at the critical Rayleigh number R_e. Both R_e and the critical wave number λ^{-1} increase with the Chandrasekhar number Q. For $\kappa > \eta$ ($p < p_m$) overstable modes appear first, at a Rayleigh number R_o,[1,5] which varies with Q as shown in

Fig. 1. As R is increased above R_o the linear modes correspond to oscillations with exponentially increasing amplitude: the two complex conjugate roots have positive real parts.

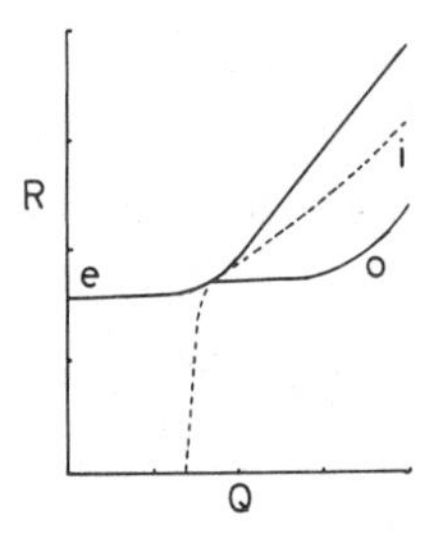

Fig. 1. Normal modes for the linearized equations. The broken line separates oscillatory from monotonic (exponential) solutions. The curves labeled o, i, e correspond to the onset of overstability, the transition from oscillatory to monotonically growing modes and the exchange of stabilities, respectively.

Danielson[4] pointed out that there was a further transition from oscillatory to monotonically growing modes. When $R=R_i$ there are two equal, real, positive roots of the characteristic equation. Finally, when R increases to R_e one of these roots becomes negative; so the exchange of stabilities corresponds to a decrease in the number of unstable modes. The value of R_i can be calculated by imposing the condition that the discriminant of the cubic equation be zero.[6] Danielson suggested that convection would become effective when monotonic modes appeared. Hence the Nusselt number should not increase appreciably above unity for configurations below the curve labeled i in Fig. 1. Analogous results hold for convection in rotating systems when $p<0.68$; experiments with mercury ($p=0.025$) are consistent with the hypothesis that N begins to increase when $R=R_i$.[7,8] Physically, convective transport requires the transfer of fluid from the bottom to the top of the layer, so that hot material from below can lose its energy at the upper boundary. Small oscillations are unable to transfer heat in this way.[9] Moreover, one might expect convection to become important when the potential energy per unit volume gained by a fluid element rising through the layer is equal to its magnetic energy density. This condition also yields a critical Rayleigh number, which is comparable to R_i.

IV. NONLINEAR RESULTS

The nonlinear equations have been integrated numerically, using finite difference methods on a staggered mesh with 24 or 48 intervals over the range $0<z<d$.[10] It is convenient to make a series of runs at a fixed Rayleigh number, decreasing the value of the imposed magnetic field. This corresponds to decreasing Q along a horizontal path in Fig. 1. Such a series was carried out with $R=10^5$, $p=1$, and $p_m=5$ for cells with $\lambda=\frac{1}{2}$. The onset of overstability occurs when $Q=63.7\times10^3$. For higher values of Q initial disturbances decayed. At $Q=60.5\times10^3$ oscillations just appeared and persisted with a small amplitude. For $Q=50.0\times10^3$, 28.1×10^3 small initial disturbances led to oscillations whose amplitude increased exponentially over several cycles until they finally settled down to steady oscillatory convection.

The transition from oscillatory to monotonically growing modes takes place when $Q=21.6\times10^3$. A numerical experiment with $Q=12.5\times10^3$ shows steady exponential growth while the amplitude is small, as predicted by linear theory. But the solution develops finite amplitude oscillations which persist indefinitely. The variation with time of the kinetic energy is shown in Fig. 2.

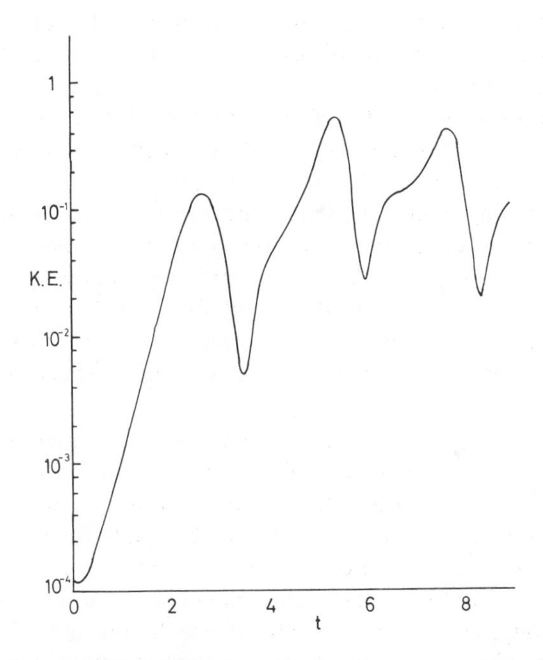

Fig. 2. Development of finite amplitude oscillations from a monotonically growing mode. The curve shows the average kinetic energy density as a function of time for a numerical experiment with $R=10^5$, $Q=12.5\times10^3$, $p=1$, $mp=5$, $\lambda=\frac{1}{2}$. The energy is measured in dimensionless units such that $B_0^2/\mu=0.5$.

This behavior persists beyond the exchange of stabilities at $Q=7.75\times 10^3$. Finite amplitude oscillations remain when $Q=6.13\times 10^3$. For $Q=3.13\times 10^3$ the flow apparently settles down slowly towards a steady state; with $Q=0.5\times 10^3$ steady flow is easily attained and when $Q=0.125\times 10^3$ the magnetic field no longer has any dynamical significance.

Similar results are obtained in a square cell, with $\lambda=1$. Even when $p=p_m=1$ (so that overstability is not predicted by the linear theory) finite amplitude oscillations develop after the onset of convection. Moreover, the Nusselt number increases gradually as Q is decreased below the critical value. Although the transition from oscillatory to monotonic modes takes place as convection is becoming important, it does not correspond to any dramatic increase in the Nusselt number.

V. DISCUSSION

These numerical experiments illustrate the dangers of extrapolating from linear to nonlinear regimes. It is not difficult to understand how finite amplitude oscillations can develop. If η is small the magnetic Reynolds number is high and the lines of force are transported with the flow.[9] The consequent distortion of the field[11] exerts a stabilizing force which tends to reverse the motion. This curvature force is underestimated by the linearized equations, since $|(\mathbf{B}\cdot\nabla)\mathbf{B}|$ becomes much greater than $|(\mathbf{B}_0\cdot\nabla)\mathbf{B}|$ when the field is distorted and enhanced. So oscillatory convection ensues unless the fluid can slip past the lines of force.

It is not surprising, therefore, that finite amplitude convection cells in a highly conducting fluid tend to oscillate. Such oscillations might actually be observed in a sunspot. There the central field allows only overstable modes (which correspond to unstable Alfvén waves[9]) in the umbra. The penumbral filaments are convection rolls aligned by the inclined magnetic field[4]; since convection is barely possible we might expect these filaments to be oscillatory.

There exist various effects which lead to cubic characteristic equations and the possibility of overstable modes: the linear theory of convection in a rotating system or of thermosolutal convection is similar to that outlined here. Once again, nonlinear behavior may be qualitatively different. But the nonlinear effects of the Lorentz force, of the Coriolis force, and of double diffusion are dissimilar. The magnetic field is solenoidal and promotes stability, giving finite amplitude oscillations even for monotonic linear solutions. On the other hand, angular momentum can be redistributed and a stabilizing solute gradient can be confined to boundary layers, allowing subcritical instabilities and permitting oscillations to develop into steady finite amplitude convection.[2,3]

Acknowledgments

I am grateful to Dr. D. R. Moore for his advice and assistance.

References

1. S. Chandrasekhar, *Hydrodynamic and Hydromagnetic Stability*, Clarendon Press, Oxford, 1961, Chap. 4.
2. E. A. Spiegel, "Convection in Stars: II Special Effects," *Ann. Rev. Astron. Astrophys.*, **10**, 269 (1972).
3. R. S. Schechter, M. G. Velarde and J. K. Plattern, "The Two-component Bénard Problem," in *Advances Chemical Physics*, Vol. 26, I. Prigogine and S. A. Rice, Eds., Interscience, New York, 1973.
4. R. E. Danielson, "The Structure of Sunspot Penumbras II," *Astrophys. J.* **134**, 289 (1961).
5. W. B. Thompson, "Thermal Convection in a Magnetic Field," *Phil. Mag.* (7) **42**, 1417 (1951).
6. N. O. Weiss, "Convection in the Presence of Restraints," *Phil. Trans. Roy. Soc. London, Ser.* **A**, **256**, 99 (1964).
7. I. R. Goroff, "An experiment on Heat Transfer by Overstable and Ordinary Convection," *Proc. Roy. Soc. London, Ser. A*, **254**, 537 (1960).
8. H. T. Rossby, "A Study of Bénard Convection With and Without Rotation," *J. Fluid Mech.*, **36**, 309 (1969).
9. T. G. Cowling, *Magnetohydrodynamics*, Interscience, New York, 1957, Chap. 4.
10. D. R. Moore, R. S. Peckover and N. O. Weiss, "Difference Methods for Time-dependent Two-dimensional convection," *Computer Phys. Commun.*, **6**, 198 (1973).
11. N. O. Weiss, "The Expulsion of Magnetic Flux by Eddies," *Proc. Roy. Soc. London, Ser. A* **293**, 310 (1966).

STABILITY OF SUPERCRITICAL BÉNARD CONVECTION AND TAYLOR VORTEX FLOW

E. L. KOSCHMIEDER

College of Engineering,
and Center of Statistics and Thermodynamics,
The University of Texas,
Austin, Texas

CONTENTS

I. INTRODUCTION

In a preceding article Koschmieder[1] has discussed various aspects of Bénard convection. This article will henceforth be referred to as I. As is well known, there is a strong correspondence between Bénard convection and the Taylor vortex instability in spite of the fundamentally different apparatus used in each of these experiments. Nevertheless, the unstable vertical distribution of density that causes Bénard convection is in many respects equivalent to the unstable radial distribution of angular momentum that is the cause of the Taylor vortices. The similarity of Bénard convection and Taylor vortex flow was first noted by Low[2] and was investigated further by Jeffreys.[3] A comprehensive presentation of the results of linear analysis of Bénard convection as well as Taylor vortex flow can be found in Chandrasekhar's book.[4] As was pointed out in I the theoretical explanation of the experimental results concerning supercritical Bénard convection poses a puzzling unsolved problem. Therefore it seemed to be helpful to obtain supplementary experimental evidence concerning supercritical Taylor vortices in order to tackle the nonlinear problem of supercritical stability from another direction. This article presents a summary of the experimental facts. For a discussion of the theoretical aspects of nonlinear stability the reader is referred to Stuart's review[5] or to the studies of Eckhaus.[6]

II. BÉNARD CONVECTION

The fundamental result of linear analysis of Bénard convection, which was started by Rayleigh,[7] is the well-known stability diagram reproduced in Fig. 1. The abscissa of Fig. 1 is the so-called wave number a, which is related to the more obvious wavelength λ through the formula

$$a = \frac{2\pi}{\lambda} \tag{1}$$

where the nondimensional wavelength λ is the horizontal extension of a pair of rolls measured in units of the fluid depth. The ordinate in Fig. 1 is the nondimensional Rayleigh number

$$R = \frac{\alpha g \Delta T d^3}{\nu \kappa} \tag{2}$$

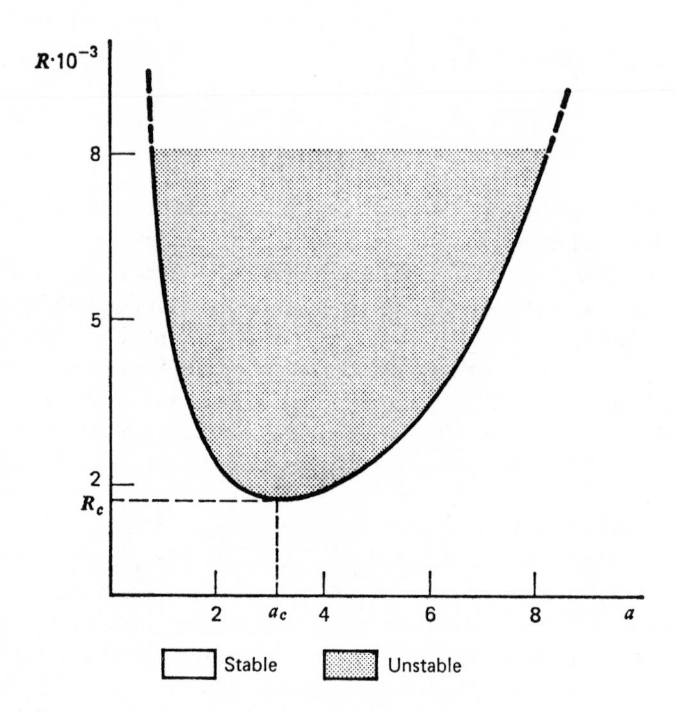

Fig. 1. Stability diagram for Bénard convection (rigid–rigid boundaries).

with the volume expansion coefficient α, the gravitational acceleration g, the applied vertical temperature difference ΔT, the fluid depth d, kinematic viscosity ν, and thermometric conductivity κ. We learn from Fig. 1 that at a minimal Rayleigh number, the so-called critical Rayleigh number R_c, disturbances with a unique wave number a_c are unstable. In the case of rigid–rigid horizontal boundaries the critical Rayleigh number is 1707 and the critical wave number $a_c = 3.12$.

As discussed in more detail in I, the eigenvalue R_c is degenerate; an infinite number of possible flow patterns with the wave number a_c is theoretically possible if, as usual, convective motions on a plane plate of infinite horizontal extent are studied. It was one of the aims of the theory of finite amplitude convection to remove this unrealistic degeneracy and to determine the proper pattern through the investigation of the flow at slightly supercritical conditions. However, as discussed in more detail in I, the hexagonal cell patterns observed by Bénard[8] and others were caused by surface tension effects, as was discovered by Block[9] and Pearson.[10] And in experiments with laterally bounded fluid layers in which surface tension effects were eliminated by a lid in touch with the fluid, the convection pattern turned out to be determined by the form of the lateral wall, as was established by Koschmieder.[11] As an example we show in Fig. 2 convective motions in a shallow fluid layer under a rigid transparent lid, bounded by a circular lateral wall. Linear theory of convective motions in a cylindrical container comes to exactly the same result, as has been shown by Charlson and Sani.[12] In square or rectangular containers flow is likewise determined by the form of the lateral boundary, as Davis[13] has shown theoretically and Koschmieder[11] and Stork and Müller[14] found experimentally.

Yet the dispute about the proper planform seems to be never ending and is nourished by experiments which produced irregular patterns in bounded containers. Charlson and Sani[12] have shown that the critical energy levels for axisymmetric and nonaxisymmetric circular fluid layers are very close together, which was to be expected from the degeneracy of the eigenvalue R_c of infinite layers. One can therefore expect that very small nonuniformities of the temperatures on the horizontal boundaries can have large effects on the selection of the cellular pattern. This can actually be demonstrated experimentally. One can, for example, deliberately introduce small deviations of the bottom temperature. This is accomplished with a bottom plate which is studded at regular intervals with small concave bumps, the bumps being of a height only a fraction of the fluid depth. This then raises the bottom isotherm to the level of the bumps at their location. A pattern of triangular cells produced artificially by this method is shown in Fig. 3*a*. This patterns is, of course, forced since we have introduced small horizontal temperature differences at the bottom plate. Any arbitrary

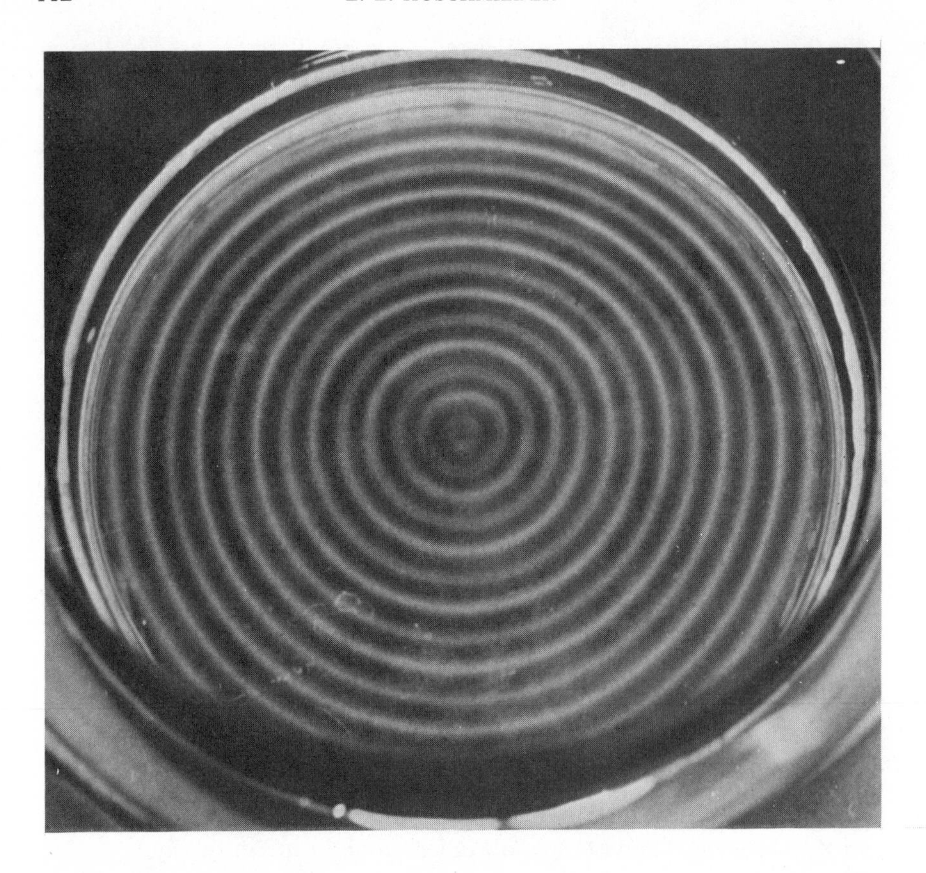

Fig. 2. Bénard convection in a cylindrical container under a rigid lid. Just critical. Visualization with aluminum powder. Dark lines indicate the location of vertical motion. Bright areas indicate predominantly horizontal motion. (After Koschmieder.[1])

pattern can be produced experimentally by analogous methods; as an example Fig. 3*b* shows a theoretically absurd pentagonal pattern. We can safely assume that the irregular pattern observed in some experiments in bounded containers were caused by nonuniformities of the temperature in the horizontal boundaries, most likely in a nonuniformly cooled top plate.

Another persistent objection to the circular pattern in a cylindrical container is the possibility of horizontal temperature gradients at the lateral wall or in the fluid. However, this objection has long since been dealt with in Koschmieder's paper.[11] Therein an experiment is described in which axisymmetry was introduced by nothing else but a step in the critical temperature difference along a circle. This step in ΔT_c was caused by a thin Bakelite plate (of 8% thickness of the fluid depth) which covered

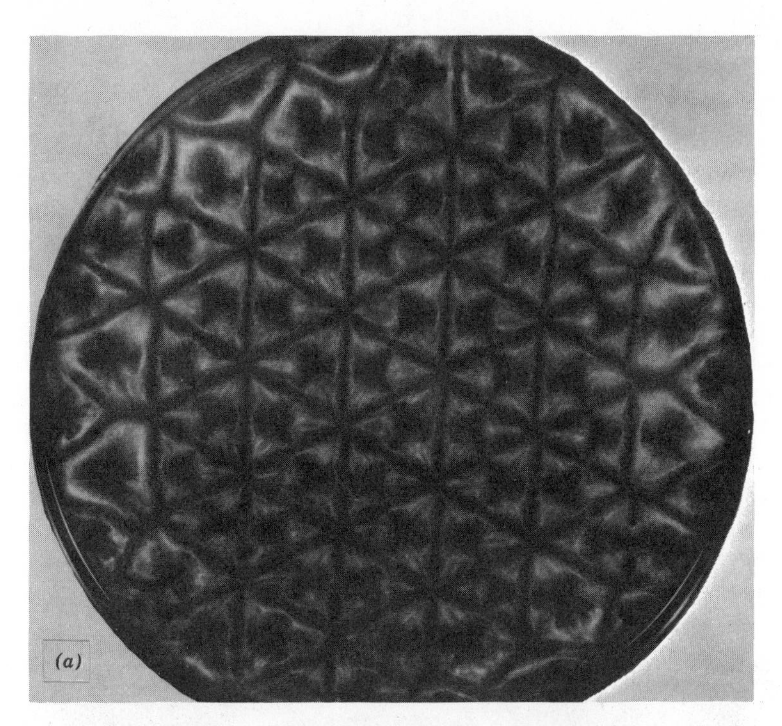

Fig. 3. (*a*) Triangular cells forced by nonuniform bottom. (Koschmieder, unpublished.)

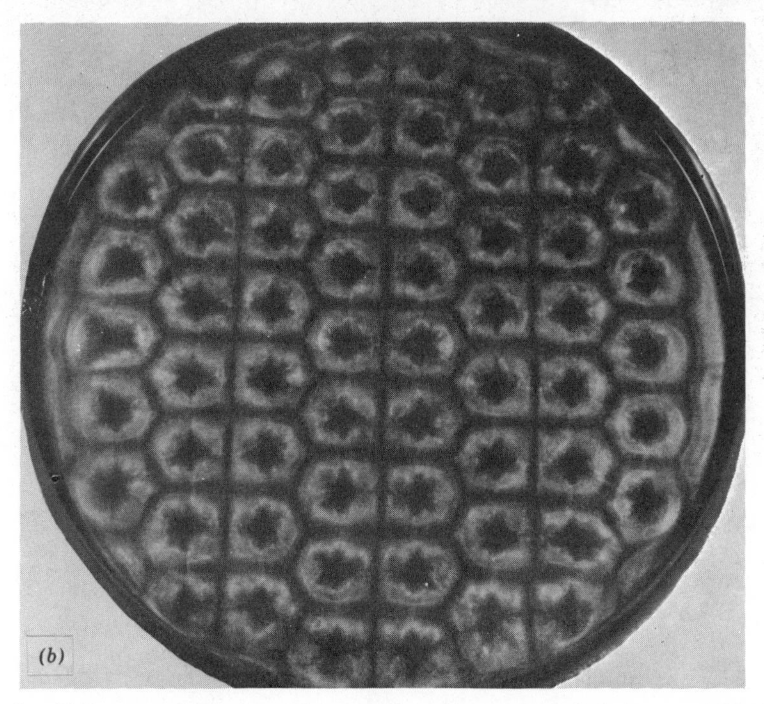

Fig. 3. (*b*) Pentagonal cells forced by nonuniform bottom. (Koschmieder, unpublished.)

the bottom copper plate except for a circular center part. Due to the decreased depth of the fluid over the bakelite plate, the critical temperature difference there was then 24% larger than over the center of the fluid layer. However, the vertical temperature gradients in the oil over the bakelite and over the center were nearly identical, since the thermal conductivity of bakelite is of the same order of magnitude as the thermal conductivity of the silicone oil which was used as fluid. When the temperature difference across the entire layer was increased slowly to the critical temperature difference of the center section, a pattern of circular concentric rolls formed at the center of the fluid, as is shown on Fig. 4. The bulk of the fluid over the dark Bakelite stayed at rest, except for a very weak roll adjacent to the step where the fluid has to move because of continuity. It is hard to see how a lateral wall can be imitated by an arrangement which disturbs the homogenuity of the temperature field less than the one just discussed. The rigid boundary condition of an actual lateral wall probably affects the fluid more than the stepwise increase of the critical temperature

Fig. 4. Convective motions with an almost free lateral boundary condition. (After Koschmieder.[11])

difference does; actually the fluid is free over more than 90% of its depth at the location of the step. Summarizing, we note that is is a *certainty* that in a shallow fluid in a bounded container with strictly uniform temperatures on top and bottom of the fluid the type of convective motion is determined by the form of the lateral wall, provided the fluid is Boussinesq. A Boussinesq fluid is, in this case, one in which the variation of the property constants with temperature are negligible, except for the variation of density.

As has been discussed in considerable detail in I, the wavelength of convective motions under supercritical conditions *increases* with increased Rayleigh number. The increase of the wavelength has been shown by 10 experiments. Further evidence has recently been added by Koschmieder and Pallas.[15] A plot showing some measured wavelengths as a function of the Rayleigh number is reproduced in Fig. 5. As has also been discussed in I, the experimental evidence contradicts seven theoretical studies, all of which predict a *decrease* of the wavelength with increased Rayleigh

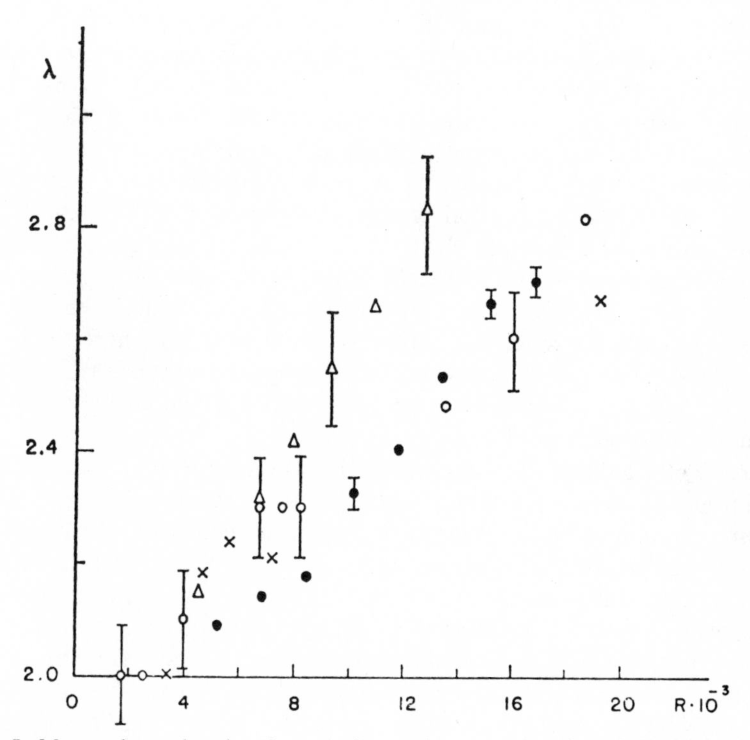

Fig. 5. Measured wavelengths of convective motions as a function of Rayleigh number. (After Koschmieder.[1])

number. Thus the variation of the wavelength as a function of the Rayleigh number remains an unexplained puzzle.

It has been suggested by Koschmieder[16] that dissipation plays an essential role in the selection of the supercritical wavelength. Since dissipation might be of importance for Taylor vortices too, the role of dissipation in Bénard convection shall be discussed here in more detail. Chandrasekhar (Ref. 4, p. 34) has shown theoretically that, according to linear theory, a *balance* is maintained at the critical Rayleigh number between the kinetic energy dissipated by viscosity and the internal energy released by buoyancy. If that balance is altered the wavelength changes. An obvious example of such a change is the variation of the critical wavelength due to different horizontal boundary conditions, as was pointed out by Davis.[13] With free-free boundary conditions the critical wavelength is $\lambda_c = 2.828$. In this case we have minimal friction at the boundaries, but also minimal release of internal energy, since ΔT_c then has the smallest value. With rigid–free boundaries $\lambda_c = 2.343$ and with rigid–rigid boundaries $\lambda_c = 2.016$. In the latter case we have maximal dissipation at the boundaries and also the largest critical temperature difference, but the shortest critical wavelength. The balance of dissipation and release of internal energy can also be altered by a poorly conducting lid on top of the fluid. It has been predicted theoretically by Hurle, Jakeman, and Pike[17] and by Nield[18] that in this case the critical wavelength should increase. This was verified experimentally by Koschmieder.[16] A poorly conducting lid necessarily reduces the release of internal energy. If balance is to be maintained, dissipation must be reduced. Hence the fluid selects a longer wavelength. On the other hand, if we artificially force an increase of the release of internal energy by the sudden application of the vertical temperature difference across the fluid, then motions with wavelengths shorter than the critical one occur, as was found experimentally by Koschmieder.[16] Again, increased internal energy release and increased dissipation due to motion in smaller rolls seem to balance.

Finally, we look at the question of the wavelength under supercritical conditions. Let us assume that a balance of dissipation and energy release is also maintained under slightly supercritical conditions. Anyhow, at $R_c + \epsilon$ the balance theorem cannot be too far from the truth. It is unlikely that an extremum principle, such as maximal heat transfer, now suddenly takes over. Actually, Chandrasekhar (Ref. 4, p. 609) has shown that the balance theorem holds also under steady supercritical conditions if the nonlinear equations and the Boussinesq approximation are used and three very plausible assumptions are made. We have found that under supercritical conditions the wavelengths are longer than λ_c. We note first that since the applied temperature difference is now larger more internal energy must be released than at the critical temperature difference. On the other hand, dissipation in the fluid has increased also, since the velocity of the fluid motion increases rapidly (exponentially according to linear theory). Apparently the increase of dissipation is larger than the increase of internal energy release. To offset the larger increase of dissipation and to maintain balance of dissipation and energy release the fluid selects a larger wavelength. Larger cells mean, under otherwise identical conditions, less dissipation, just as under an insulating lid larger cells reduce dissipation. The reduced dissipation of larger supercritical cells seems to be able to balance the only moderately increased internal energy release of supercritical flow.

We shall now discuss the effect of initial conditions on the wavelength of convective motions. Since the initial conditions are very important for Taylor vortices, we shall be more specific here than we were in I with regard to this question. As follows from the stability diagram in Fig. 1, the wavelength of supercritical convective motions is nonunique according to linear theory. There is an interval of possible wavelengths, ranging from values shorter than the critical wavelength to values longer than the critical wavelength, if the Rayleigh number is larger than R_c. To prove nonuniqueness experimentally, one would have to observe steady convective motions with different wavelengths at one given particular Rayleigh number. Some experimental evidence concerning this point has been provided by Willis, Deardorff, and Somerville.[19] These authors heated shallow layers of air, with Prandtl number $\sigma = 0.71$, water ($\sigma = 6.7$), and silicone oil ($\sigma = 450$) up to about $10R_c$ and measured the wavelength of a fairly irregular pattern. When $10R_c$ was reached the heating and, consequently, the Rayleigh number were reduced and the wavelength was measured again. The results of the measurements are shown in Fig. 6. No hysteresis of the wavelength was observed with air; however, the wavelength in water and in particular in silicone oil depended on the initial conditions and had a hysteresis. As Willis, Deardorff, and Somerville note, they are not certain how long one must wait before the observed average wavelength obtained from one set of initial conditions agrees with the average wavelength obtained from different initial conditions. Since the viscous diffusion time d^2/ν and the thermal diffusion time d^2/κ of water and silicone oil are one to two orders of magnitude larger than the corresponding times in the air layer, it may be possible that the hysteresis of the wavelength in water and silicone shown in Fig. 6 originates from time-dependent conditions rather than from a true nonuniqueness. As a matter of fact, Koschmieder[20] has observed that the wavelength of convective motions in a silicone oil layer ($\sigma = 450$) which is first heated suddenly from below changes, if the applied supercritical temperature difference is maintained, in the course of hours from its original value

$\lambda < \lambda_c$ to exactly the value $\lambda > \lambda_c$ which one obtains if the fluid is heated very slowly from rest to the same supercritical Rayleigh number. To summarize, the evidence concerning the effect of the initial conditions on the wavelength of convective motions is, as yet, inconclusive. However, it appears to this author that it is likely that the wavelength of convective motions is independent of initial conditions.

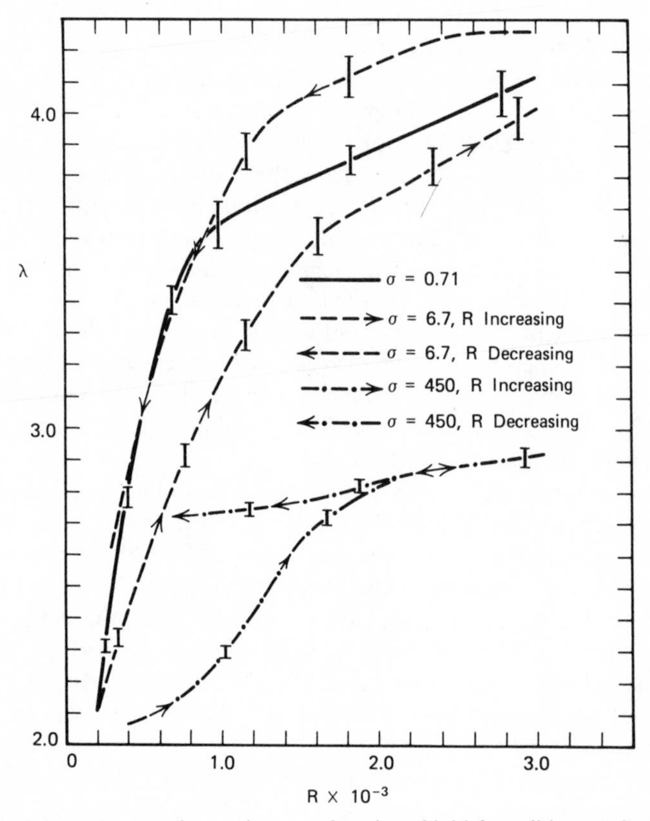

Fig. 6. Wavelength of convective motions as a function of initial conditions. (After Willis and Deardorff.)

Concluding this section we cite a recent elegant measurement of the velocity of convective motions made by Bergé and Dubois.[21] Quantitative knowledge of the velocity field is a necessity, if we want to be able to verify future theoretical analyses of supercritical convection.

III. TAYLOR VORTEX FLOW

Since the classical investigation of Taylor[22] the stability of the shear flow between two vertical concentric cylinders which rotate around their common axis has been the subject of many theoretical and experimental studies. For simplicity we are here concerned mainly with the flow between two cylinders, the inner one of which rotates, and the outer one which is at rest. As was shown by Rayleigh,[23] the shear flow, commonly called Couette flow, of an inviscid fluid between both cylinders is unstable if the angular momentum of the fluid decreases outwards anywhere in the fluid. Taylor then investigated the stability of viscous flow between the cylinders. He found that viscosity postpones the onset of the instability. He introduced as the characteristic parameter a nondimensional number, which is now referred to as the Taylor number and is, in the case of infinitely long cylinders, given by the formula[24]

$$T_N = \frac{2\eta^2}{1-\eta^2} \cdot \frac{\Omega^2 d^4}{\nu^2} \tag{3}$$

where η is the ratio of the radii of the cylinders, Ω is the angular velocity of the inner cylinder, d is the gap width between the cylinders, and ν is the kinematic viscosity of the fluid. Taylor predicted theoretically and verified experimentally that the instability sets in at a certain critical Taylor number ($T_{Nc} = 1723.9$ for $\eta = 0.975$) in the form of axisymmetric toroidal rolls, which have a size given by the critical wave number $a_c = 3.127$. A picture of Taylor vortices is shown in Fig. 7. The dark heavy lines on this photograph indicate the sinks, the location where the flow at the outer cylinder is directed radially inwards; the fine dark lines are at the location of the sources, where the flow has an outward radial component. It is obvious that in neighboring rolls the flow has opposite circulation. The rolls or vortices have a characteristic wavelength λ. The nondimensional wavelength is the ratio of the vertical extension of a pair of vortices divided by the gap width, or in a formula $\lambda = l/d$. The critical wavelength λ_c is $\lambda_c = 2.01$ in a fluid column of infinite length and with a resting outer cylinder. As we see, the critical Taylor number has almost the same value as the critical Rayleigh number and, likewise, the critical wavelength of Taylor vortices has almost the same value as the critical wavelength of Bénard convection. A critical wavelength of $\lambda_c = 2.01$ means in practical terms that a single Taylor vortex is as long in vertical direction as it is deep in radial direction, as convective rolls, at the onset of convection, are as wide horizontally as the fluid is deep, provided that both boundaries are rigid. Taylor vortices can, however, have a critical wavelength much shorter than $\lambda = 2$. This happens when both cylinders rotate and rotate in

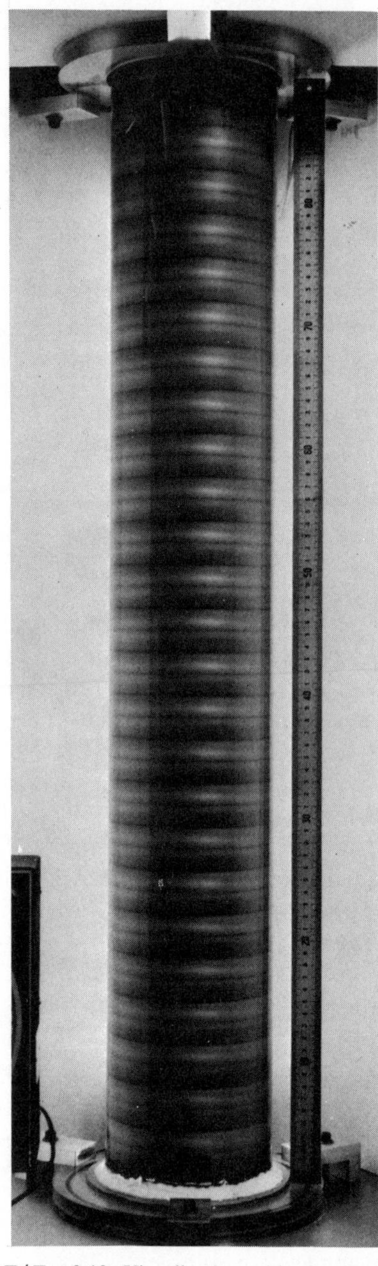

Fig. 7. Taylor vortices. $T/T_c = 9.10$. Visualization with aluminum powder. Dark lines are location of radial motion. Bright areas indicate predominantly tangential motion. (After Burkhalter and Koschmieder.[35])

opposite directions. Remember that the critical wavelength of Bénard convection has different values too, depending on the type of horizontal boundaries.

Unlike the convection problem, the critical Taylor number is not a degenerated eigenvalue; the toroidal vortices are the only form of flow that occurs. For a while it appeared that a second kind of flow exists, namely, the so called spiral flow. The most convincing picture of spiral flow can be found in Taylor's original paper and is reproduced in Fig. 8. As can be seen, the vortices are then wrapped around the inner cylinder in a continuous spiral. Note the very marked difference in the size of the vortices—they are alternately larger and smaller. Later experimental work has shown that spiral flow does not appear at the same critical Taylor number as the regular Taylor vortices. Spiral flow can be produced in two ways. One method involves the help of an axial flow with a parabolic flow profile which runs through the cylinder gap while the inner cylinder rotates. In this case spiral flow forms if the Reynolds number of the flow is sufficiently large and the standard Taylor number is exceeded. The spiral

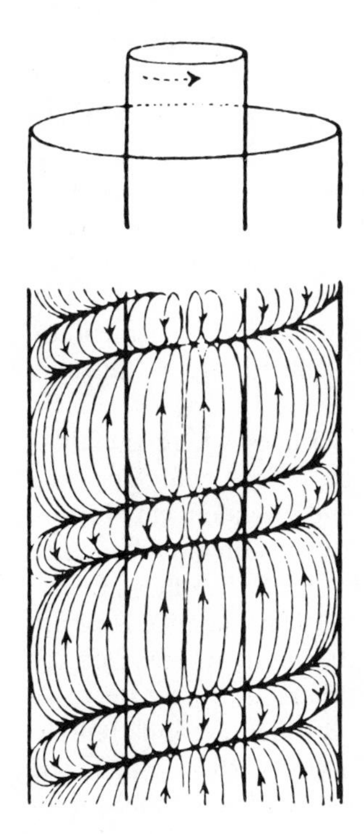

Fig. 8. Spiral flow. (After Taylor.[22])

vortices are, however, all of the same size. This type of spiral flow has been investigated experimentally by, among others, Schwarz, Springett, and Donnelly[25] and theoretically by Krueger and DiPrima[26] in the case of low Reynolds number flow and by Hughes and Reid[27] for larger Reynolds numbers.

The other way to create spiral flow is through the application of a radial temperature gradient to the fluid in the gap between both cylinders. As Snyder[28] has shown experimentally spiral flow forms under these circumstances at Taylor numbers larger than the standard critical Taylor number if the radial temperature difference is sufficiently large or, at even larger temperature differences, at subcritical Taylor numbers. Spiral flow caused by a radial temperature has vortices of alternately smaller and larger size, as does the spiral flow that Taylor observed. The appearance of vortices of different size has not been explained theoretically but can be easily understood. With a radial temperature difference and in the absence of rotation the fluid in the gap rises along the warm cylinder and sinks along the cold cylinder, forming one large loop. Rotating the inner cylinder at near critical Taylor numbers superposes Taylor vortices onto the overall thermal circulation. Since every second vortex has a circulation opposite to the overall thermal circulation, every second vortex is smaller than its neighboring vortex which has a circulation of the same direction as the thermal circulation. Exactly the same kind of superposition with alternately smaller and larger rolls occurs in the case of convection on a nonuniformly heated plane, as was shown experimentally by Koschmieder.[29] No explanation is available for the fact that the flow with a radial temperature gradient does not stay in axisymmetric toroidal rings but rather changes into spiral flow if the radial temperature difference is sufficiently large. Menzel and Roesner[30] have recently confirmed theoretically the initial stabilizing effects of the radial temperature differences which Snyder observed.

We shall now return to the standard Taylor vortices where the fluid motions are caused solely by differential rotation of the cylinders. The investigation of the wavelength of the vortices should provide a clue for the understanding of supercritical Taylor vortices. Early experimental evidence concerning the wavelength has been provided by Pai,[31] Schultz-Grunow and Hein,[32] Donnelly and Schwarz,[33] and Snyder.[34] The evidence is partly contradictory in so far as some experiments indicate an increase of the wavelength with increased Taylor number while others indicate a decrease of the wavelength. Recent measurements by Burkhalter and Koschmieder[35] show that actually the wavelength of Taylor vortices which one would observe in an infinitely long cylinder is *independent* of the Taylor number as long as the flow is axisymmetric and as long as the Taylor number is increased quasisteadily. The cautious reference to infinitely long cylinders is necessary since end effects can have an effect on the wavelength. End effects can also influence the critical Taylor number in the vortices near the cylinder end, just as is the case with the critical Rayleigh number of convective rolls near the lateral walls. We note, however, that there is a marked contrast in the behavior of the wavelengths of Taylor vortices and Bénard convection under supercritical condition. While the wavelength of Bénard convection increases, the wavelength of Taylor vortices remains constant. If, on the other hand, at some supercritical T_N the flow loses its axisymmetry and becomes doubly periodic with tangential waves (for an example of such flow see Fig. 9*c*), then the wavelength can increase, as discovered by Coles.[36] In doubly periodic flow the relation among Taylor number, number of tangential waves, and axial wavelength is very complex. If the Taylor number is increased much further, so that the flow between the cylinders becomes turbulent, then the tangential waves in the vortices fade and turbulent Taylor vortices remain with axial wavelengths larger than the critical wavelength (see Burkhalter and Koschmieder[35]). Such wavelengths were actually observed by Pai[31] and Schultz-Grunow.[32] A set of pictures showing the transition from laminar axisymmetric vortices to fully turbulent vortices with longer wavelengths is shown in Fig. 9. To summarize, the wavelength of supercritical Taylor vortices remains constant, according to the experimental evidence, as long as the flow is axisymmetric, and the wavelength increases with higher Taylor numbers if the flow is doubly periodic or turbulent.

Some theoretical studies have been concerned with the question of the wavelength of the vortices under supercritical conditions. Stuart[37] has introduced the so-called "shape assumption" which restricts the flow to the critical wavelength under slightly supercritical conditions. The shape assumption has been used to compute the amplitude of the Taylor vortices. Knowledge of the amplitude is essential for understanding of the torque measurements of Taylor vortex flow, a topic that cannot be pursued here. Kirchgässner and Sorger[38] have shown rigorously that the critical wavelength is the only stable wavelength in an interval just above critical. Stuart's shape assumption and the study of Kirchgässner and Sorger are in agreement with the experimental facts, however, the critical wavelength is maintained over a much larger interval in Taylor number (up to 80 T_{Nc}) than both studies envisioned. Further information is available concerning the possible interval of wavelengths for a given supercritical Taylor number. Chandrasekhar (Ref. 4, p. 303), using linear theory and the narrow gap approximation gives a formula for the wavelength interval. In a first approximation the wavelength interval for Taylor vortices is the same as that for the possible wavelengths of Bénard convection. Thus Fig.

1 also shows the stability diagram of Taylor vortex flow if the Rayleigh number in Fig. 1 is replaced by the Taylor number. Figure 1 does then also indicate the possibility that, according to linear theory, supercritical Taylor vortex flow is nonunique. Kogelman and DiPrima[39] have made a nonlinear stability analysis and find another somewhat smaller interval of possible wavelengths which is shown in Fig. 11.

Fig. 9. Development of turbulent Taylor vortices: $r_i = 10.36$ cm, $\eta = r_i / r_0 = 0.912$. (Koschmieder, unpublished.) (*a*) $T/T_c = 1.10$, fluid oil, $\nu = 0.44\ \text{cm}^2/\text{sec}$, column length 55.5 cm. (*b*) $T/T_c = 2.22$, same oil, same column. (*c*) $T/T_c = 3.16$, fluid water, $\nu = 1.0 \cdot 10^{-2}\ \text{cm}^2/\text{sec}$, column length 52.8 cm. (*d*) $T/T_c = 14.73$, water, same column. (*e*) $T/T_c = 45.0$, water, same column. (*f*) $T/T_c = 93.1$, water, same column.

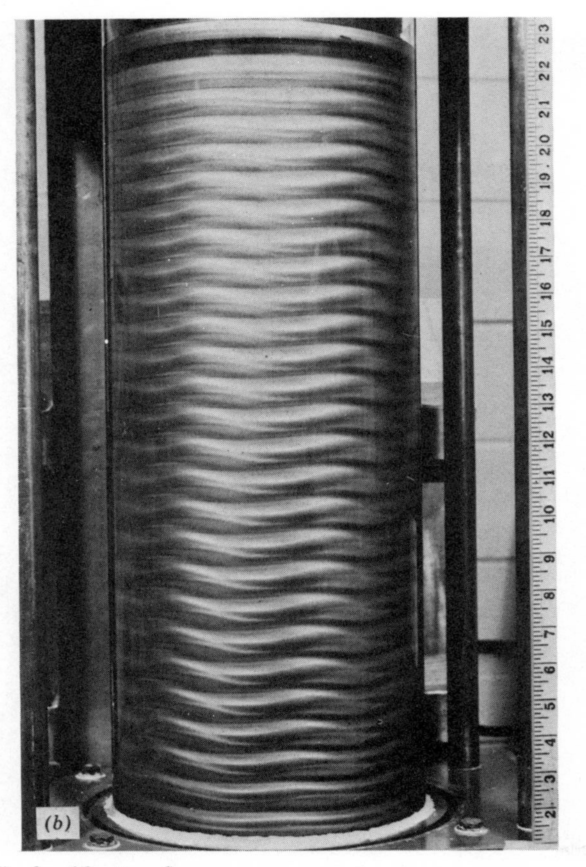

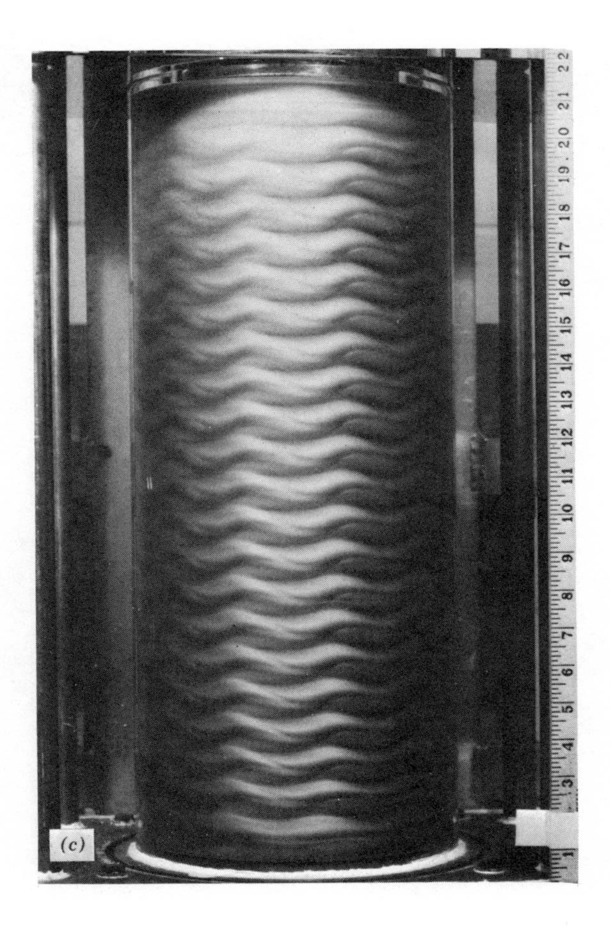

Fig. 9. (*Continued*)

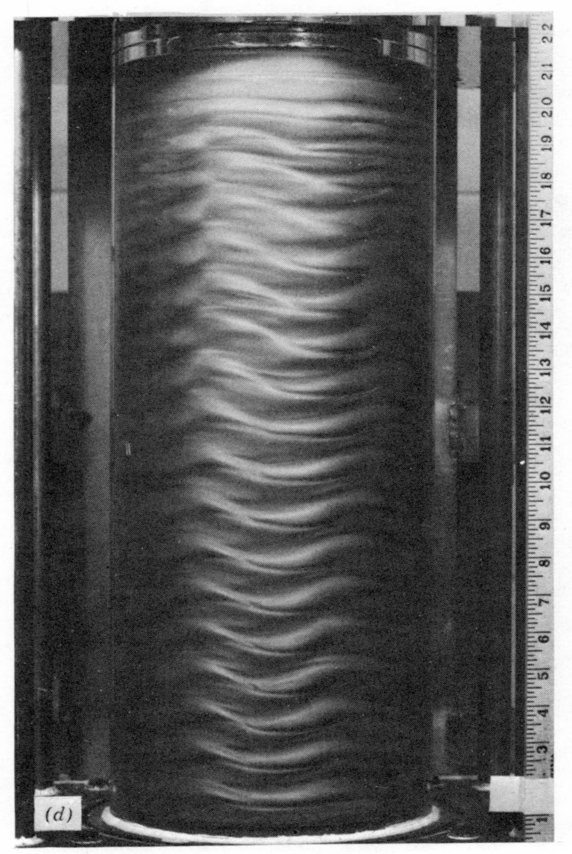

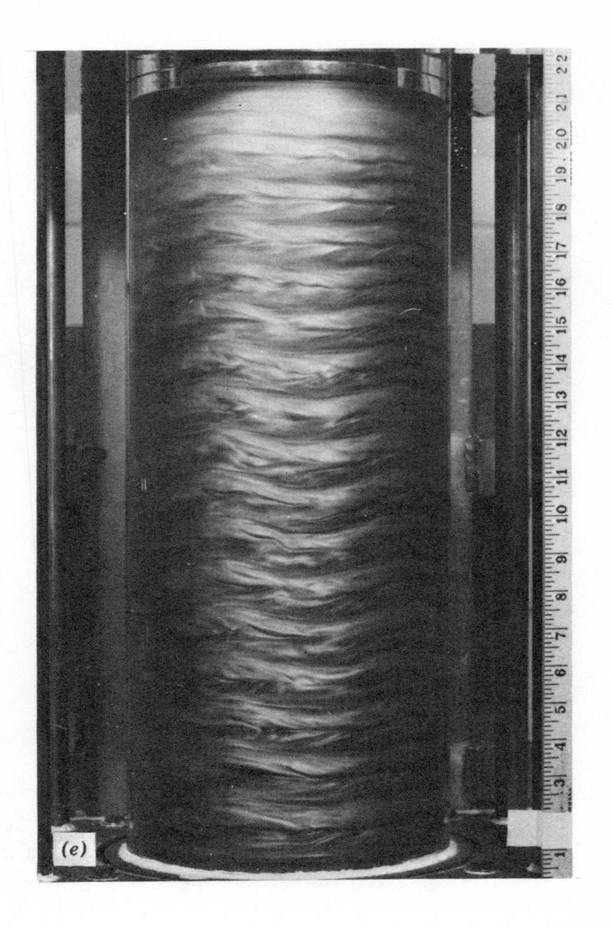

Fig. 9. (*Continued*)

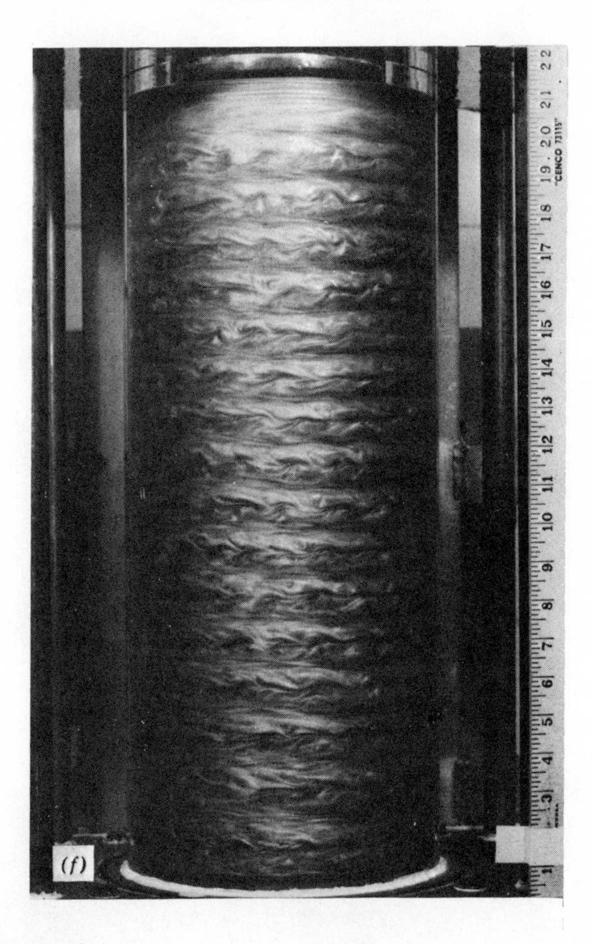

Fig. 9. (*Continued*)

The nonuniqueness of supercritical Taylor vortex flow has been confirmed by unambiguous experimental evidence. Coles[36] has observed some degree of nonuniqueness in doubly periodic flow. Snyder[34] has shown that the critical wavelength of Taylor vortices between two rotating cylinders is maintained even if, for example, the outer cylinder is gradually brought to rest after the vortices are formed at T_{Nc}. Burkhalter and Koschmieder[40] have shown that Taylor vortices with an entire range of wavelengths shorter than λ_c can be produced through sudden starts of the inner cylinder to supercritical Taylor numbers. Vortices produced by this method maintain their wavelength for an indefinite time if the Taylor number is maintained or even if the Taylor number is varied slowly over large intervals. Burkhalter and Koschmieder[40] also found that wavelengths longer than λ_c can be established by filling the gap between the cylinders with fluid, while the inner cylinder rotates at a constant supercritical angular velocity. A picture showing Taylor vortices of different λ in the same apparatus with the same fluid at the same Taylor number is shown in Fig. 10. Figure 11 shows the stability diagram of Taylor vortex flow

Fig. 10. Nonuniqueness of Taylor vortex flow. $T/T_c = 9.10$. Sudden start experiment at left, quasisteady experiment center, filling experiment at right. (After Burkhalter and Koschmieder.[40])

according to Chandrasekhar[4] and according to Kogelman and DiPrima[39] together with some experimentally observed steady-state wavelengths. As we see, Taylor vortices can, dependent on the initial conditions, utilize a wide range of wavelengths, whereas we have seen that convective motions apparently are not able to do so. In this regard Bénard convection and Taylor vortex flow are strikingly different. However, the nonuniqueness of Taylor vortices does not seem to occur if the Taylor vortices are turbulent. As Burkhalter and Koschmieder[40] find, turbulent Taylor vortices caused by sudden starts have, within the range of the statistical scatter, the same wavelengths longer than λ_c as turbulent Taylor vortices which have been created by slow increases of the Taylor number.

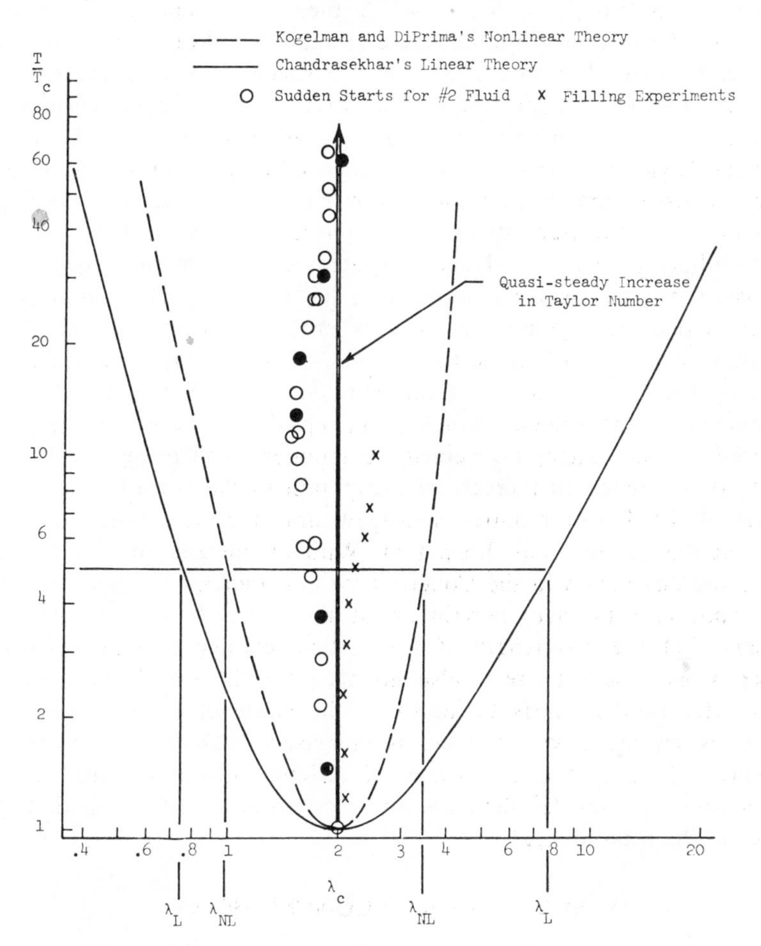

Fig. 11. Stability diagram for Taylor vortex flow with some measured wavelengths. (After Burkhalter and Koschmieder.[40])

We would like to suggest that dissipation plays an essential role in the selection of the wavelength of the vortices, just as in the case of Bénard convection. Little is known though about dissipation in Taylor vortices. Stuart[37] has suggested that it may be that in the equilibrium state the rate of transfer of energy from the mean flow to the disturbance precisely balances the rate of viscous dissipation of energy of the disturbance. Actually some information about the effect of dissipation on the wavelength could be obtained with linear theory through the investigation of the variation of the critical wavelength between two rotating cylinders. As mentioned before, the critical wavelength decreases substantially if both cylinders rotate in opposite directions. One would expect a substantial increase of dissipation in this case. A theoretical explanation of the behavior of the wavelength under these circumstances would be a significant step forward. We would also like to speculate that dissipation is responsible for the fact that the critical wavelength of Taylor vortices is maintained if the Taylor number is slowly increased to fairly high supercritical Taylor numbers. There is, under subcritical conditions, a fair amount of dissipation in the Couette flow on which the vortices develop. This is in contrast to Bénard convection where the subcritical state is one of rest without dissipation. Hence any increase in dissipation due to the disturbance, the convective motions, is much more important in Bénard convection than in Taylor vortex flow, where the ratio of the amount of dissipation in the disturbances, the vortices, to the amount of dissipation caused by the Couette flow is small. This must be so since the r and z components of the velocity which make up the vortices are small as compared to the tangential velocity component which represents the Couette flow. Hence an increase of dissipation in the vortices due to an increase of the Taylor number is insignificant since it concerns only a small fraction of the total dissipation. With an increase of the Taylor number the dissipation in the Couette flow also increases, of course, and the dissipation ratio may remain about the same. It is therefore not necessary that the wavelength of the vortices change when the Taylor number is increased. There is also no need for the wavelengths which appear after sudden starts to adjust to the value of $\lambda = \lambda_c$ which one observes when the Taylor number is increased quasisteadily, since the differences of dissipation in vortices of different λ at the same Taylor number are even smaller than the small contribution of dissipation by vortices to the total dissipation.

IV. SUMMARY AND CONCLUSIONS

There is an almost perfect correspondence between Bénard convection and Taylor vortex flow as far as the onset of the instabilities is concerned, or as long as linear theory is the appropriate way to describe both phenomena. This similarity is most clearly expressed by the fact that the stability diagram (Fig. 1) which follows from linear theory is the same for Bénard convection and for Taylor vortices, which means that the values of the critical Rayleigh number and the critical Taylor number are practically the same, provided corresponding boundary conditions are considered. The values of the critical wavelengths are also the same, provided again that the boundary conditions match. In essence, the results of linear theory for both problems apparently differ only in one aspect, namely, while there appears to be an infinite number of possible flow patterns at the critical Rayleigh number in the case of convection, there is only one possible flow in the case of the Taylor vortex instability. However, the degeneracy of the solutions in the Bénard convection problem originates from the simplifying but unrealistic assumption of a fluid layer of infinite horizontal extent. There is, therefore, no specification of the geometry of the flow at all, while the geometry of the apparatus in the Taylor vortex case is clearly defined, although the cylinders are supposed to be of infinite length. There are also some characteristic similarities between both problems which become apparent in experiments. The transport properties, torque in the case of Taylor vortices and heat in the case of Bénard convection, are described by similar curves with the "breaks" at the critical values of Taylor number or Rayleigh number. Lateral boundaries in Bénard convection have the same consequences as the horizontal top and bottom boundaries of the fluid column in a Taylor vortex apparatus. In either case the instabilities invariably develop under subcritical conditions at first at these boundaries, which are necessarily finite disturbances in an otherwise homogeneous temperature or velocity field. With increased Rayleigh or Taylor number the instability progresses, in both cases, steadily into the fluid layer until, at the critical values, the instability extends throughout the entire fluid.

The correspondence between Bénard convection and Taylor vortices ends abruptly under supercritical conditions when linear theory is no longer sufficient for a description of the phenomena. There is a spectacular difference in the variation of the wavelength of supercritical Bénard convection and of supercritical Taylor vortices. Measurements show that the wavelength of supercritical axisymmetric Taylor vortices is equal to the critical wavelength and *independent* of the Taylor number, provided the Taylor number is increased quasisteadily and that the measured wavelength is extrapolated to the wavelength which one would observe between cylinders of infinite length. On the other hand, the wavelength of supercritical Bénard convection definitely *increases* with increased Rayleigh number, assuming again that the increase in R is slow. This observation is, in all likelihood, true also for convective motions on a plane plate of infinite horizontal extent. There is an equally striking discrepancy in the response of both instabilities to initial conditions. The wavelength of supercritical Taylor vortices is clearly *dependent* on initial conditions. On the other hand, the wavelength of supercritical Bénard convection seems to be *independent* of initial conditions, although this fact has as not yet been established unambiguously. However, nobody has ever contended that the wavelength of supercritical Bénard convection, established by one means or another, does not change at all even if the Rayleigh number is changed. But just that is the case with Taylor vortices—a wavelength once established by some means does not change, even if the Taylor number is changed.

There is as yet no explanation of the differences in the behavior of the supercritical flows. There is no theoretical explanation for the increase of the wavelength of supercritical convection nor is there an explanation for the constant value of the wavelength of axisymmetric supercritical Taylor vortex flow. Both questions can only be explained by nonlinear stability theory. It is, however, suggested here that the reason for the different behavior of both instabilities does not necessarily lie in the nonlinear nature of the governing equations. There is the possibility that *dissipation*, or better the differences in dissipation, of Bénard convection and Taylor vortex flow may be the cause of the different behavior of both flows.

Acknowledgments

The author would like to express his appreciation to Professor I. Prigogine for the invitation to take part in this symposium. This paper was written while the author enjoyed the hospitality of the Advanced Study Program of the National Center for Atmospheric Research, Boulder, Colorado.

References

1. E. L. Koschmieder, *Advances in Chemical Physics*, Vol. 26, I. Prigogine and S. A. Rice, Eds., Interscience, New York, p. 177 (1973).
2. A. R. Low, *Nature*, **115**, 299 (1925).
3. H. Jeffreys, *Proc. Roy. Soc. (London) Ser. A*, **118**, 195 (1928).
4. S. Chandrasekhar, *Hydrodynamic and Hydromagnetic Stability*, Oxford University Press, Oxford, 1961.
5. J. T. Stuart, *Ann. Rev. Fluid Mech.*, **3**, 347 (1971).
6. W. Eckhaus, *Studies in Non-linear Stability Theory*, Springer-Verlag, Berlin, 1965.
7. Lord Rayleigh, *Phil. Mag.*, **32**, 529 (1916).
8. H. Bénard, *Rev. Gen. Sci. Pure Appl.*, **11**, 1261, 1309 (1900).
9. M. J. Block, *Nature*, **178**, 650 (1956).
10. J. R. Pearson, *J. Fluid Mech.*, **4**, 489 (1958).
11. E. L. Koschmieder, *Beitr. Phys. Atmos.*, **39**, 1 (1966).
12. G. S. Charlson and R. L. Sani, *Int. J. Heat Mass Transfer*, **13**, 1479 (1970).
13. S. H. Davis, *J. Fluid Mech.*, **30**, 465 (1967).
14. K. Stork and U. Müller, *J. Fluid Mech.*, **54**, 599 (1972).
15. E. L. Koschmieder and S. G. Pallas, *Int. J. Heat Mass Transfer*, **17**, 991 (1974).

16. E. L. Koschmieder, *J. Fluid Mech.*, **35**, 527 (1969).
17. D. T. Hurle, E. Jakeman, and E. R. Pike, *Proc. Roy. Soc. (London), Ser. A*, **296**, 469 (1967).
18. D. A. Nield, *J. Fluid Mech.*, **32**, 393 (1968).
19. G. E. Willis, J. E. Deardorff, and R. C. J. Somerville, *J. Fluid Mech.*, **54**, 351 (1972).
20. E. L. Koschmieder, *Bull. Am. Phys. Soc.*, **16**, 1309 (1971).
21. P. Bergé and M. Dubois, *Phys. Rev. Lett.*, **23**, 1041(974).
22. G. I. Taylor, *Phil. Trans. Roy. Soc., A*, **223**, 289 (1923).
23. Lord Rayleigh, *Proc. Roy. Soc., A*, **93**, 148 (1916).
24. P. H. Roberts, *Proc. Roy. Soc., A*, **283**, 550 (1965).
25. K. W. Schwartz, B. E. Springett, and R. J. Donnelly, *J. Fluid Mech.*, **20**, 281 (1964).
26. E. R. Krueger and R. C. DiPrima, *J. Fluid Mech.*, **24**, 521 (1964).
27. T. H. Hughes and W. H. Reid, *Phil. Trans. Roy. Soc. (London), Ser. A*, **263**, 57 (1968).
28. H. A. Snyder and S. K. F. Karlson, *Phys. Fluids*, **7**, 1696 (1964).
29. E. L. Koschmieder, *Beitr. Phys. Atmos.*, **39**, 208 (1966).
30. K. Menzel and K. Roesner, to appear.
31. S. I. Pai, NACA Tech. Note #892, 1943.
32. F. Schultz-Grunow and H. Hein, *Z. Flugwiss.*, **4**, 28 (1956).
33. R. J. Donnelly and K. W. Schwarz, *Proc. Roy. Soc. (London), Ser. A*, **283**, 531 (1965).
34. H. A. Snyder, *J. Fluid Mech.*, **35**, 273 (1969).
35. J. E. Burkhalter and E. L. Koschmieder, *J. Fluid Mech.*, **58**, 547 (1973).
36. D. Coles, *J. Fluid Mech.*, **21**, 385 (1965).
37. J. T. Stuart, *J. Fluid Mech.*, **4**, 1 (1958).
38. K. Kirchgässner and P. Sorger, *Quart. J. Mech. Appl. Math.*, **22**, 183 (1969).
39. S. Kogelman and R. C. DiPrima, *Phys. Fluids*, **13**, 1 (1970).
40. J. E. Burkhalter and E. L. Koschmieder, *Phys. Fluids*, **17**, 1929 (1974).

LABORATORY EXPERIMENTS ON DOUBLE-DIFFUSIVE INSTABILITIES

J. S. TURNER

Department of Applied Mathematics and Theoretical Physics,
University of Cambridge,
Cambridge, England

Abstract

The convective motions which can arise in fluids containing two components with different molecular diffusivities have recently been studied using a variety of laboratory experiments. This paper is a summary of a wide range of experimental techniques and results, and was originally prepared as an introduction to a time lapse ciné film of our current experiments. Many of the phenomena observed remain to be explained in detail, since they lie well outside the range of conditions which can be described by linear stability theory.

I. INTRODUCTION

There has been an increasing interest during the last few years in double-diffusive phenomena, the convective motions driven by the differential diffusion of two properties in a stratified fluid. The relation between the linear stability of this case and that in the single-component Bénard problem has been thoroughly studied, and the theoretical work has been the subject of a recent review by Schechter, Velarde, and Platten[8] in this Series. It is enough to say here that necessary conditions for the instability of a double-diffusive system are that the two substances stratifying the fluid have different molecular diffusivities and opposing effects on the vertical density gradient. A striking feature of the theoretical predictions is that instabilities can arise even when the net density distribution is decreasing upwards. All the experiments described below have this property, but nevertheless finite amplitude or even turbulent convective motions can develop. It is implied throughout that the fluids under discussion are completely miscible, so that motions due to interfacial surface tension variations (the Marangoni effect) are excluded.

These ideas about instabilities driven by double-diffusion mechanisms have proved very relevant for the ocean, where salt content and heat are the properties of interest, and in fact much of the recent work has been developed with this application in mind. (For reviews which emphasize the geophysical applications, see Refs. 19 and 20.) Laboratory experiments have been used extensively, however, to explore the various double-diffusive processes which could occur in the ocean and to interpret the observations as they have become available. It is the purpose of the present paper to make these experimental results and methods known to those more familiar with such problems in a different context.

It seems particularly appropriate to describe these experiments to this audience, since many of these studies have been carried out (following the suggestion of Stern and Turner[14]) using two solutes with comparable diffusivities instead of salt and heat. Usually sugar and sodium chloride have been used (with a ratio of diffusivities of about three), but similar effects have been observed with even closer diffusivities, and other quite different applications of the results may occur to physical chemists. Many of the techniques to be described are essentially visual, and cannot be appreciated fully without several photographs, or even better, ciné films, as in the original presentation of this paper. Reference can be made to Turner and Chen[21] for more photographs of the phenomena described in Sections IV to VII. But many of the experiments, especially those involving two layers, are very easy to set up and to observe with elementary equipment, and enough details are given here to allow interested readers to try them in their own laboratory (or kitchen).

II. SIMPLE ONE-DIMENSIONAL EXPERIMENTS

The form of the motion set up by differential diffusion depends on the relation between the two diffusivities and the density gradients, that is, on which component provides the driving energy. The simplest example of a system in which this energy comes from the component having the *larger* diffusivity is a stable salinity gradient, heated uniformly from below. Initially a temperature boundary layer develops at the bottom; this breaks down through an overstable oscillation[9] to form a thin convecting layer. As heating is continued, the depth of this well-mixed layer increases in time (like $t^{1/2}$) until the thermal boundary layer ahead of the convecting region reaches a critical Rayleigh number and a second layer forms.[18] In this way many layers can be built up; the lowest layers merge in succession and more appear at the top.[22] Strong turbulent convection persists in each layer, driven by the heat flux across each interface, while enough salt is left behind to maintain a hydrostatically stable density distribution.

In the opposite situation, when a small amount of hct salty water is poured carefully on top of a stable temperature gradient, so that now the component (salt) with the *smaller* diffusivity provides the potential energy to drive the motion, a very different effect is observed. Long narrow convection cells, called "salt fingers," are formed, due to the more rapid sideways diffusion of heat relative to salt, as first demonstrated by Stern.[13] It is not essential to have a large ratio of diffusivities, and similar phenomena are easily observed using two solutes such as sugar and salt. Stern and Turner[14] showed that there is not as big a difference between the two cases as appears at first sight. Convecting layers can also be formed

Fig. 1. A thick "finger" interface, formed between two fluid layers about 10 cm deep. The lower layer contains sodium chloride of specific gravity about 1.10 and the upper is sugar solution, sp. gr. 1.08. A shadowgraph has been used in this and later experiments to reveal the variations of refractive index.

from a gradient in the "finger" configuration, with small-scale fingers in each of the interfaces supplying the unstable buoyancy flux which drives the motion in the layers.

For many experimental purposes, it is convenient to set up two or more layers directly by pouring one fluid on top of another (using some kind of porous float to spread the lighter layer gently over the heavier). This is very easy to do using sugar and salt solutions, and typical specific gravities of 1.09 and 1.10 for the upper and lower layers. With salt solution (having the higher diffusivity) on top, a very thin "diffusive" interface is formed. With sugar solution on top, salt fingers appear in the interface; this is at first very thin, but gradually spreads out to reveal a regular array of fingers. Fingers can form readily with very small concentrations of the destabilizing component, and this process has been investigated quantitatively by Huppert and Manins.[4]

The large refractive index difference between these solutions also makes visualization very straightforward. "Moiré" patterns can be seen by viewing the experimental tank against an illuminated grid of inclined, regu-

Fig. 2. Shadowgraph picture of a thin interface between a layer of salt solution sp. gr. 1.092 above sugar solution (sp. gr. 1.098), in a container with a sloping bottom boundary. A circulation has been set up in both layers, as shown by the distortion of an initially vertical dye streak. The flow down the slope to the bottom has put dyed salt under the sugar solution and led to the formation of fingers, as shown by the rapid vertical diffusion of the dye.

larly spaced lines. Simpler still, and very satisfactory for photographic purposes, is a shadowgraph technique. A small source of light placed about 3 m behind a tank 5 cm thick forms a pattern of light and dark areas on a sheet of tracing paper fixed to the front of the tank. The distribution of illumination is related to the density gradients, and both horizontal layering and salt fingers can be seen by this means (see Figs. 1 and 2). Finger interfaces have been observed in plan using a related method, which shows that the fingers have a square cross section, with upward and downward motions alternating in a close-packed array.[12]

It is also of interest to mention an elegant color Schlieren technique devised by Shirtcliffe,[10] which gives a characteristic color to the image of a region according to the direction of the refractive index gradient through it. This shows up the motions at both kinds of interface much more clearly than does the simple shadowgraph. At a "finger" interface, for example, it reveals a stable density gradient through the center, with a pattern of fingers superimposed on it, and thin unstable regions at the ends of the fingers which break down to drive convection in the layers on each side of the interface.

III. MEASUREMENT OF LAYER DEPTHS AND INTERFACIAL FLUXES

To correct the impression which may have been given so far that laboratory work on layered systems has been entirely visual and qualitative, a brief reference must be made to some of the quantitative results. (See the review articles cited above for a fuller discussion.) For the case of a salinity gradient heated from below, Turner[18] obtained a relation between the original linear salinity gradient S_z, the rate of heating H, and the depth h of the *first* layer at the time it stops growing and a second layer forms above it; this can be expressed as

$$h \propto H^{3/4} S_z^{-1} \tag{1}$$

Measurements have also been made[16] of the fluxes across an established interface between a layer of hot salty water below colder fresh water. These have shown that the heat flux is well described by the form of relation used for a single-component fluid at high Rayleigh number, with an additional factor which depends on the relative contributions of S and T to the density difference. Huppert[3] showed that the measurements may be conveniently summed up by the expression

$$H = A\left(\frac{\alpha\Delta T}{\beta\Delta S}\right)^2 (\Delta T)^{4/3} \tag{2}$$

where α and β are "coefficients of expansion" for heat and salt, required to convert temperature and concentration differences to density differences. The ratio of the salt to the heat fluxes falls rapidly in the range $1<\beta\Delta S/\alpha\Delta T<2$, but above about $\beta\Delta S/\alpha\Delta T=2$ it becomes constant and independent of this ratio. Using this result with (2), Huppert showed theoretically than an intermediate layer, or series of layers, is stable if the overall S and T differences lie in the "constant" range, and unstable if the system is in the variable range. This is being tested in current experiments and is referred to again in Section VII.

Corresponding measurements of the two fluxes have been made for "finger" interfaces, which transport salt vertically faster than heat.[6,7,17] Again there is a strong dependence on the density ratio across the interface, and the ratio of heat to salt fluxes is constant over a considerable range; in fact, in this case no "variable" regime has yet been found directly. There are also some comparable measurements of fluxes and flux ratios across sugar–salt interfaces of both kinds.[11,14]

IV. SLOPING BOUNDARIES AND SIDE-WALL HEATING

With the geophysical applications in mind, recent experimenters have begun to take into account horizontal nonuniformities of various kinds. It is unreasonable to think of the formation of layers in the ocean as a purely one-dimensional process,[5] since observations of layers attributed to double-diffusion are often associated with the intrusion of one water mass into another. (For example, such layers are prominent above and below the core of relatively warm salty water flowing from the Mediterranean out into the Atlantic.)

One simple type of nonuniformity which has been investigated is a variation of depth of a convecting layer. Gill and Turner[2] have shown how large-scale quasihorizontal motions can be set up in such a region even when the buoyancy flux across the horizontal boundary is uniform (see Fig. 2). This effect is a purely geometrical consequence of the sloping boundary. The net result of the double-diffusive transports across the interface is to provide an unstable buoyancy flux which makes the bottom layer heavier. A given flux causes the density of the lower layer to increase most rapidly in its shallowest part where there is less dilution, and this sets up a horizontal pressure gradient to drive a circulation in the sense which includes a flow down the slope. This downslope flow can reverse the relative concentrations of sugar and salt in a laboratory experiment, so that, for example, salt fingers are formed at the bottom of a tank which originally contained a layer of salt solution above sugar solution. Similar effects are observed when a sloping interface, rather than a solid sloping boundary, produces the nonuniformity of depth.[21]

A different effect produced by a sloping boundary is shown in Fig. 3. In this experiment, smooth opposing gradients of salt and sugar solutions were set up, with a maximum of salt concentration at the top and a maximum of sugar at the bottom. With vertical side walls, the surfaces of constant concentration were normal to the boundaries. The condition of no flux through these walls was thus automatically satisfied, and there was no tendency to set up horizontal density gradients or to produce any instability. When a sloping boundary was inserted, as shown, the density surfaces were distorted so that they could remain normal to the slope. This upset the hydrostatic equilibrium, and flows parallel to the slope developed; just above the boundary a thin downflow formed, and above it a thicker upslope counterflow. This flow did not continue indefinitely, but it rose until it reached the level where it was neutrally buoyant, and then turned outwards to form a series of approximately equally spaced layers, which developed nearly uniformly all along the sloping plate.

This kind of instability is characteristic of double-diffusive systems in which the side-wall boundary conditions do not match the conditions in

Fig. 3. Layers forming at an inclined boundary, in a fluid containing opposing gradients of two components with different diffusivity. Total depth 18 cm; (top) NaCl solution, sp. gr. 1.126; (bottom) sugar solution sp. gr. 1.137, with linear variations of properties between. This system was stable until the sloping plate was inserted, and diffusion distorted the horizontal surfaces of constant concentration.

the interior. More familiar examples are the convecting layers formed when a tank containing a stable gradient of a solute is heated or cooled from the sides. Thorpe, Hutt, and Soulsby[15] concentrated on the case of layers formed in a narrow gap, and compared their experiments with the results of a linear stability analysis. Chen, Briggs, and Wirtz[1] studied the initiation of instability along a single heated wall in a wide tank, and showed that the layer thickness is close to the lengthscale

$$l_1 = \frac{\alpha \Delta T}{\beta \, dS/dz} \tag{3}$$

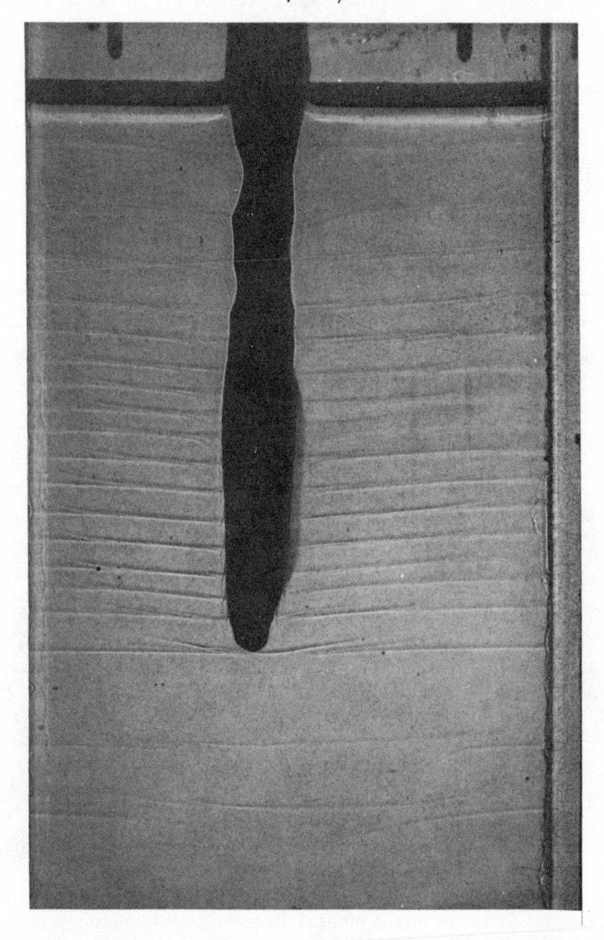

Fig. 4. A shadowgraph picture of layers formed in a weak salinity gradient by the insertion of block of ice, here viewed edge on. At the edge of the ice both the salinity and temperature fields are distorted relative to the conditions in the interior.

the height to which a fluid element with temperature difference ΔT would rise in the initial density gradient.

Another amusing example of the same effect is illustrated in Fig. 4. A weak gradient of common salt was set up by floating a layer of fresh water over salt solution and stirring gently to spread out the interface. A thin block of ice, here viewed edge on, was then inserted and layers quickly formed and spread right across the tank. In this case, of course, neither the salinity nor the temperature at the edge of the ice matched those at the same horizontal level in the interior of the fluid.

V. INSTABILITY OF LONG FINGERS

When the experimental tank is filled with opposing gradients of sugar and salt solution, with the sugar concentration (the component of lower diffusivity) a maximum at the top, fingers form rapidly and extend from top to bottom of the fluid. When there are large individual concentrations and a small net density difference, the fingers appear less stable and regular than those shown in Fig. 1, for example. Prominent in the early stages of such experiments are small-scale instabilities, which lead to a thickening of the fingers at random positions along their length. These suggest that the vertical motion driven by the potential energy in the sugar distribution is too rapid to be balanced by laminar viscosity and diffusion alone, as implied by theories such as that of Huppert and Manins.[4]

When horizontal nonuniformities are also introduced by filling the tank from one end, another kind of motion is observed. This consists of unsteady alternating motions in layers, rather like those produced by side-wall heating.[1] It leads to the distortion of the fingers away from the vertical (which is most readily seen on time lapse films), but these wave motions usually become stabilized at a finite amplitude, and the net density distribution remains smooth.

However, when a large amplitude perturbation is introduced, either by altering the flow rate during the filling process or by making a mechanical disturbance at one end of the tank, a large-scale instability can develop. In the latter case, for example, the local disturbance produced by lifting a flap on the end wall propagates rapidly right along the tank in the form of nearly horizontal wave motions. Then there is a rapid overturning, leading to a breakdown of the fingers and the formation of a convecting layer, which occurs almost simultaneously along the whole length of the tank.

A curious feature of this experiment is that a convecting layer can only be produced during a very limited time. When the fingers are allowed to "run down" for as little as 5 minutes after the tank is filled, even vigorous stirring does not lead to the instability. The two effects are shown in Fig. 5. At the bottom of the tank is a convecting layer, formed earlier in the experiment by the disturbance due to lifting the flap into a downward sloping position. The flap was later raised to the position shown, but no further breakdown of the fingers was produced at this higher level. The instability therefore depends on having both a finite initial disturbance and a large enough destabilizing sugar gradient in relation to the net density gradient. One approach to the finite amplitude stability problem posed by these experiments was suggested by Stern and Turner,[14] but there is as yet no detailed theory.

Fig. 5. The tank was filled with opposing linear gradients of sugar above salt solution (sp. gr. of sugar at top 1.130, sp. gr. of salt at bottom 1.141; depth 25 cm), and fingers extended from top to bottom. The convecting layer near the bottom of the tank was generated by setting the flap in a downward sloping position; the local disturbance propagated right along the tank, about three times as far as the width shown here. The later movement of the flap to the upward slanting position shown has produced no further breakdown of the fingers.

VI. SOURCES OF ONE FLUID IN ANOTHER OF DIFFERENT DIFFUSIVITY

When a small (2 mm diam.) source of solution is released into a tank of homogeneous fluid of nearly the same density, its behavior is quite different depending on whether the tank fluid has the same or a different composition. When both fluids are sodium chloride solution, for example, a thin plume is formed, which for small density differences remains laminar until it reaches the top or bottom of the tank. If, on the other hand, sugar solution flows at the same slow rate into homogeneous sodium chloride, the resulting flow is much more violent (see Fig. 6). Because of the different diffusion rates across the plume boundary, more salt is added

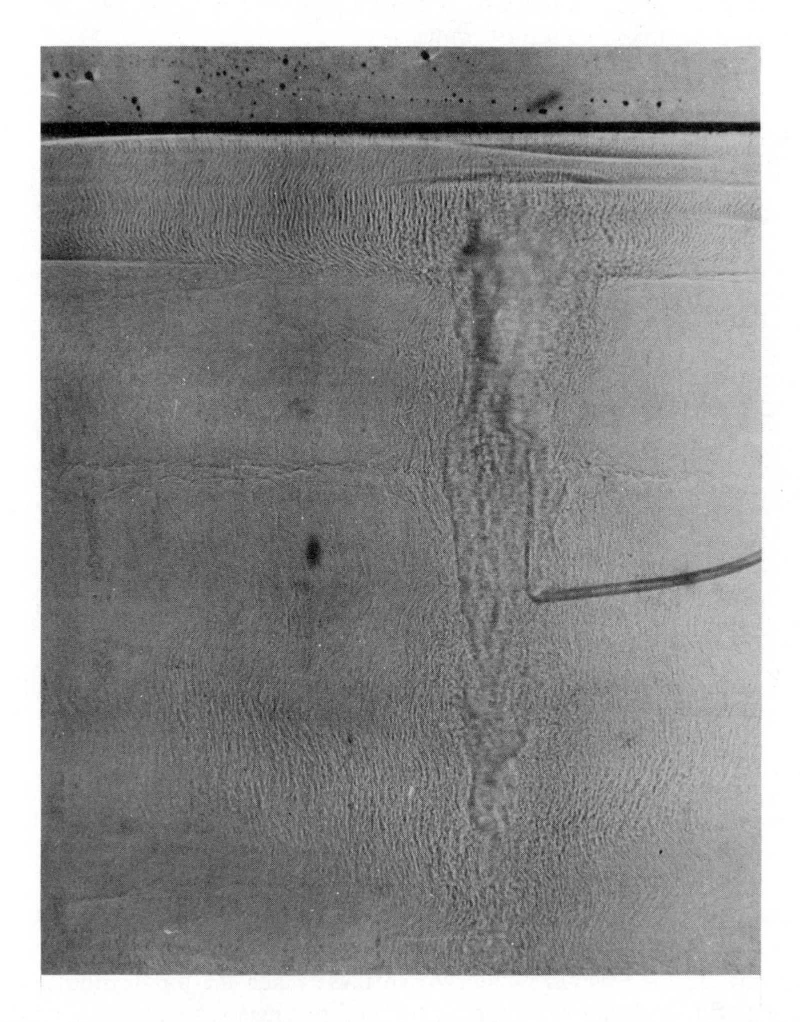

Fig. 6. The strong vertical motions which develop when sugar solution is released slowly from a 2 mm-diameter tube into homogeneous salt solution of nearly the same specific gravity (1.188). Differential diffusion creates buoyancy anomalies in both senses, forming upward and downward moving plumes which eventually lead to vertical stratification and the formation of layers.

than sugar is removed, and the plume fluid becomes heavier and its immediate surroundings lighter. This generates more rapid vertical motions, both upwards and downwards, and the separation rate is enhanced by turbulence developing in the plumes, which increases the concentration gradients and the rate of diffusion. With careful adjustment of the densities and flow rates, the plume appears to split near the orifice, with part going up and part down. Similar effects can be produced with hot salt solution released in a tank of cold fresh water.

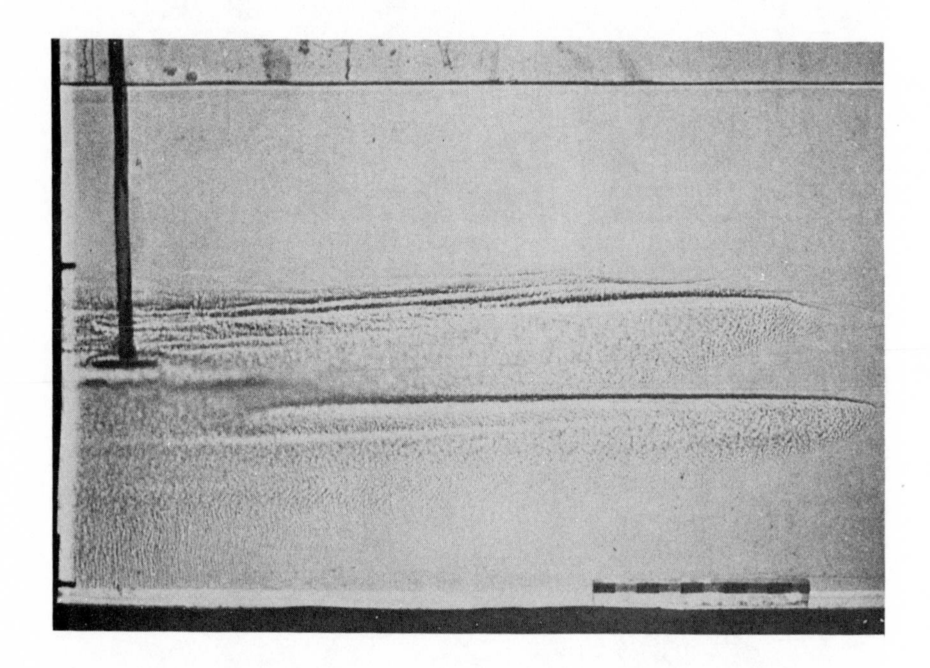

Fig. 7. The effect of a small source of sugar solution (sp. gr. 1.100) flowing at mid-depth into a linear gradient of NaCl (sp. gr. at top 1.070, sp. gr. at bottom 1.130, depth 22 cm). There is strong vertical convection in both senses near the source, followed by spreading at two levels, but not at the level of the source.

The final result of this process is to generate relatively strong vertical plumes, in both senses, in which the buoyancy fluxes are much greater than that in the original plume. When these reach the top and bottom of the experimental tank they spread out and produce strong vertical density and composition gradients. In a long tank, horizontal gradients of the two components can also be set up in this way; these have nearly compensating effects on the density so that the net horizontal density gradient is small.

When the tank of salt solution into which the small source of fluid flows is stratified at the beginning of the experiment, another surprising effect is observed. Injected fluid of the same composition and intermediate density just spreads out as a thin intrusion at its own density level. Source fluid of different composition, however, first produces strong vertical convection in both senses, as described above. This settles out at levels above and below the source and then intrudes, more rapidly than in the former case because of the larger mass fluxes in the two separated plumes. An example of such an experiment, with sugar added at its level of neutral buoyancy to a gradient of sodium chloride, is shown in Fig. 7; no intruding fluid is visible at all at the level of the source.

VII. THE TIME-HISTORY OF A SERIES OF CONVECTING LAYERS

A different kind of instability of a double-diffusive fluid is illustrated in Fig. 8. In this experiment, opposing gradients of salt and sugar were set up, with a maximum salinity at the top, decreasing linearly to zero at the bottom, and a maximum sugar concentration at the bottom going to zero at the top. The individual concentrations were large, corresponding to specific gravities of about 1.131 and 1.136 at the top and bottom, while the net density difference was small but hydrostatically stable. An initial disturbance was produced by making the tank fluid a few degrees cooler than the room, and a series of thin (about 1 cm) layers and sharper interfaces became visible on the shadowgraph. These are believed to arise

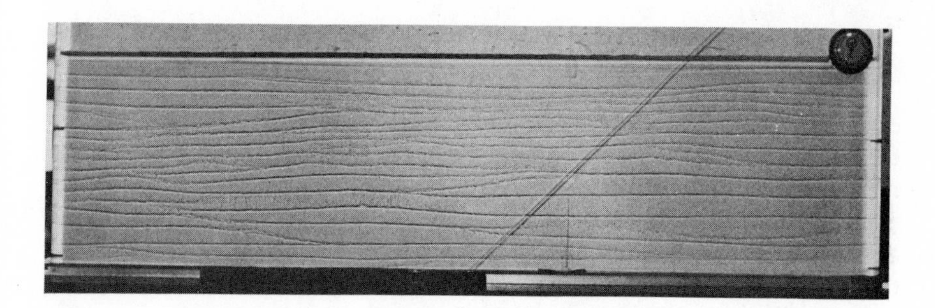

Fig. 8. The merging of thin layers formed in a double-diffusive fluid with opposing gradients. Total depth 21 cm; (top), NaCl solution, sp. gr. 1.131; (bottom), sugar solution, sp. gr. 1.136, with linear variations between. The initial disturbance was due to side-wall heating. Differential diffusion of sugar and salt has caused the densities of neighboring layers to approach the same value so that they merge. This happens at different rates in different horizontal positions, giving the appearance of waves on the interfaces.

from an inherent instability of the opposing solute distributions, rather than directly as a result of the heating, since their scale was not sensitive to the imposed temperature difference.

As the experiment continued, the convective velocities increased while the layers became more uniformly mixed. The system of layers then became unstable in another way, due to the mechanism described by Huppert[3] and referred to earlier. If the buoyancy fluxes across the interfaces bounding a layer above and below are unequal, the density changes gradually until it approaches that of one of the neighboring layers, and the two then merge. Figure 8 shows a stage of this process at which the merging has taken place at some horizontal positions but not at others, which makes it appear that there are waves on the interfaces; the effect is purely kinematic, however. After about 4 hours (in this experiment) the interfaces became nearly horizontal, but further merging took place overnight. The final layer depth was about 4 cm, indicating that there had been two successive mergers of the original 1-cm deep layers.

Quantitative experiments aimed at comparing these observations with Huppert's predictions are in progress, but the qualitative results have already shown why one must be cautious in interpreting oceanographic observations of layers. Comparing measured layer depths with the relation (1) may be unrealistic for two reasons. First, the formation process is unlikely to be one dimensional, except in very special circumstances. Secondly, several stages of merging could have occurred subsequently, thus making it meaningless to compare measurements with a theory based on the initial formation alone.

VIII. FINAL REMARKS

Many of the experiments described here are still at the qualitative, exploratory stage, and much remains to be done before all the phenomena they have revealed can be regarded as properly understood. A fundamental difficulty is that there is a big gap between the experiments and the detailed theoretical work in this field. It is very easy to produce large-amplitude, turbulent convection in the laboratory, whereas the bulk of the theory relates to the initial, infinitesimal disturbances which can be treated by linear theory. A few finite-amplitude numerical calculations have taken a step beyond these linear results, but there is a great need for more such work and for more phenomenological theories related to particular experiments.

Many applications of these ideas have already been suggested[20] in diverse fields, apart from the geophysical context in which much of the recent work has been done. It seems certain that other important double-diffusive effects and further applications remain to be discovered.

References

1. C. F. Chen, D. G. Briggs and R. A. Wirtz, "Stability of Thermal Convection in a Salinity Gradient due to Lateral Heating," *Int. J. Heat and Mass Trans.*, **14**, 57 (1971).
2. A. E. Gill and J. S. Turner, "Some New Ideas about the Formation of Antarctic Bottom Water," *Nature*, **224**, 1287 (1969).
3. H. E. Huppert, "On the Stability of a Series of Double Diffusive Layers," *Deep-Sea Res.*, **18**, 1005 (1971).
4. H. E. Huppert and P. C. Manins, "Limiting Conditions for Salt-Fingering at an Interface," *Deep-Sea Res.*, **20**, 315 (1973).
5. H. E. Huppert and J. S. Turner, "Double-diffusive convection and its Implications for the Temperature and Salinity Structure of the Ocean and Lake Vanda," *J. Phys. Oceanog.*, **2**, 456 (1972).
6. P. F. Linden, "Salt Fingers in the Presence of Grid-Generated Turbulence," *J, Fluid Mech.*, **49**, 611 (1971).
7. P. F. Linden, "On the Structure of Salt Fingers," *Deep-Sea Res.*, **20**, 325 (1973).
8. R. S. Schechter, M. G. Verlarde and J. K. Platten, "The Two-Component Bénard Problem." *Advances in Chemical Physics*, Vol. 26, I. Prigogine and S. A. Rice, Eds., Interscience, New York, 1974, p. 265.
9. T. G. L. Shirtcliffe, "Thermosolutal Convection: Observation of an Overstable Mode," *Nature*, **213**, 489 (1967).
10. T. G. L. Shirtcliffe, "Colour-Schlieren Observations of Double-Diffusive Interfaces," unpublished manuscript, 1972.
11. T. G. L. Shirtcliffe, "Transport and Profile Measurements of the Diffusive Interface in Double-Diffusive Convection with Similar Diffusivities," *J. Fluid Mech.*, **57**, 27 (1973).
12. T. G. L. Shirtcliffe and J. S. Turner, "Observations of the Cell Structure of Salt Fingers," *J. Fluid Mech.*, **41**, 707 (1970).
13. M. E. Stern, " The 'Salt Fountain' and Thermohaline Convection," *Tellus*, **12**, 172 (1960).
14. M. E. Stern and J. S. Turner, "Salt Fingers and Convecting Layers," *Deep-Sea Res.*, **16**, 497 (1969).
15. S. A. Thorpe, P. K. Hutt and R. Soulsby, " The Effect of Horizontal Gradients on Thermohaline Convection," *J. Fluid Mech.*, **38**, 375 (1969).
16. J. S. Turner, "The Coupled Turbulent Transports of Salt and Heat across a Sharp Density Interface." *Int. J. Heat and Mass Trans.*, **8**, 759 (1965).
17. J. S. Turner, "Salt Fingers across a Density Interface," *Deep-Sea Res.*, **14**, 599 (1967).
18. J.S. Turner, "The Behavior of a Stable Salinity Gradient Heated from Below," *J. Fluid Mech.*, **33**, 183 (1968).
19. J. S. Turner, *Buoyancy Effects in Fluids*, Cambridge University Press, Cambridge, 1973.
20. J. S. Turner, "Double-Diffusive Phenomena," *Ann. Rev. Fluid Mech.*, **6**, 37 (1974).
21. J. S. Turner and C. F. Chen, "Two-Dimensional Effects in Double-Diffusive Convection," *J. Fluid Mech.*, **63**, 577 (1974).
22. J. S. Turner and H. Stommel, "A New Case of Convection in the Presence of Combined Vertical Salinity and Temperature Gradients," *Proc. Natl. Acad. Sci. U.S.*, **52**, 49 (1964).

CYLINDRICAL COUETTE FLOW INSTABILITIES IN NEMATIC LIQUID CRYSTALS

PAWEL PIERANSKI

Physique des Solides,
Université Paris, Sud,
Orsay, France and

ETIENNE GUYON*

Physics Department,
University of California,
Los Angeles, California

CONTENTS

I. INTRODUCTION

We have recently presented a simplified comparative description of some convective linear instabilities leading in particular to the evaluation of thresholds.[1] The experimental data used were essentially results obtained on nematic liquid crystals where the anisotropic properties lead to a larger variety of effects than in ordinary isotropic fluids. In particular, linear (infinitesimal) instabilities can be obtained in plane Poiseuille and shear flow experiments whereas it is found that they are of the "snap

*Present address: Physique des Solides, Université Paris Sud, Orsay, France.
Supported by NSF grant 38687.

through" type in ordinary fluids. The need for a simplified description comes from the fact that the hydrodynamic behavior of liquid crystals is far more complex than that of isotopic liquids. In addition to the knowledge of the velocity field, $\vec{v}$, it requires that of a director field, $\vec{n}$, where $\vec{n}$ is a unit vector directed along the local direction of the optical axis of the uniaxial nematic fluid. We also assume that an initial uniform state of the director $\vec{n}$ is obtained by a suitable treatment of the limiting plates containing the material. Otherwise the role of the defects would play a role similar to that of dislocations in the mechanics of solids. Five coefficients having the dimensions of a viscosity are introduced in the hydrodynamic description.[2–4] Their meaning is easily seen if one considers the three shear flow geometries (*a*), (*b*), (*c*) of Fig. 1. In addition to the three viscosities, η_a, η_b, η_c measured when $\vec{n}$ remains fixed along one of the three orthogonal directions with respect to $\vec{v}$, in geometries (*b*) and (*c*) the torques α_2 and α_3 tend to distort the molecular order. The sign of α_2 is unambiguous but α_3 can be of either sign. For simplicity we will consider here only the case of Fig. 1 where α_3 is also negative.[5] In large enough shears the molecules would align at a small angle θ_0 with the direction of $\vec{v}$, such that $\tan^2\theta_0 = \alpha_3/\alpha_2$.[3]

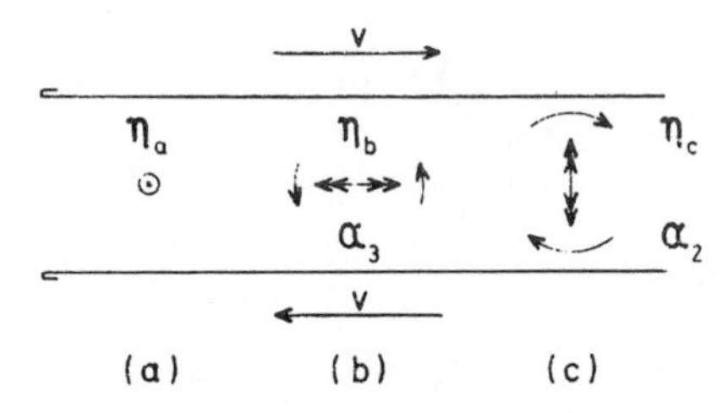

Fig. 1. The five coefficients of the hydrodynamics of nematics: three viscosity coefficients $\eta_{a,b,c}$ and two torques proportional to α_2 and α_3 ($|\alpha_3| \ll |\alpha_2|$). We consider here the case where α_3 has the sign given in the figure.

Rather than again giving a summary of the nematic liquid crystal properties, we refer to recent reviews on the subject.[6,7] We also choose not to reproduce the comparative discussion presented in Ref. 1 but will rather get into the spirit of this approach by giving here an original theoretical discussion of the Taylor instability in nematics. This simplified description leads to new qualitative predictions in the evaluation of the instability thresholds. We base it on a preliminary naive one-dimensional description on the Taylor problem in isotropic fluids.

II. TAYLOR INSTABILITY IN AN ISOTROPIC FLUID

We consider the cylindrical Couette flow of a fluid of kinematic viscosity $\nu=\eta/\rho$ (η=viscosity, ρ=specific mass) contained between two concentrical cylinders of axis z and radii R_1 and R_2 ($>R_1$), rotating at different angular velocities Ω_1 and Ω_2. When Ω_1 is larger than a critical value $\Omega_{1\,\mathrm{cr}}$ (Ω_2), a convective instability develops which is formed by toroidal rolls of axis z. The physical origin of the instability has been clearly connected, from the original work of Lord Rayleigh,[8] to the destabilizing role of the centrifugal forces (corresponding to that of the buoyancy force in the Bénard-Rayleigh heat convection problem) and the threshold expressed in terms of a dimensionless number which represents the ratio of the centrifugal force to the stabilizing viscous force. The description proposed here gives an explicit account of the coupling between the acting destabilizing variables.

We describe the instability using a one-dimensional model[9] where only the z dependence of the velocity and pressure fluctuations are retained. The finite thickness is taken into consideration by expressing that the wave vector of the modulation, $|\vec{k}|$, is of the order of π/h where $h=R_2-R_1$. Close to the threshold, the shape of the rolls does not depend on θ (rotational symmetry) and neglecting the θ variation is perfectly legitimate. Neglecting the radial dependence is a very crude simplification. However the use of such an approach has given a reasonable semiquantitative description of other convective phenomena.[10,11] We will see that it gives here the essential characteristics of the Taylor instability as long as Ω_1 and Ω_2 are close enough and are of the same sign. The term involving the fluctuation of velocity along z, v_z, does not appear in this model, as it should depend on r, nor does the fluctuating pressure term $p'(z)$, which plays a role only in the equation for v_z, appear.

We take the perturbed fluid velocity in cylindrical coordinates

$$v_r(z)$$

$$V_\theta(r)+v_\theta(z)=\omega r+\frac{C}{r}+v_\theta(z)$$

$$\omega=\frac{\Omega_2R_2^2-\Omega_1R_1^2}{R_2^2-R_1^2}\qquad C=\frac{(\Omega_1-\Omega_2)R_2^2R_1^2}{R_2^2-R_1^2}$$

The term $V_\theta(r)$ gives the unperturbed flow and satisfies to the boundary conditions for $r=R_1$ and R_2. The two components v_r and v_θ ($\ll V_\theta$) are fluctuating infinitesimal disturbances.

We look for solutions of the form

$$\frac{v_r(z)}{v_{r_0}} = \frac{v_\theta(z)}{v_{\theta_0}} = e^{st}\cos kz \tag{1}$$

The Navier Stokes equations for v_{r_0} and v_{θ_0} are

$$\frac{\partial v_{r_0}}{\partial t} - 2\left(\omega + \frac{C}{r^2}\right)v_{\theta_0} = -\nu\left(k^2 + \frac{1}{r^2}\right)v_{r_0} = -\frac{v_{r_0}}{T_v} \tag{2}$$

$$\frac{\partial v_{\theta_0}}{\partial t} + 2\omega v_{r_0} = -\frac{v_{\theta_0}}{T_v}$$

This set of coupled equations has the general form with which we have comparatively analyzed some linear instability problems (see the table in Ref. 1). The two velocity components v_{r_0} and v_{θ_0} are coupled by the two positive force terms:

$$F_{ir} = -2\left(\omega + \frac{C}{r^2}\right)v_{\theta_0}$$

represents the radial centrifugal force due to a fluctuation of the tangential velocity (i refers to isotropic) and

$$F_{i\theta} = 2\omega v_{r_0}$$

gives the contribution of the tangential Coriolis force under the influence of a fluctuation of the radial velocity. The time constants $T_v \sim (\nu k^2)^{-1}$ give the exponential damping of the velocity v when no coupling is present.

The threshold obtained by inserting (1) in (2) and putting $s = 0^+$ (principle of exchange of stabilities) is expressed by the following implicit relation

$$\frac{1}{T_v^2} + 4\omega\left(\omega + \frac{C}{r^2}\right) = 0 \tag{3}$$

This expression reproduces the qualitative features of the exact solution. In particular it leads to an infinite threshold when the necessary condition is reached $\Omega_2 R_2^2 = \Omega_1 R_1^2$ which expresses the equality of the circulations at the inner and outer rim, and is also a sufficient condition for an inviscid fluid.

Assuming a distance between cylinders $h \ll R_1$, the threshold variation obtained (Fig. 2, curve I) agrees qualitatively with that obtained with a detailed numerical calculation.[8] For example, for $\Omega_2 = 0$ we get

$$\Omega_1 = A\frac{\nu}{h(Rh)^{1/2}} \qquad \text{with} \qquad A = \frac{\pi^2}{(2)^{1/2}}$$

instead of 41.3.

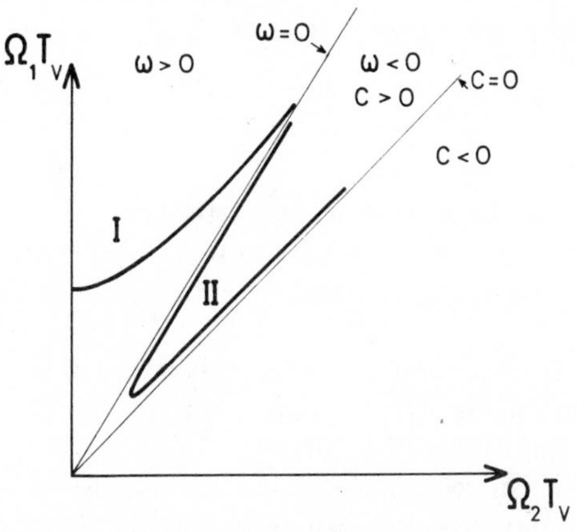

Fig. 2. Instability threshold in the isotropic (Curve I) and nematic geometry b (Curve II) problems. The line $\omega = 0$ corresponds to the change of sign of the Coriolis force; the line $C = 0$ corresponds to the change of sign of the centrifugal force. The minimum value of $\Omega_1 T_v$ is of the order of 1 in the isotropic case and $\lambda^{-1/2}$ in the nematic one.

The large difference in the numerical factors is clearly related to the crudeness of the model and the direction of this error is understandable: In a one-dimensional description we overestimate the value of the time constant T_V as the effect of viscosity comes from the variation of the velocity along r as well as along z. But we insist again that the purpose of this simplified view is to give enough physical insight to approach the more complex nematic problem.

III. TAYLOR INSTABILITY IN NEMATICS

The nematic film is aligned uniformly under the influence of the limiting cylindrical surfaces.

A. Case of Geometry b
(The alignment corresponds to that of Fig. 1b in the plane case.)

We consider first the case where the molecules are parallel to the velocity $V_\theta(r)$ in the unperturbed state. We take an infinitesimal distorsion of the director, such that $\vec{n}=n_r(z)$, 1, $n_z(z)$ (with $|n_r|$ and $|n_z|\ll 1$), periodic along z with a wave vector $\vec{k}$. In the hydrodynamic equations of nematics, we consider the contribution due to first-order perturbation terms depending on z. (We do not consider the small uniform distorsion due to the shear which would saturate to a small angle θ_0 for the large shears.) A detailed but standard calculation of the viscous stress tensor,[2-4] which there is no need to reproduce here, leads to a new term of radial force depending on z and connected with an anisotropic viscosity

$$F_{ar}=(\eta_b-\eta_c-\alpha_3)\frac{C}{r^2}\frac{\partial n_z}{\partial z} \tag{4}$$

The term α_3 is small compared to the difference of viscosities $\eta_b-\eta_c$. The force F_{ar} can be understood as a hydrodynamic focusing term related to the anisotropy of the viscosity when the molecular orientation varies in the θz plane.

We also consider the equation giving the relaxation of the director under the effect of a destabilizing hydrodynamic torque due to the gradient of the velocity $v_\theta(z)$ (similar to that of Fig. 1*b*) and of the elastic force characterized by an elastic coefficient K (in this one-dimensional model we neglect the anisotropy of the elasticity as well as the distortion along r

$$(\alpha_3-\alpha_2)\frac{\partial n_z}{\partial t}+\alpha_3\frac{\partial v_\theta}{\partial z}-Kk^2n_z=0 \tag{5}$$

At threshold the compatibility equation between equations (2) modified by (4), (2′), and (5) can be written as

$$\frac{1}{T_v^2}+4\omega\left(\omega+\frac{C}{r^2}\right)-\frac{2\omega C}{r^2}\frac{\alpha_3(\eta_b-\eta_c-\alpha_3)}{\rho K}=0 \tag{6}$$

Because of the anisotropic behavior of the liquid crystal, the last term of this equation is in practice very much larger than the second term which gives the coupling in the isotropic case and can be neglected. (For MBBA which is nematic at 20°C, $\lambda=\alpha_3(\eta_b-\eta_c-\alpha_3)/\rho K\sim 10^3$.) This is a large factor because the relaxation time of a fluctuation of the director is very slow compared to the diffusion time of a velocity fluctuation, T_v.[13] A similar effect also explains why the thermal convection threshold could be lower by a factor 10^3 in nematics compared with isotropic liquids of comparable average properties.[12]

Let us note that the domain of existence of this instability is given by the necessary condition $\omega C<0$ or $\omega>0$. This implies a change of sign of the Coriolis force with respect to the isotropic case, as can be seen by comparing Fig. 3*a* and 3*b*. The variation of threshold is given schematically as a function of Ω_1 and Ω_2 in Fig. 2, curve *II*. We also note that in the domain of existence of the isotropic instability ($\omega>0$) the anisotropic force F_{ar} is in opposition to the centrifugal one. The convective threshold should then be much larger than in an isotropic medium.

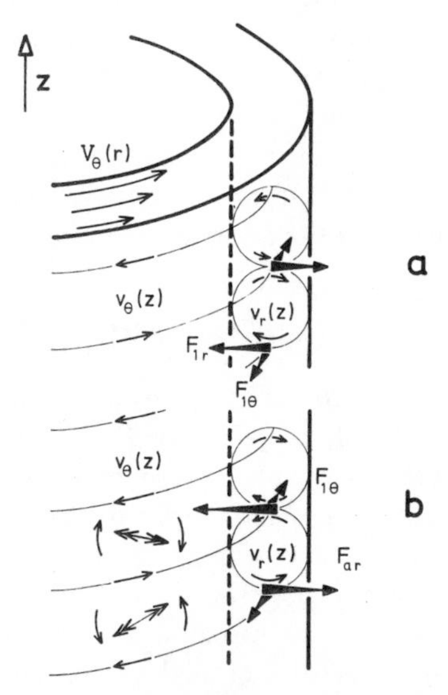

Fig. 3. Taylor instability of a cylindrical Couette flow. The variation along z of the infinitesimal disturbances v_r and v_θ is considered. (*a*) Isotropic case: the coupling between v_r and v_θ is due to the action of the centrifugal $F_{ir}(\alpha v_\theta)$ and Coriolis $F_{i\theta}(\alpha v_r)$ forces. (*b*) Nematic case: the unperturbed direction of the director (shown as ←←——→→) is along V_θ. The tangential disturbance $v_\theta(z)$ induces a destabilizing torque on n proportional to α_3 $(\partial v_\theta/\partial z)$. In the velocity field $V_\theta(r)$, the distortion space $n_z(z)$ induces a hydrodynamic focusing force which couples with the isotropic Coriolis force. (Note the difference of sign of this force from that of Fig. 3*a*, related to the different domain of existence of the instability on the diagram of Fig. 2.)

B. Case of Geometry *c*

Here the molecules are perpendicular to the cylinder walls. Fluctuation of the director such that $\vec{\Omega}=(1,n_\theta(z),n_z(z))$ are considered. The explicit expression of the stress tensor leads to a new first-order term in force depending on z:

$$F_{a\theta}=(\eta_a-\eta_b)\frac{C}{r^2}\frac{\partial n_z}{\partial z} \tag{7}$$

A description of this force is similar to that of the hydrodynamic focusing force (4). The coupled play of this tangential force and of the isotropic centrifugal term should give an instability when $\Omega_2>\Omega_1$[14] ($C<0$) by a mechanism similar to that described in Ref. 1. However, we cannot neglect here the effect of the uniform distortion of $\vec{n}$ in the $r\theta$ plane under the influence of the velocity gradient $dV_\theta(r)/dr$. It turns out that, for typical shears at thresholds, estimated without considering this distortion, the distortion should practically have reached the saturation angle θ_0. Our solution is not consistent and the problem in geometry c reduces practically to that in geometry b as the angle θ_0 is very small.

C. Case of Geometry *a*

In a previous work[15,16] we have given a detailed description, both experimental and theoretical, of a shear flow linear instability in the corresponding plane geometry with $\vec{n}$ perpendicular both to the velocity and velocity gradient. In a plane geometry this was the only configuration among the three considered here which could give rise to an observable linear instability threshold. Two types of instabilities were obtained. In a first homogeneous regime[15] the director distorts uniformly across the sample when the shear exceeds a critical value. The mechanism for the instability is given in the caption of Fig. 4. Another type of instability falls in the category of rolls having their axis parallel to the velocity.[16] This instability involves the action of the hydrodynamic focusing terms as the director is distorted from point to point along the direction of the unperturbed axis of the liquid crystal.[17] The calculations leading to these modes can be extended easily to the case of a cylindrical geometry where the perturbed director $\vec{n}=(n_r(z), n_\theta(z), 1)$. The threshold for both homogeneous and roll instabilities is much smaller than for the geometry b. In the latter problem we study the coupled relaxation of the two components n_r and n_θ which both relax very slowly, whereas the problem b still considers the coupling between a fluctuating component of $\vec{n}$ and one velocity component which relaxes more rapidly.

Experiments are in progress at Orsay to study these different Couette flow problems.

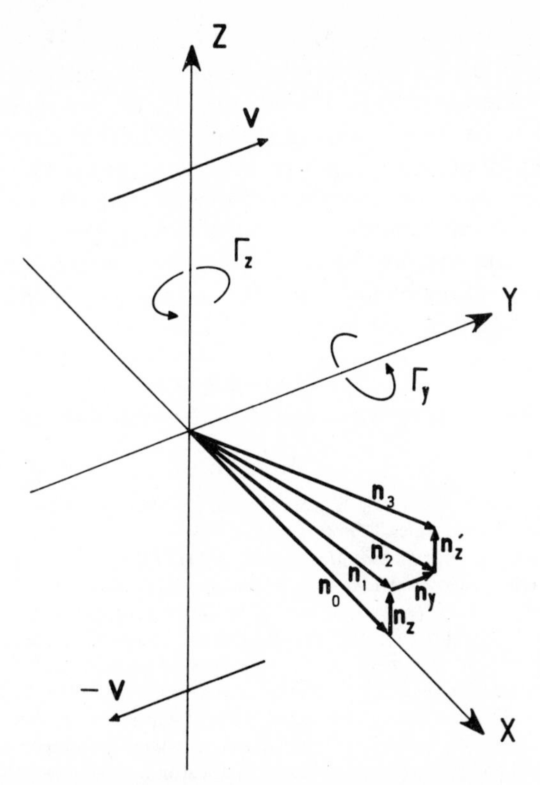

Fig. 4. The mechanism for the development of an instability in the case of Fig. 1*a*. We start with an initial fluctuation $\mathbf{n}_z$. The flow creates a torque Γ_z which tends to displace the extremity of $\mathbf{n}_1$ along the flow line and brings $\mathbf{n}_1$ to $\mathbf{n}_2=\mathbf{n}_1+\mathbf{n}_y(|\mathbf{n}_y|\ll\mathbf{n}_0|)$. At this stage we look at the projection of $\mathbf{n}$ on the yz-plane. We are in the situation described by Fig. 1*b* and, according to the sign of the torque Γ_y acting on it, $\mathbf{n}_2$ moves towards $\mathbf{n}_3=\mathbf{n}_2+\mathbf{n}_z$, thus increasing $\mathbf{n}_z$. This leads to instability. (See Ref. 15).

IV. CONCLUSION

We have presented the simplified solution of the Couette flow instabilities in nematics as a model problem to review the typical original features which have been met in the various convective instability problems in these materials: electrohydrodynamic,[10,11,18] thermal,[12] and hydrodynamic[15,16] problems. In the various problems the role of a focusing term, reinforcing an initial fluctuation, was found. The convective flow itself couples with the molecular orientation and reinforces the focusing effect. The low threshold values are associated with the slow relaxation of the director under the restoring action of the elastic constant.

In addition to providing new types of instabilities, the sensitivity to fields and the versatility of liquid crystals offers new tools for these studies. The relaxation time constant and, consequently, the instability thresholds can be varied over a large range by the application of electric and magnetic fields. The strong anisotropic properties of the materials allow a very accurate control of the molecular distortion. Finally the rapid change of properties with temperatures (viscosities varying typically by a factor of 10 over 20°C) and the possibility of obtaining several liquid phases in the presence of a strong enough temperature gradient open new doors in the physics of convective phenomena.

Acknowledgments

A critical discussion of this work with P. G. de Gennes is acknowledged.

References

1. E. Guyon and P. Pieranski, "Proceedings of the Van der Waals Centennial Conference on Statistical Mechanics," *Physica*, 185 (1974).
2. J. L. Ericksen, *Arch. Ration. Mech. Anal.* **4**, 231 (1960).
3. F. M. Leslie, *Quart. J. Mech. Appl. Math.*, **19**, 357 (1966).
4. O. Parodi, *J. Phys. (Paris)*, **31** 581 (1970) has obtained a linear relation between the coefficients by expressing the Onsager relations. However a sixth coefficient is obtained in a flow with an alignment intermediate between (b) and (c) of Fig. 1.
5. We have analyzed recently (P. Pieranski and E. Guyon, *Phys. Rev. Lett.*, **32**, 924 1974) the hydrodynamic shear flow instabilities, obtained in the complementary problem $\alpha_3>0$ where no equilibrium angle is found, which are quite different from the one discussed here. In particular, we have obtained in a given case a cascade of instabilities of the molecular orientation.
6. A. Rapini, *Progr. Solid State Chem.*, **8**, 337 (1973).
7. P. G. de Gennes, *The physics of Liquid Crystals* Oxford Press (1974).
8. For a detailed discussion on the subject and references we refer to the work of S. Chandrasekhar, *Hydrodynamic and Hydromagnetic Stability*, Clarendon Press, London, 1961.
9. See also "Hydrodynamic Stability" by J. T. Stuart in *Laminar Boundary Layers*, L. Rosenhead, Ed., Clarendon Press, Oxford, 1963.
10. W. Helfrich, *J. Chem. Phys.*, **51**, 2755 (1969).
11. E. Violette Dubois, P. G. de Gennes, and O. Parodi, *J. Phys.* (*Paris*), **32**, 305 (1971).
12. E. Violette Dubois, P. Pieranski, and E. Guyon, *Molecular Crystals and Liquid Crystals*, **26**, 193 (1974).
13. In the expression of the time constant T_V, the viscosity η_b related to the geometry has to be used. In practice we feel that a more exact evaluation, including in particular the anisotropic elastic constants, would be very heavy and cumbersome in the absence of experimental data.
14. The change of sign on $\Omega_1-\Omega_2$ in the necessary condition for instability, if one could neglect the uniform distortion, is very analogous to the problem of heat convection in nematics when the molecules are perpendicular to the walls. We have found that in this case thermal convection was obtained by heating from above. (P. Pieranski, E. Dubois Violette and E. Guyon, *Phys. Rev. Lett.*, **30**, 736, 1973).

15. P. Pieranski and E. Guyon, *Solid-State Commun.*, **13**, 435 (1973).
16. P. Pieranski and E. Guyon, *Phys. Rev. A.*, **9**, 404 (1974).
17. A comparison between the threshold of the homogeneous and roll instabilities involves a detailed numerical calculation. However in the case $\alpha_3 > 0$ (see Ref. 5) only the roll instability can be obtained.
18. J. Williams, *J. Chem. Phys.*, **39**, 384 (1963); W. Helfrich, *J. Chem. Phys.*, **51**, 2755 (1969); Orsay Liquid Crystal Group, *Mol. Cryst. Liquid Cryst.* **13**, 187 (1971).

THEORETICAL AND EXPERIMENTAL STUDY OF STATIONARY PROFILES OF A WATER–ICE MOBILE SOLIDIFICATION INTERFACE

PIERRE R. VIAUD

Laboratoire d'Aérothermique du C.N.R.S., Meudon, France

CONTENTS

I. INTRODUCTION

The essential object of this work is the study, experimental and theoretical, of the stationary geometrical forms adopted by a water–ice solidification front progressing in a liquid phase (freezing) under varied external conditions.

From a general point of view, phase-change phenomena cover a vast and varied domain, embracing equally the physics of solids, minerology, crystallography as physical chemistry, mathematics, and metallurgy. Such a subject can therefore be approached from many points of view, differing one from another, according to the particular branch of physics concerned.

However, if one wishes to follow a logical scheme and look for a coherent representation of the question, it is indispensable to start from very wide concepts, such as thermodynamics offers, before going into more specific domains, such as those concerning mass transfer or dislocations. It seems necessary also, to analyze relatively simple situations with elementary conditions first before approaching more complicated situations with more elaborate conditions.

Crystallization processes have long been the object of numerous experimental, applied, and theoretical studies.[1] Nevertheless, a certain number of important questions remain concerning this problem, and our inability to treat crystallization from a global point of view shows that considerable research is still indispensable in this domain.

In particular, as Parker has shown in *Crystal Growth Mechanisms*,[2] one of the trickiest points is the introduction of the discontinuity which appears during phase changes.

One could imagine that, as the temperature decreases, the molecules composing a liquid gradually modify their liquid character (disorder at long range) in favor of the solid character (order at long range). But in fact, every experiment on phase changes proves that the process appears only in a minimal fraction of the volume occupied by the system considered: the interface separating the liquid from the solid.

Thermodynamics describes these phenomena as transitions of the first order. The latent heat corresponds to a rapid decrease of the entropy S of the system and to an abrupt change of slope of the curve representing, for example, the variation at constant pressure, of the free energy G of the system as a function of temperature T.

This spatial discontinuity, due to the fact that the system is composed of two phases having different properties, is defined mathematically on the curve $G(T)$ by the intersection of the two branches $G_s(T)$ and $G_l(T)$ characterizing each phase.

Nevertheless, Parker[2] remarks that, if thermodynamics can describe a transformation of this type, it is utterly incapable of predicting or demonstrating the appearance of such a transformation. Furthermore, as phase changes are spatially discontinuous phenomena, occuring only in certain local regions of the system, transport phenomena will necessarily appear which, together with interfacial phenomena, will control the phase change.

The transport phenomena are generally treated from the macroscopic point of view using the equations of mass and heat diffusion. These studies can be roughly grouped in the literature under the heading "The problem of Stefan".[3,4]

Interfacial phenomena demand, for their part, an examination at the molecular level.

Consequently, according to the system considered and the boundary conditions imposed, the relative importance of transport and interfacial phenomena is very variable; thus the geometric form of a solidification interface, for example, is in fact controlled by different processes, either cooperative or antagonistic. It is at this level, in fact, that the principal difficulties of the problem occur.

The equilibrium conditions of two-phase system can be described. One has sufficient elements to allow an approach to the problem of interfacial thermodynamics.[5,6,7] It is known that the phenomena studied are irreversible and that the liquid–solid transformation concerns the thermodynamics of systems out of equilibrium[8,9]; but the prediction possibility and the complete description of such phenomena have never been proposed.

Of course, the work presented here does not pretend to fill this gap, but is proposed simply as a formulation of the stability problem based on the thermodynamics of irreversible processes and illustrates several aspects from a qualitative point of view.

II. THE STABILITY OF AN INTERFACE

The geometric form of a moving solidification interface is of a capital importance in the quality of the solid material obtained. It is because of this that numerous studies have been made on metals, alloys, eutectics, and so on.[10,11]

Twenty years ago, Rutter and Chalmers[12,13] proposed a criterion predicting the instability of an initially plane interface the moment that solidification conditions are such that a boundary layer appears at the interface in the liquid phase, the Constitutional Supercooling.

This criterion gave interesting results[12] and was revised later by numerous authors who analyzed crystalline growth and stability of interfaces of varied geometric forms:

- Spherical or cylindrical[14–17]
- Ellipsoidal[18,19]
- Parabolic with a circular or elliptic base[20]
- Polygonal[21,22]
- Varied forms (needles, flakes)[23]

Ten years ago, in a series of very well known articles[24,25] Mullins and Sekerka developed for the analysis of crystalline growth a qualitative criterion for the stability of form. These authors have since added numerous improvements and extensions to their initial work.[26–28]

One of the most important justifications for these studies is that, in many natural and industrial processes, a solidification interface occurs with forms, perturbed and irregular to a greater or lower extent, having nothing in common with a planar, cylindrical, or spherical form.

In most of the work cited above, the diffusion of heat and the diffusion of mass are treated simultaneously. The problem, without difficulty for a planar interface, rapidly becomes very complex, insoluble even for non-planar interfaces. As Mullins and Sekerka[25] note, concerning the work of Temkin[29] on a hemispherical perturbation, the solution of the problem in not unique and does not fulfill all the boundary conditions.

However, before pursuing this study, it is perhaps necessary to note that the phrase "stable interface" encompasses two different concepts:

1. To verify if a geometric form of an interface, chosen as fulfilling the boundary conditions, leads to a solution of the diffusion equations when the crystallization occurs.
2. To analyze the behavior of small perturbations imposed on a chosen interfacial form (generally plane, cylindrical, or spherical).

Concerning this, one should note certain interpretations of the theory of small perturbations which are too wide.

Indeed, let us imagine any stationary system, characterized by certain parameters unchanging in the time. In the precise case that concerns us here, supposing that the velocity of the liquid–solid interface is uniform is the same as saying that the form of the interface is invariant during the evolution of the crystallization process.

This remark shows a notable difference which exists between a process of solidification in one direction where effectively a stationary state can be defined, and a radial or spherical solidification process where the radius of curvature of the interface, and its area, are increasing functions of time.

It is therefore necessary to precise with respect to which variable the stationary state is defined. Herein, where we will consider only one-dimensional solidification processes, an interface is stationary if its velocity of displacement and its geometrical form are invariant in time. On the other hand, the classic approach of these phenomena consists of taking a reference frame attached to the interface for which a uniform displacement velocity is supposed; thus one defines, explicitly or implicitly, a control volume linked to the interface, which is a closed system (the difference of the liquid and solid densities is neglected in general).

Within these hypotheses, the only irreversible processes that can exist in the system are diffusive ones; the equations describing the phenomena being linear (Fourier's equation for heat diffusion and that of Fick for the mass), Glansdorff and Prigogine[30] have shown that no instability phenomenon can occur.

In particular, a small perturbation of form imposed on a stationary interface can only regress and can never lead the system to a stationary state different from the initial one, unless one of the initial hypotheses is not respected.

This point can be illustrated with the criteria of Mullins and Sekerka,[25] for example. For a pure material, of thermal conductivity λ, labeling solid and liquid phases s and l and calling the thermal gradient at the interface G, the criteria of Mullins and Sekerka can be written:

$\lambda_s G_s + \lambda_l G_l > 0$ stable planar interface (perturbations regress)

$\lambda_s G_s + \lambda_l G_l < 0$ unstable interface (perturbations are amplified)

The second inequality (criterion of instability) implies:

$$G_l < 0 \qquad \text{and} \qquad |\lambda_l G_l| > |\lambda_s G_s|$$

Now, it is true that a negative gradient in the liquid provokes the development of a sawtooth interface, in a closed system; but it is impossible to define a stationary state with a constant negative gradient in the liquid phase, which was one of the hypothesis defining the initial state.

Moreover, the existence of any stable interfacial form other than plane is excluded, while experience shows that stable macroscopically perturbed interfaces can be obtained in nature as in the laboratory.

To overcome this difficulty, we examine the question in a different manner. First of all, we construct a model sufficiently simple so as to obtain easily handled analytic solutions. Solidification processes in an open system are then considered; in this way thermal conduction and convection phenomena can be introduced naturally, which is a necessary condition for obtaining several stationary states, depending on the external conditions imposed on the system.

In such systems, the entropy production $P(S)$ is evaluated. As it is known that $P(S)$ is a definite positive (or zero) function,[30] one possesses therefore a general evolution criterion for these system in the neighborhood of equilibrium.

Indeed, with these hypotheses, if one supposes, *a priori*, the existence of two different stationary states, ① and ②, the system occupies state ① or ② according to the sign of $\int_{①}^{②} P(S)\,dt$. The system will remain in ① or evolve spontaneously from ② to ① if $\int_{②}^{①} P(S)\,dt > 0$ and vice versa if $\int_{②}^{①} P(S)\,dt < 0$.

Moreover, as states ① and ② are supposed stationary [$P_{①}(S)$ and $P_{②}(S)$ independent of t] the preceding criterion can be written:

$$\begin{cases} \dfrac{P_{②}(S)}{P_{①}(S)} < 1 & \text{state ① stable with respect to ②} \\ \dfrac{P_{②}(S)}{P_{①}(S)} > 1 & \text{state ② stable with respect to ①} \end{cases} \quad (1)$$

With (1), therefore, one possesses a criterion allowing us to predict the state of a system the moment that $P(S)$ is known. Two remarks must nevertheless accompany this scheme:

- Local equilibrium must be supposed, which assumes sufficient collisional effects to exclude large deviations from statistical equilibrium. For the solid–liquid systems considered herein, this hypothesis is not very restrictive and is supposed to be always verified.
- One cannot hope to analyze the time evolution of the transition from the states ① ⇄ ② since the hypothesis concerning the stationary of the states is no longer respected.

III. MATHEMATICAL MODEL

The model used is based on the principle of the zone melting process,[31] well known in metallurgy. It permits the study of phase changes on a band of material of known physical characteristics that can be solidified or melted by moving a thermal device parallel to this band. The device comprises a heat source and sink separated by unit distance and is called the oven. All space coordinates are given with respect to this distance and in consequence are dimensionless.

The side walls limiting the system are adiabatic. A two-dimensional model with well-defined boundary conditions (Fig. 1) is thus defined.

The following hypothesis are supposed true:

- The difference of density between the liquid and solid phases is negligible.
- The material is pure (no mass flux).
- The physical characteristics of the liquid and solid phases are constant and independent of temperature.
- The liquid and solid phases are isotropic and homogeneous.
- The system is stationary.
- The pressure is constant and equal to normal atmospheric pressure.
- The effects of the kinetics of crystallization are, for the moment, negligible; that is to say that the interface temperature T_i remains equal to

T_0 (the solid–liquid equilibrium temperature) whatever the velocity V of the oven.

- There is no thermal resistance between the material and the oven.

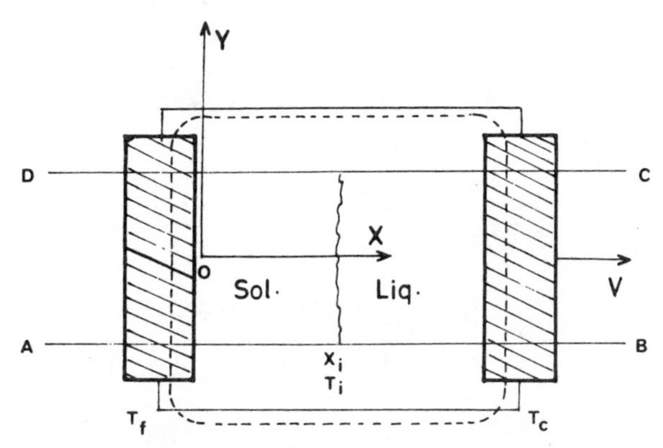

Fig. 1. Model used for the study of phase change. A heating system, maintained at T_C and a cooling system at T_F move together at speed V along a band of matter $ABCD$ centered on $0X$. Whatever the values of T_C, T_F, and V constants, the system attains after a certain time, a stationary state with a solid–liquid interface at the abscissa X_i.

The general equation for such a system is:

$$\frac{\partial T}{\partial t} - \alpha \nabla^2 T = 0$$

In the reference frame of the cold source, in which the system is stationary, one has simply:

$$\nabla^2 T + V/\alpha \nabla T = 0 \tag{2}$$

where: T is the temperature, V the driving velocity of the oven, $\alpha = \lambda/\rho c$ the thermal diffusivity, λ the thermal conductivity, c the specific heat at constant pressure, and ρ the density.

Without restricting the generality of the problem, it can be supposed that a solidification is performed, which means that the source–sink device moves at velocity V in the direction of increasing x.

If it is supposed that the interface is plane, (2) can be solved without difficulty for the liquid phase (index L) and solid phase (S) with the boundary conditions:

$$\begin{cases} x=0 & T_S=T_F \\ x=X_i & T_S=T_L=T_0 \quad \text{and} \quad \lambda_S\left(\dfrac{\partial T_S}{\partial x}\right)_{X_i}-\lambda_L\left(\dfrac{\partial T_L}{\partial x}\right)_{X_i}=\rho_S \mathfrak{L} V \\ x=1 & T_L=T_C \end{cases} \tag{3}$$

where: $\mathfrak{L}$ is the latent heat of solidification per unit mass and X_i is the position of the stationary interface, one finds:

$$T_S=T_F+\frac{\Delta T_F\{\exp[-(V/\alpha_S)x]-1\}}{\exp[-(V/\alpha_S)X_i]-1} \tag{4}$$

$$T_L=T_C-\frac{\Delta T_C\{\exp[-(V/\alpha_L)(x-1)]-1\}}{\exp[-(V/\alpha_L)(X_i-1)]-1} \tag{5}$$

$$\partial T_S/\partial x=-\frac{V}{\alpha_S}\frac{\Delta T_F\{\exp(-V/\alpha_S)x\}}{\exp[-(V/\alpha_S)X_i]-1} \tag{6}$$

$$\partial T_L/\partial x=\frac{V}{\alpha_L}\frac{\Delta T_C\{\exp-(V/\alpha_L)(x-1)\}}{_L\exp[-(V/\alpha_L)(X_i-1)]-1} \tag{7}$$

where X_i is the solution of:

$$\frac{(\rho c)_S\exp[-(V/\alpha_S)X_i]\Delta T_F}{\exp[-(V/\alpha_S)X_i]-1}+\frac{(\rho c)_L\exp[-(V/\alpha_L)(X_i-1)]\Delta T_C}{\exp[-(V/\alpha_L)(X_i-1)-1]}+\rho_S\mathfrak{L}=0 \tag{8}$$

and $\Delta T_F=T_0-T_F$ and $\Delta T_C=T_C-T_0$.

To simplify the notation let us introduce:

$$\exp[-(V/\alpha_S)x]=x_S\leqslant 1 \qquad \exp[-(V/\alpha_L)(x-1)]=x_L\geqslant 1$$

$$\exp[-(V/\alpha_S)X_i]=x_{iS}<1 \qquad \exp[-(V/\alpha_L)(X_i-1)]=x_{iL}>1$$

A. Closed System

Let us examine first the trivial problem where $V=0$, corresponding to a plane interface in a closed system. The temperatures T_S and T_L are linear functions of x and the thermal gradients are constants in each phase.

The position X_0 of the interface is given by:

$$X_0 = \frac{1}{1+(\lambda_L \Delta T_C / \lambda_S \Delta T_F)} \tag{9}$$

One notices that in (9), for a given material, the position of the interface X_0 depends only on the ratio $\Delta T_C/\Delta T_F$; thus, there exists an infinity of coupled values T_C, T_F imposing the same position X_0 for the solidification front.

From the point of view of equilibrium thermodynamics this result would seem paradoxal; but in reality, even though the interface is stationary in a closed system, the system is not in thermodynamic equilibrium because it is subject to a constant nonzero heat flux.

One can nevertheless differentiate solutions giving the same X_0 by evaluating their relative entropy productions.

In this particular system, the entropy production $P_0(S)$ can be written easily[30] since the only irreversible phenomenon is the heat diffusion. If $\sigma_0(S)$ is the entropy production per unit time per unit volume or the entropy source, one has for any system of volume v containing a pure material of thermal conductivity λ,

$$P_0(S) = \lambda \int_v \sigma_0(S)\, dv \geqslant 0 \tag{10}$$

where $\sigma(S)$ takes the particularly simple form:

$$\sigma_0(S) = \lambda \left(\frac{\nabla T}{T} \right)^2 \tag{11}$$

and for the two-phase system considered here, one has:

$$P_0(S) = \lambda_S \int_0^{X_0} \left(\frac{\nabla T_S}{T_S} \right)^2 dx + \lambda_L \int_{X_0}^{1} \left(\frac{\nabla T_L}{T_L} \right)^2 dx \tag{12}$$

Together with (7) one can deduce that:

$$P_0(S) = \frac{(T_C - T_F)}{T_L T_F} (\lambda_S \Delta T_F + \lambda_L \Delta T_C) \tag{13}$$

It is immediately clear that the absolute minimum of P_0, that is $P_0 = 0$, is obtained for $T_C = T_F$ which corresponds to a closed isothermal system, while if $T_F < T_0 < T_C$, P_0 is strictly positive. Figure 2 shows a network of curves giving the variation of $P_0(S)$ as a function of $\Delta T_C/\Delta T_F$ versus ΔT_F and we can see that the nonunivocality $X_0 \rightarrow (T_C, T_F)$ seen above vanishes.

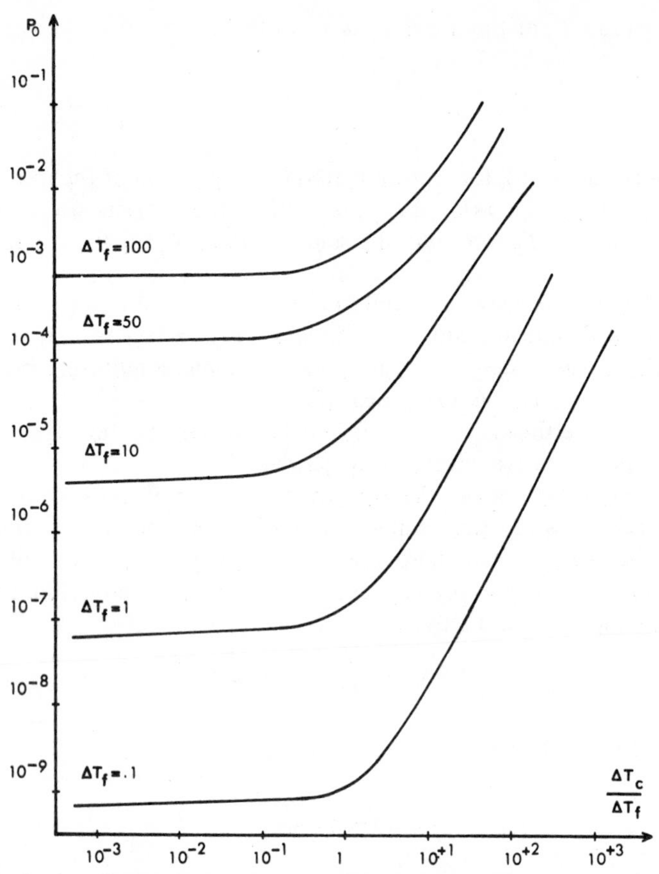

Fig. 2. Production of entropy as a function of the conditions at the limits T_F and T_C when the motion V is zero (closed system).

The results given in Fig. 2, as well as all those which are presented later on, correspond to a water–ice interface with the following numerical values:

$$\rho_S = 0.92 \text{ g/cm}^3 \qquad \rho_L = 1 \text{ g/cm}^3$$

$$(\rho c)_S = 0.46 \text{ cal/(°C) (cm}^3) \qquad (\rho c)_L = 1 \text{ cal/(°C) (cm}^3)$$

$$\lambda_S = 5.23 \times 10^{-3} \text{ cal/(°C) (sec)/(cm)}$$

$$\lambda_L = 1.35 \times 10^{-3} \text{ cal/(°C) (sec)/(cm)}$$

$$\alpha_S = 11.4 \times 10^{-3} \text{ cm}^2\text{/sec} \qquad \alpha_L = 1.35 \times 10^{-3} \text{ cm}^2\text{/sec}$$

$$\mathfrak{L} = 80 \text{ cal/g} \qquad \gamma_{eg} = 20 \text{ ergs/cm}^2 \text{ (surface energy)}$$ [46]

B. Open System

Let us now consider the case where V, oven-moving velocity is positive; this describes a solidification process. Relations (4) to (7) which are necessary to compute $P(S)$ as given by (12) can be used only if one uses the explicit form (8) of the function $X_i(V)$. But from (8), one cannot obtain the analytical expression for the function $X_i(V)$ because the arguments of the exponentials are very different ($\alpha_S \cong 10\alpha_L$) and, moreover, not necessarily small. One must then use together both limited expansions (with caution, V very small) and numerical calculation. Let us first assume that the arguments $x_{iS} = \exp[-(V/\alpha_S)X_i]$ and $x_{iL} = \exp[-(V/\alpha_L)(X_i - 1)]$ are much smaller than unity; expanding (8) up to first order, we obtain for the function $X_i(V)$ the explicit form:

$$X_i = \frac{\lambda_S \Delta T_F}{\rho_S \mathfrak{L} V + \lambda_S \Delta T_F + \lambda_L \Delta T_C} \tag{14}$$

Expression (14) allows us to specify the analytical structure for small V of the functions (4) to (7) defined previously.

For instance, with (14), relation (7) for the thermal gradient in the liquid at the interface becomes after expansion up to first order:

$$\left(\frac{\partial T_L}{\partial x}\right)_{X_i} = \frac{\Delta T_c(\rho_S \mathfrak{L} V + \lambda_S \Delta T_F + \lambda_L \Delta T_C)}{(\rho_S \mathfrak{L} V + \lambda_L \Delta T_C)} \tag{15}$$

One then has

$$\frac{d(\partial T_L/\partial x)_{X_i}}{dV} = -\frac{\rho_S \mathfrak{L} \lambda_S \Delta T_F \Delta T_C}{(\rho_S \mathrm{L} V + \lambda_L \Delta T_C)^2} \tag{16}$$

Therefore, for small V, the gradient in the liquid decreases starting from its initial value (17) for $V=0$:

$$\left(\frac{\partial T_L}{\partial x}\right)_{X_i,\, V=0} = \frac{\lambda_S}{\lambda_L} \Delta T_F + \Delta T_C \tag{17}$$

On the other hand, starting from (7), one can look for the limit of $(\partial T_L/\partial x)_{X_i}$ when V tends to infinity because one has

$$\lim_{v \to \infty} X_i = 0$$

and

$$\lim_{v \to \infty} \frac{(V/\alpha_L)\Delta T_C}{1 - x_{iL}} = +\infty$$

As $(\partial T_L/\partial x)_{X_i}$ is a continuous differentiable function for every X_i in the interval $]0,1[$, for every $V \geqslant 0$ it has thus at least one minimum.

Figure 3 gives the variations for the water–ice system, computed numerically from (6), (7) and (8), of the gradients in the liquid and in the solid as a function of V for $\Delta T_C = \Delta T_F = 5°C$ and $\Delta T_F = 10°C$, $\Delta T_C = 5°C$. As has just been proved, one sees that the function G_L goes through a minimum. However, it is not possible to obtain an explicit expression for this minimum because it corresponds to values of V such that the first-order expansion which has been used is no longer valid.

Figure 4 shows a series of curves $V(X_i)$ computed numerically from (8).

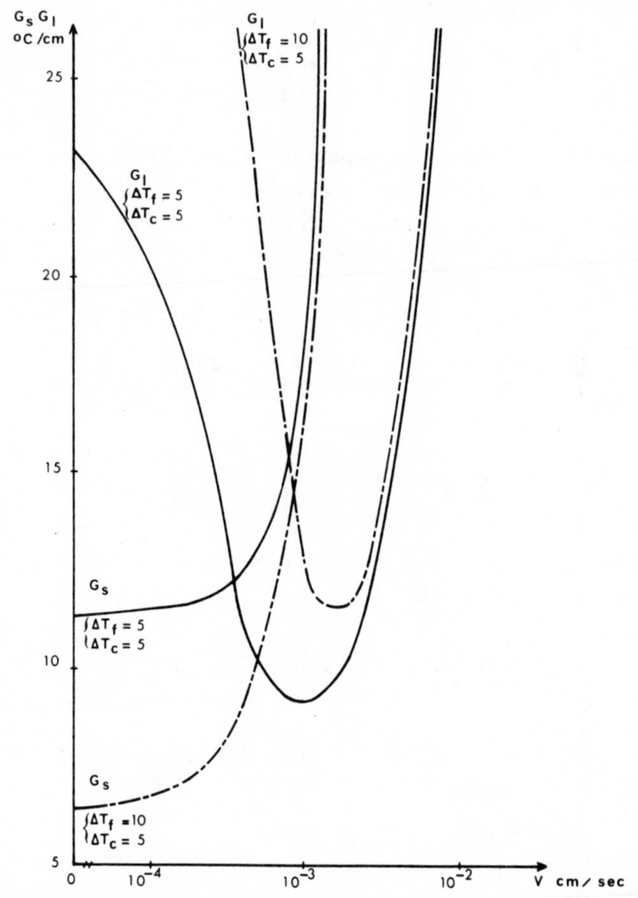

Fig. 3. Gradients in the liquid at the interface G_L and in the solid G_S as functions of V for $T_F = 268°K$, $T_C = 278°K$ (solid line) and $T_F = 263°K$, $T_C = 278°K$ (dashed).

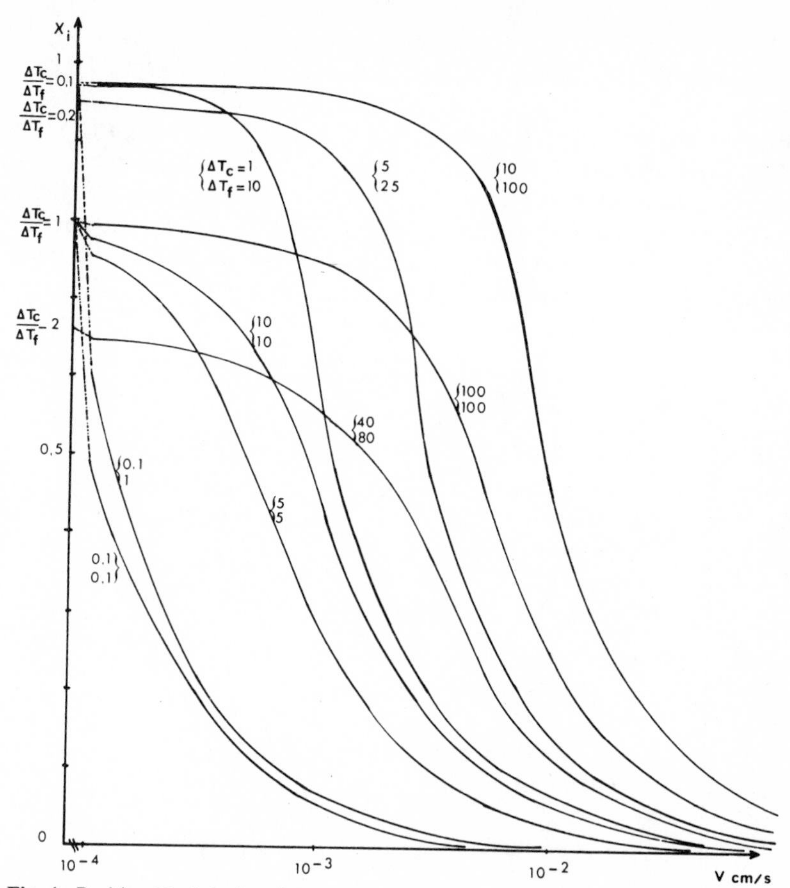

Fig. 4. Position X_i of the interface as a function of V for diverse thermal conditions.

For each value of $\Delta T_C/\Delta T_F$ determining uniquely the position of the interface X_0 for $V=0$, one obtains a set of curves $X_i(V, \Delta T_F)$.

As in the preceding section, we see again that a stationary state determined by a fixed X_i corresponds to several possible boundary conditions; we shall thus range them according to their respective entropy productions.

When $V>0$ the entropy production (12) can be written:

$$P(S) = V(\rho c)_S \left\{ \log\left(\frac{T_F}{T_0}\right) + \frac{x_{iS}}{(x_{iS}-1)} \frac{\Delta T_F}{T_0} - \frac{1}{(x_{iS}-1)} \frac{\Delta T_F}{T_F} \right\}$$

$$+ V(\rho c)_L \left\{ \log\left(\frac{T_0}{T_C}\right) + \frac{x_{iL}}{(x_{iL}-1)} \frac{\Delta T_C}{T_0} - \frac{1}{(x_{iL}-1)} \frac{\Delta T_C}{T_C} \right\} \quad (18)$$

where

$$x_{iS} = \exp[-(V/\alpha_S)X_i] \qquad \text{and} \qquad x_{iL} = \exp[-(V/\alpha_L)(X_i - 1)]$$

As easily verified, when V goes to zero in the expression (18), $P(S)$ tends to P_0 as given by (13).

In Fig. 5, we have drawn a number of curves $P(V)/P_0$ as a function of V for the parameters $\Delta T_F = \Delta T_C = 5°C$; $\Delta T_F = 10°C$, $\Delta T_C = 5°C$; and $\Delta T_F = 25°C$, $\Delta T_C = 5°C$. One of the most interesting features about these curves is the existence of a minimum of the function $P(V)/P_0$, which is considered in more details later.

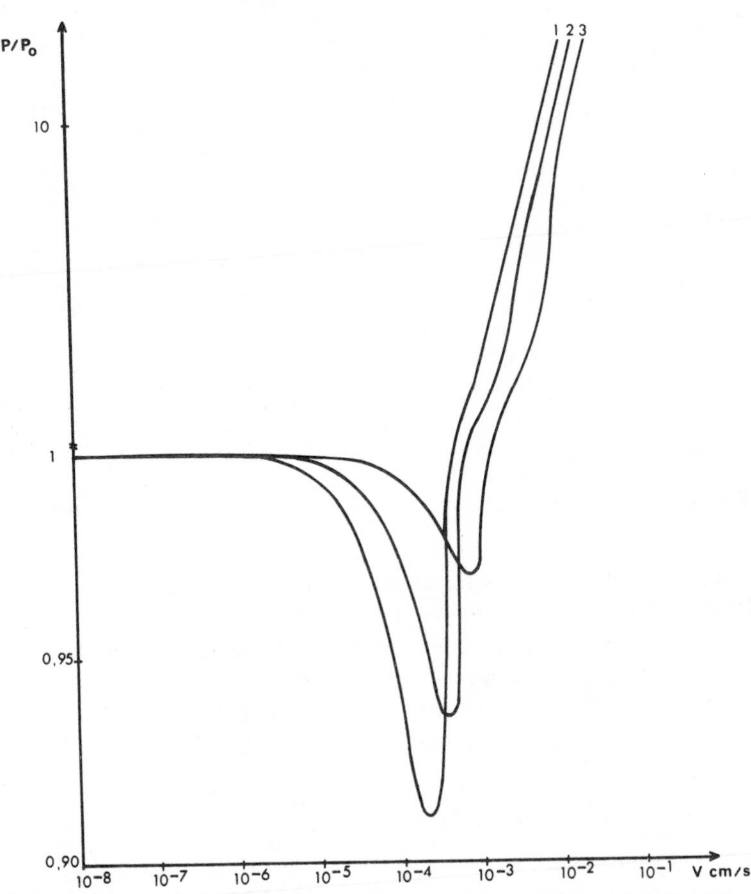

Fig. 5. Production of relative entropy P/P_0 as a function of V for (1) $\Delta T_F = \Delta T_C = 5°C$; (2) $\Delta T_F = 10°C$, $\Delta T_C = 5°C$; (3) $\Delta T_F = 25°C$, $\Delta T_C = 5°C$.

When the system under study is closed, that is, the driving velocity is zero, the entropy production is due solely to the effects of heat diffusion and depends only on the imposed boundary conditions T_F and T_C; from (9) and (13), we obtain:

$$P_0 = \frac{\lambda_S \Delta T_F^2}{T_0 T_F} \cdot \frac{1}{X_0} + \frac{\lambda_L \Delta T_C^2}{T_0 T_C} \cdot \frac{1}{1 - X_0}$$

Let us first assume that the motion of the system at a velocity $V > 0$ induces only a simple variation of the position of the interface; or equivalently, let us assume that V is sufficiently small for the entropy production due to convective phenomena to be negligible. With this assumption, we have $dP_0/dX_0 > 0$ for most of the domain of variation of T_C, T_F which is physically conceivable.

Indeed:

$$\frac{dP_0}{dX_0} = -\frac{\lambda_S \Delta T_F^2}{T_0 T_F} \cdot \frac{1}{X_0^2} + \frac{\lambda_L \Delta T_L^2}{T_0 T_L} \cdot \frac{1}{(1 - X_0)^2}$$

that is, with (9):

$$\frac{dP_0}{dX_0} = \frac{(\lambda_S \Delta T_F + \lambda_L \Delta T_C)^2}{T_0} \cdot \frac{\lambda_S T_F - \lambda_L T_C}{\lambda_S \lambda_L T_F T_C}$$

The sign of dP_0/dX_0 is then the sign of $(\lambda_S T_F - \lambda_L T_C)$ and we have

$$\frac{dP_0}{dX_0} \geqslant 0 \quad \text{if} \quad \frac{\lambda_S}{\lambda_L} \geqslant \frac{T_C}{T_F}$$

For our present study, this last inequality is not very restrictive. Indeed, for the system water–ice, we have $\lambda_S/\lambda_L = 3.8$ and $T_F < 273°K$, $T_C < 373°K$ at atmospheric pressure.

In Fig. 6, we have drawn the domain $dP_0/dX_0 \geqslant 0$. One notices that dP_0/dX_0 is positive unless one imposes a T_F smaller than 80°K (liquid nitrogen) and a T_C larger than 300°K.

Therefore, for most thermal conditions one expect the entropy production due to diffusive phenomena decrease when V increases because X decreases when V increases.

On the contrary, the entropy production due to convective phenomena increases with V.

Under these conditions, it is interesting to investigate the monotonic character of the function $P(V)/P_0$ when V increases from zero to infinity and to look for the possible existence of a minimum. For this, it is

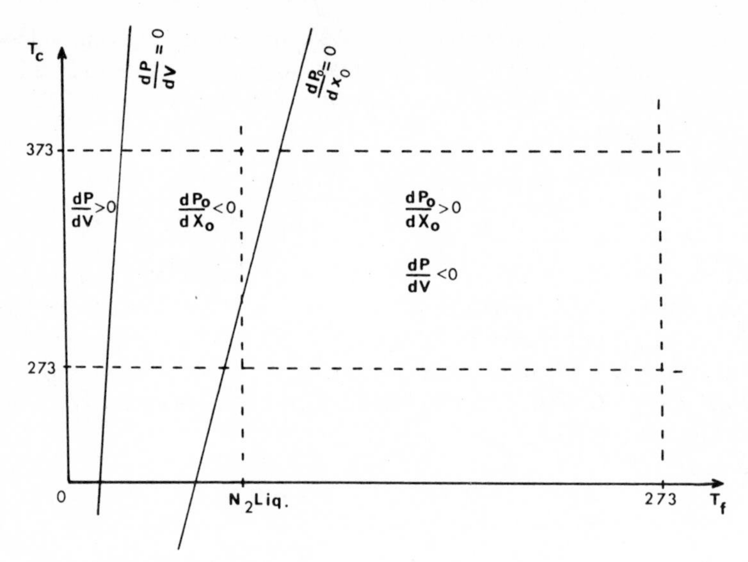

Fig. 6. Sign of dP/dV and dP_0/dX_0 in the plane T_F, T_C.

sufficient to show that $\partial P/\partial V$ is a negative function of V in the neighborhood of zero.

Assuming that the arguments x_{iS} and x_{iL} are small compared to unity and also that $\Delta T_F/T_F$ and $\Delta T_L/T_C$ are small compared to unity, we obtain after first-order expansion of (18):

$$\frac{\partial P}{\partial V} = \rho_S \mathfrak{L} \frac{\Delta T_F}{T_0}\left(\frac{1}{T_F} - \left(\frac{\lambda_S}{\lambda_L}\right)^2 \cdot \frac{1}{T_C}\right) - (\rho c)_S \frac{\Delta T_F}{T_F} - (\rho c)_L \frac{\Delta T_C}{T_C} \quad (19)$$

As $(\lambda_S/\lambda_L)^2 \cong 14$, the term $1/T_F - 1/T_C(\lambda_S/\lambda_L)^2$ is negative in most of the domain (T_C, T_F) which can be considered and the entropy is indeed a decreasing function of V in the neighborhood of zero and has at least one minimum for a strictly positive value $\tilde{V}$ of V. The domains where $\partial P/\partial V$ is positive, zero, and negative are indicated in Fig. 6.

This analytical study of the function $P(V)$ exhibits a fundamental feature, namely:

- For certain positive values of the velocity, the entropy production of the stationary open system under study is smaller than that of the equivalent closed system ($V=0$).
- The function P/P_0 goes through at least one minimum for a nonvanishing value of V.

From the point of view of applications, the existence of minima for the curves P/P_0 allows us to define an optimalization of the process of zone melting developed by Viaud.[32]

These conclusions, although surprising from certain points of view, can be explained. Indeed, even if the two stationary systems we have compared [the first being a closed system (index zero) and the second an open system] seem identical, they are basically different because the first depends on two independent variables (T_F and T_C) while the second depends on three independent variables (T_F, T_C, and V); it is therefore not possible to deduce the behavior of the second from that of the first. This exposes one of the difficulties which is often glossed over in studies of moving interfaces when, for instance, one rejects the thermal boundary conditions to infinity or assumes the monotonic behavior of certain functions (thermal gradients as function of V). In the study of a simple model for which one can determine the temperature field and the thermal gradients analytically such a behavior of these systems is not evident *a priori.*

C. Effect of a Perturbation of Shape

So far we have determined the entropy production for systems in which the solidification interface was assumed to be planar.

Using the same method, one can also consider the case of a nonplanar interface and establish the corresponding entropy balance.

In particular, it will certainly be very interesting to compare the productions for planar and nonplane interfaces, all external variables being identical, and to look for the case which corresponds to the smaller entropy production.

Let us admit that, for fixed values of T_F, T_C, and V, we can define two stationary systems:

- One system is characterized by a plane interface with abscisse X^* and at temperature T_0^* (Fig. 7*a*) (all quantities relative to this interface will be denoted with the symbol *).
- The other is characterized by an interface with regularly distributed protuberances (Fig. 7*b*).

Later on, we examine the exact nature of the deformations of the interface; for the moment, we shall simply suppose a parabolic protuberence with wavelength λ and amplitude a (Fig. 8) according to equation:

$$y = a\left[1 - \left(\frac{2x}{\lambda}\right)^2\right]$$

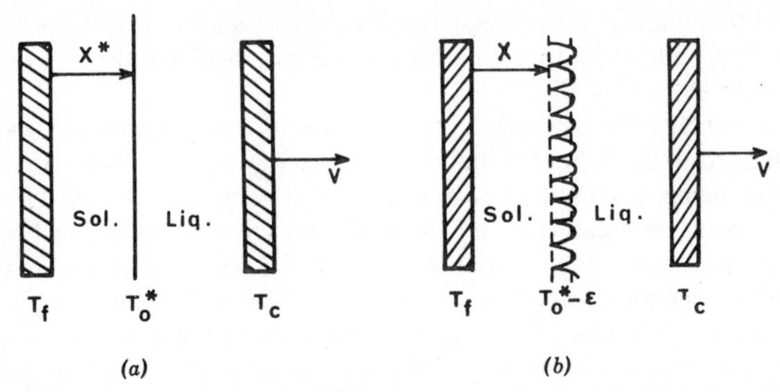

Fig. 7. Schematic representation of solidification process, (*a*) for a plane interface; (*b*) for a perturbed interface.

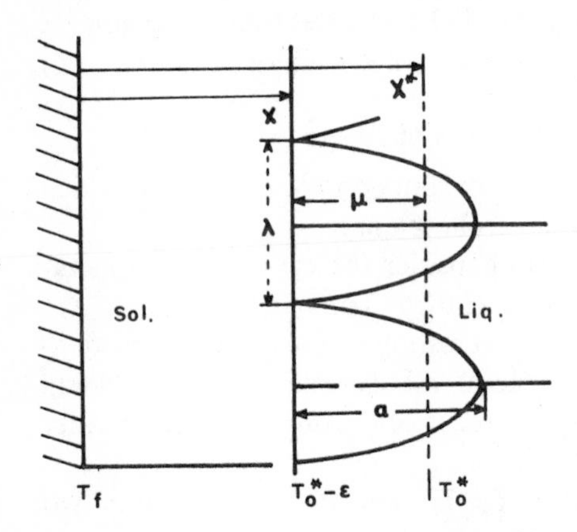

Fig. 8. Characteristic of the perturbed interface and of the equivalent plane interface.

This interface is defined as follows:

- It is located by its abscissa X corresponding to the last straight section entirely contained in the solid phase (Fig. 8).
- We shall assume that this section is characterized by a temperature $(T_0^* - \varepsilon)$; indeed, it is impossible to attribute to the section with abscisse X the temperature T_0^* because this would imply the existence of solid phase at a temperature $T > T_0^*$.
- The equivalence of both interfaces X and X^* is then defined by writing the equality of masses for both configurations, that is,

$$\frac{(X^* - X)\lambda}{2} = \int_0^{\lambda/2} a\left[1-(2x-\lambda)^2\right]dx$$

and

$$X^* - X = \mu = \tfrac{2}{3} \cdot a$$

• We also assume that the temperature field remains one dimensional despite the existence of perturbations. We examine later on the validity of this assumption.

The two systems we shall compare are thus respectively defined by:

$$[T_F, T_C, T_0^*, X^*, V] \qquad \text{plane interface}$$

$$[T_F, T_C, T_i = T_0^* - \varepsilon, X = X^* - \mu, V] \qquad \text{perturbed interface}$$

If one writes the entropy production for each, one has:

$$\frac{P}{P^*} = \frac{(\rho c)_S \left[\log \dfrac{T_F}{T_i} + \dfrac{(\Delta T_F - \varepsilon)^2}{T_i T_F} \cdot \dfrac{T_i - T_F x_{iS}}{1 - x_{iS}} \right] + (\rho c)_L \left[\log \dfrac{T_i}{T_L} + \dfrac{(\Delta T_C - \varepsilon)^2}{T_i T_C} \cdot \dfrac{T_C - T_i x_{iL}}{1 - x_{iL}} \right]}{(\rho c)_S \left[\log \dfrac{T_F}{T_0^*} + \dfrac{\Delta T_F^2}{T_0^* T_F} \dfrac{T_0^* - T_F x_{iS}^*}{1 - x_{iS}^*} \right] + (\rho c)_L \left[\log \dfrac{T_0^*}{T_c} + \dfrac{\Delta T_C^2}{T_0^* T_C} \cdot \dfrac{T_C - T_0^* x_{iL}^*}{1 - x_{iL}^*} \right]} \tag{20}$$

where

$$T_i = T_0^* - \varepsilon$$

$$x_{iS} = \exp\left[-\frac{V}{\alpha_S} X \right] \qquad x_{iS}^* = \exp\left[-\frac{V}{\alpha_S} X^* \right]$$

$$x_{iL} = \exp\left[-\frac{V}{\alpha_L} (X - 1) \right] \qquad x_{iL}^* = \exp\left[-\frac{V}{\alpha_L} (X^* - 1) \right]$$

Figure 9 shows a certain number of curves P/P^* as a function ε, with μ and V as parameters. They are obtained by numerical computation of (20), using arbitrarily chosen set of values of T_F, T_C, V, ε, and μ. One notices that for each (T_F, T_C, μ) there exists a value of ε such that the entropy production is minimum.

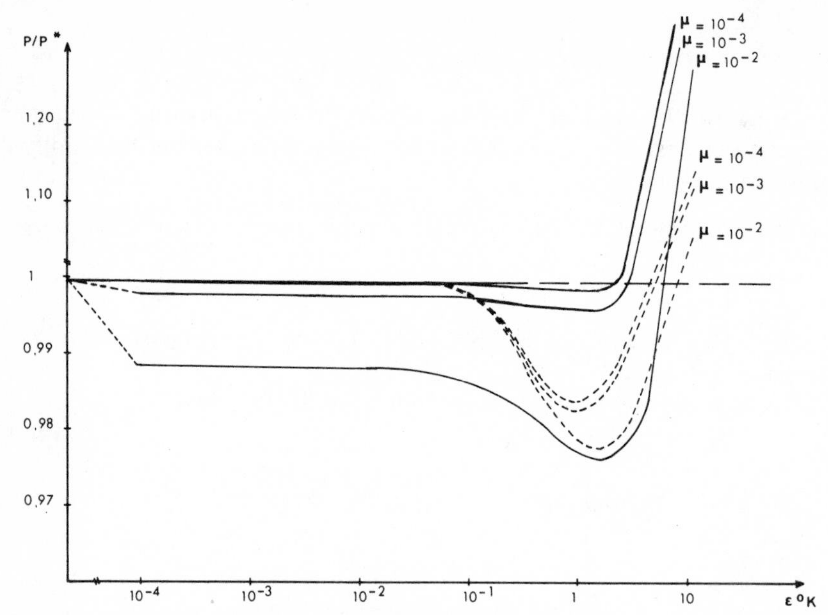

Fig. 9. Curves P/P^* as a function of undercooling ε for $V = 10^{-4}$ cm/sec (solid line) and $V = 10^{-3}$ cm/sec (dashed) with $\Delta T_F = 40°C$, $\Delta T_C = 5°C$.

One also notices that there are (ε, μ) pairs corresponding to a perturbed interface $(\mu \neq 0)$ such that the entropy production is smaller than that of the equivalent plane interface, as defined above $(P/P^* > 1)$. Criterium (1):

$$(1) \quad \begin{cases} \dfrac{P}{P^*} > 1 & \text{perturbed state stable with respect to *state} \\ \dfrac{P}{P^*} < 1 & \text{* state stable with respect to perturbed state} \end{cases}$$

allows us then to say that the stationary state of the perturbed interface is more stable than that of the plane interface as soon as $P/P^* > 1$.

Indeed, let us assume that we have a perturbed stationary interface $(\mu = 10^{-4})$ moving with $V = 10^{-4}$, and that we increase ε (Fig. 10); the representative point of the system is on AB. If a local temperature fluctuation $\Delta\varepsilon$ puts the system locally in C, then there is an increase of the amplitude of the perturbation $\mu = 10^{-4} \rightarrow \mu = 10^{-3}$. Now, if the perturbation increases, the local velocity increases and the system will have to move towards E, but when it reaches D, it reaches the stationary state defined by the curve $(\mu = 10^{-4}, V = 10^{-3})$; thus the perturbation regresses, and V local decreases bringing the system back to B. In this way, one exhibits the stability of the perturbed interface starting from a certain value of ε for given V.

In fact, this reasoning is only approximate because it is established using variables whose values have been taken discrete for the needs of numerical computation, but *a fortiori* it is valid for continuous variations of V, μ because then the trajectory BCD is reduced to the neighborhood of a point.

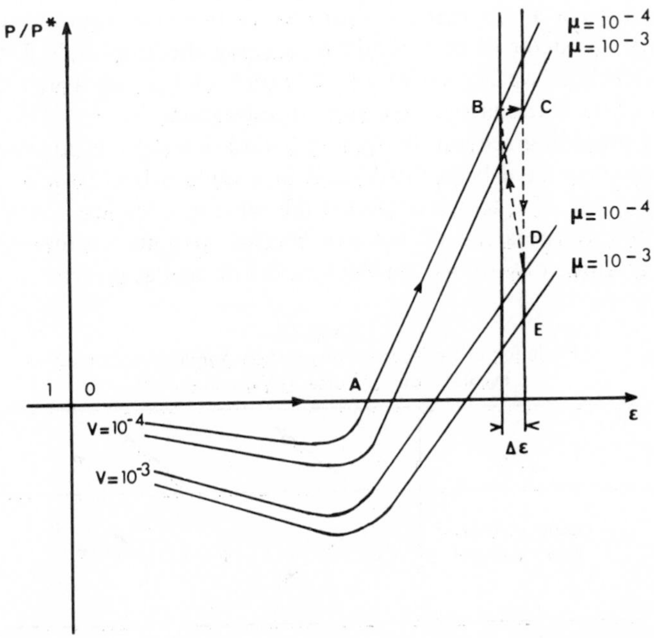

Fig. 10. Cycle showing the stability of a perturbation of the form of given amplitude.

In fact, there is another explanation for the greater stability of a perturbed interface with respect to a plane interface submitted to a set of determined constraints.

For this, one must first notice that the amplitude of the temperature fluctuations increases with the temperature gradients in the liquid and solid phases.

On the other hand, the amplitude of a perturbation decreases when the gradients increase. We thus obtain here the necessary conditions for a stable structure, that is, two opposite constraints which increase simultaneously under the effect of the generalized forces (thermal gradients): one which generates perturbations, the other which destroys them. But we have seen (see Fig. 3) that the temperature gradient in the liquid is a decreasing function of V for small values of V.

Let $\tilde{V}>0$ be a value of V such that $G_L(V)$ is minimum; the probability of appearance of a thermal fluctuation with large amplitude follows a law of the same kind as $G_L(V)$ and thus goes through a minimum for $\tilde{V}$.

On the contrary, the probability of development of a large perturbation follows a law in $1/G_L(V)$ and thus has a maximum for $\tilde{V}$.

Table I gives a schematic description of this situation and shows the stationary character of perturbations reaching the amplitude $\tilde{\mu}$. Once this state is reached, we are on the V, ε branch of Fig. 10 which shows the stability of the stationary state under consideration.

In the preceding section, temperature variations at the interface as well as deformation have been introduced in a quite arbitrary way; however, the assumption of a temperature for the moving interface, lower than the equilibrium temperature of the two phases, is quite correct and is the subject of numerous studies on the kinetics of crystal growth.

TABLE I.
Evolution of the Thermal Fluctuation Amplitudes and the Perturbations of Form as Functions of V.

	V					
	V	0	↗	$\tilde{V}$	↗	∞
G_l	Liquid Tempera -ture gradient	$+G_0$	↘	$\tilde{G}$	↗	∞
ε	Thermal fluctua tions amplitude	ε	↘	$\tilde{\varepsilon}$	↗	∞
μ	Shape pertuba-tions amplitude	μ	↗	$\tilde{\mu}$	↘	0

D. Kinetics of Crystal Growth

The word solidification is used to describe the transformation of a liquid phase to a solid phase. The phase change is due to extraction of heat from the system but the transformation process itself must be analyzed in two steps:

- Crystal nucleation.
- Growth of the nucleus fed by atoms or molecules from the liquid phase.

There are numerous studies concerning the conditions for nucleation at the origin of a phase transformation[33] but they do not fail within the context of this work.

Moreover, for the same reasons, we shall not discuss the details of crystal growth and shall just say a few words about the atomistic theories of crystalline interfaces.

There exists indeed several theories describing the mechanism of the molecular kinetics of solidification.[34,35] Generally, one distinguishes two different modes of crystal growth:

- Step or discontinuous growth
- Continuous growth

At the present time, only the model of diffuse interface proposed by Cahn and coworkers[36,37] seems compatible with experimental observations. It predicts that for weak driving forces (weak undercooling) the variations of surface free energy necessary for a uniform propagation of a crystalline face are greater than these forces. Then the solid surface can advance only through lateral motion of steps and kinks whose origin is a screw dislocation or a two-dimensional nucleation site. Then, one observes a step growth.

For large driving forces, the surface can move normally to itself because all molecules which hit it remain captured. One then observes continuous growth. Under these conditions, Cahn, Hillig, and Sears[37] have proposed a criterion which allows the prediction of the transition between these two mechanisms of crystal growth.

With the first mechanism, the lateral growth of an elementary step whose height is the lattice distance a (or a multiple of it) is such that the interface does not move except during a transition from one step to the next.

The proposed theory is based on the evaluation of the variation of free energy per unit volume, $-\Delta F_V$, which acts as the driving force.

In the case of solidification, ΔF_V is proportional to the undercooling ΔT and can be written, in first approximation:

$$-\Delta F_V = \frac{\rho_S \mathfrak{L} \Delta T}{T_0}$$

This reasoning, however, is based on an important assumption: the growth process starts on a perfectly planar crystalline surface. Now, in the neighborhood of the equilibrium temperature T_0, this assumption is very questionable and it is more plausible to suppose a crystalline growth from a molecularly rough surface.

Cahn[36] has attempted to take this fact into account through the introduction of a "diffuseness" parameter g which depends on the number of molecular layers involved in the liquid–solid phase transition.[37]

He proposes to separate the various growth regimes in the following way:

- Step growth
- Transition
- Continuous growth

1. Step Growth

When the undercooling is very weak, crystallization on the surfaces of small index occurs laterally. At every point of the surface the speed of the advance is given by:

$$V = av$$

where a is the height of the step and ν the step frequency at this point.

Clearly, this process is closely dependent on the probability of the formation of new steps.[54]

For a perfect crystallographic surface, sites of two-dimensional nucleation are necessary for the creation of each new step.

On the contrary, for an imperfect crystallographic surface with impurities, dislocations, and grain boundaries, each imperfection will be a permanent source for the creation of new steps.

2. Continuous Growth

When undercooling of the interface is large, as has been mentioned, the solidification front advances in a continuous manner, in a direction normal to itself without needing the step by step filling processes of monomolecular layers.[41,42]

In this case, the kinetics of growth have been established[41] by supposing that the transfer processes due to diffusion in the liquid are equal to those of molecular capture and fixation at the solid surface.

3. Transitory Regime

For undercoolings obeying certain conditions, the two preceding mechanisms can operate simultaneously.

Using the results of Cahn, Hillig, and Sears,[37] Tarshis and O'Hara[40] have given a particularly simple form for the kinetic laws governing the diverse solidification processes examined above.

In the following relations [(21) to (25)] V is the velocity of crystal growth, ΔT the undercooling of the interface, and μ the kinetic constant.

For step growth, as indicated above, several kinetics are possible according to the origin and the nature of the sites producing the steps. Tarshis examines four of these:

- Normal step growth

$$V = \mu_{\infty} \Delta T \tag{21}$$

- Screw dislocation growth

$$V = \mu_1 \Delta T^2 \tag{22}$$

- Generation of a new layer limited by the passage of one layer to the next

$$V = \mu_4 \exp \frac{\mu_3}{3\Delta T} \tag{23}$$

- Generation of a new layer is limited by the rate of two-dimensional nucleation

$$V = a_0 A \mu_2 \exp \frac{\mu_3}{\Delta T} \tag{24}$$

where a_0 is the step height and A the area of the crystal.

- Continuous growth

$$V = \mu_0 \Delta T \tag{25}$$

As this law (25) is the simplest, Tarshis expresses all the constants in terms of μ_0.

The determination of μ_0 is based on more or less empirical relations.[36,39,56] Nevertheless, by choosing judiciously the diverse parameters for a given material, one can calculate or at least evaluate, the order of magnitude of kinetic constants μ concerning the different modes of crystalline growth cited above.

For water–ice, which interests us here, Tarshis and O'Hara[38] have proposed explicitly the kinetic constants based on the results of Hillig[45]:

- Continuous growth

$$V = 0.12\, \Delta T \tag{26}$$

- Step growth

$$V = 2.26\, \Delta T \tag{27}$$

- Screw dislocation mechanism

$$V = 3 \times 10^{-2} (\Delta T^2) \tag{28}$$

- Growth limited by the passage from one layer to the next

$$V = 3\times 10^{-2}\left(\exp\frac{-1.05}{3\Delta T}\right) \tag{29}$$

with V in cm/sec and ΔT in C°.

Returning to relation (18) for the entropy production $P(S)$, our calculations can be further refined by introducing an interface temperature which takes account of the crystallisation kinetics (26) to (29), and they can be classified from criterion (1). One finds[43] that the classification corresponds to that of Cahn et al.[17,40] and it can be shown, that for small values of $V (V < 10^{-4}$ cm/sec), it is the mechanism (29) (bidimension nucleation) that corresponds to the smallest entropy production, followed by mechanism (28) (screw dislocation) then (25) (continuous growth) when V increases. It is of interest to note the convergence of two absolutely independent methods:

- The work of Cahn and coworkers based on a discussion of molecular phenomena at the interface and on experimental results.
- The present study, based on an evaluation of entropy in the model, in fact arbitrary, defined at the beginning of the chapter.

E. The Coupling of Curvature and Kinetics

In the two preceding paragraphs, we have examined separately:

1. The effect of an arbitrary undercooling and perturbation of the interface.
2. The undercooling effects due to the crystallization kinetics.

For these two situations, we have shown that the entropy production has a minimum value.

In a similar manner one can examine the most general case taking into account not only the crystallization kinetics but also the geometrical form of the interface.

We must compare two systems supporting the same external conditions, T_F, T_c, V (Fig. 7):

- One characterized by a plane interface whose temperature T_i depends on one of the kinetic laws given above, ①.
- The second characterized by a deformed interface whose temperature T_i depends both on the kinetic law of crystallization and on the form of the perturbations ②. One could attribute different kinetics to these two systems.

The description of the first system poses no special problems and the results obtained in the preceding paragraph can be used.

The description of the second, however, requires a number of hypotheses and a preliminary study (Fig. 7).

Let us suppose that the perturbations are of a parabolic type, wavelength λ and amplitude a. The exact form of the perturbations has in fact little importance and one can choose, without greatly changing the solution, parabolic, circular, eroded saw-tooth forms, the main restriction being not to choose a form such that the mean radius of curvature is zero as would be the case, for example,with a sinusoidal perturbation. The rightness of this restriction will clearly appear in the following chapter in which all experimentally observed perturbations presents a very marked cusp on the side of the solid phase. The possible choice of a semicircular form presents the drawback that the number of degrees of freedom of the form is reduced by the relationship between a and λ ($\lambda=2a$).

For these reasons we preferred to introduce a parabolic perturbation, function of two independent parameters a and λ which permit us to approximate to first- or second-order perturbation of very varied forms.

With this, it is possible to define global equivalence of the system ① and ②:

- The conservation of mass in ① and ② which determines the position of the interface X_i in the system ②.
- The parameter of interfacial curvature of ② (mean value of the radius of curvature in terms of the arc):

$$\left\langle\frac{1}{r}\right\rangle=\frac{1}{\operatorname{arc} 0\lambda}\int_0^{\lambda}\frac{dx}{r(x)}=\frac{\int_0^{\lambda}-y''/(1+y'^2)^{3/2}\,dx}{\int_0^{\lambda}(1+y'^2)^{1/2}\,dx} \tag{30}$$

where $y(x)$ is the analytical representation of the perturbation and y',y'' are the first and second derivatives, respectively,

- The undercooling of the curvature effect:

$$\Delta T=\left\langle\frac{1}{r}\right\rangle\cdot\frac{\gamma_{eg}}{\rho_S} \tag{31}$$

Here it is supposed that Kelvin's relation[31] remains valid out of thermodynamic equilibrium and that the free surface energy is isotropic and independent of the curvature.[44,45]

For a parabolic curvature of wavelength λ and amplitude a:

$$y=a\left[1-\left(\frac{2x}{\lambda}\right)^2\right]$$

Equation (31) can be written

$$\Delta T=\frac{\gamma_{eg}}{\rho_S}\frac{16a}{\lambda^2}\cdot\frac{th(X)}{X+\frac{1}{2}sh(2X)} \tag{32}$$

where

$$X=\log\left\{\frac{4a}{\lambda}+\left[1+\left(\frac{4a}{\lambda}\right)^2\right]^{1/2}\right\}$$

For the numerical calculations we have taken $\gamma_{eg}=20$ dynes/cm.[46] Using this together with (18), the relative entropy production of the two systems can be evoluated.

One finds for system ①:

$$P_1=(\rho c)_S V\left\{\log\frac{T_F}{T_0-f_i(V)}+\frac{x_{iS}[T_0-f_i(V)-T_F]}{(x_{iS}-1)[T_0-f_i(V)]}-\frac{T_0-f_i(V)-T_F}{(x_{iS}-1)T_F}\right\}$$
$$+(\rho c)_L V\left\{\log\frac{T_0-f_i(V)}{T_C}+\frac{x_{iL}[T_C-T_0+f_i(V)]}{(x_{iL}-1)[T_0-f_i(V)]}-\frac{T_C-T_0+f_i(V)}{(x_{iL}-1)T_C}\right\} \tag{33}$$

where $f_i(V)$ is one of the crystallization kinetics defined in the preceding paragraph. Similarly, for ②:

$$P_2=(\rho c)_S V\left\{\log\frac{T_F}{T_0-f_j(V)-\Delta T}+\frac{x_{iS}(T_0-f_j(V)-\Delta T-T_F)}{(x_{iS}-1)[T_0-f_j(V)-\Delta T]}\right.$$
$$\left.-\frac{T_0-f_j(V)-\Delta T-T_F}{(x_{iS}-1)T_F}\right\}$$
$$+(\rho c)_L V\left\{\log\frac{T_0-f_j(V)-\Delta T}{T_C}\right.$$
$$\left.+\frac{x_{iL}(T_C+f_j(V)+\Delta T-T_0)}{(x_{iL}-1)[T_0-f_j(V)-\Delta T]}-\frac{T_C+f_j(V)+\Delta T-T_0}{(x_{iL}-1)T_C}\right\} \tag{34}$$

where ΔT is the undercooling given by (32) and $f_i(V)$, $f_j(V)$ are each one of the crystallization kinetics (26) to (29).

If one supposes, for the sake of simplicity, that the kinetics are identical in the two systems, the organigram of the numerical calculations allowing us to obtain P_2/P_1 appears as follows:

- One takes a set of discrete values of λ and of a which, using (32), determine ΔT.
- One takes for (26) to (29) a kinetic law defining T_i ① and T_i ② as functions of V.
- P_1, P_2 and P_2/P_1, which are functions of V for the T_F, T_C, $T_i(V)$, λ and a given, are determined with (33) and (34).

Figure 11 shows several curves P_2/P_1 calculated numerically with the crystallization kinetics where $\Delta T = V/2.26$ (normal step growth).

Here again, one finds two domains according to the value of P_2/P_1:

- $P_2/P_1(V) < 1$ in which the stationary planar interface is stable.
- $P_2/P_1(V) > 1$ in which the stationary parabolically deformed interface (a, λ) is stable.

Under these conditions, it is not surprising that an interface of solidification can take on very diverse geometrical forms as we will see in the following chapter.

This work, simultaneously taking account of the two phenomena analyzed separately above, that is interfacial kinetic effects and effects of curvature leads us again to the characteristics already noted, namely, the existence of a minimum and the definition of two stable stationary regimes, one for a plane interface, the other having periodic perturbations.

One can however question the rigor of such results and ask if they are not a consequence of one of the hypotheses adopted for the analysis.

F. Discussion of the Hypothesis Adopted

A priori, the most questionable hypothesis is perhaps that of the existence of an unperturbed thermal field even with a perturbed interface. It can be shown, however, that the above results are greater than those that would be obtained by introducing a two-dimensional thermal field; that is to say, in reality, the stable region of the perturbed interfaces determined by $V > V_c$ (Fig. 11) for fixed thermal constraints is greater than that proposed: V_c (two-dimensional) $<$ V_c (one-dimensional)

To show this, it suffices that

$$P_2(x, y) \geqslant P_2(x) \tag{35}$$

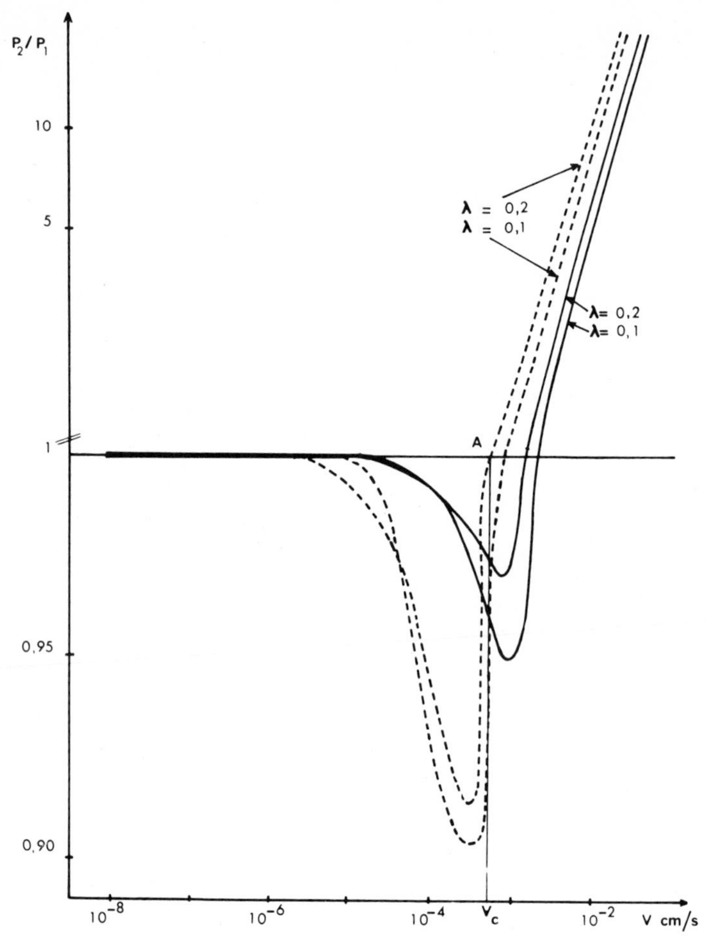

Fig. 11. Curves P_2/P_1 as functions of V for different parameters λ; ΔT_C; ΔT_F for $a = 10^{-2}$ cm: $\Delta T_C = 5°C$, $\Delta T_F = 25°C$ (solid line); $\Delta T_C = 5°C$, $\Delta T_F = 5°C$ (dashed).

Performing a development of $P_2(x,y)$ for an interface with periodic perturbations of wavelength λ, one has

$$P_2(x,y) = P_2(x) + \eta P_{2,1}(x,y) + \cdots \tag{36}$$

where $P_2(x)$ is the solution corresponding to the unperturbed stationary interface given by (34) and η, a positive small parameter of the form $\lambda V/\alpha$ (α is the thermal diffusion). Because by definition $P_{2,1}(x,y)$ describes two-dimensional diffusion phenomena around the interface, it is positive and (35) is always true.

Concerning this problem, one should perhaps note that the determination of an exact two-dimensional solution is difficult in any very general form. It depends on the scale used; whatever the real dimensions of the interfacial perturbations, a scale can always be defined such that the interface can be considered to be planar, but equally, one can use scales of analysis such that the hypothesis of a planar interface is not true.

Rigorously, once the size of a crystalline mass grows with any appreciable speed, the planar interface no longer exists.

According to the requirements, one can examine the interface:

- At the molecular scale as Cahn, Hillig, and Sears did[37] where the characteristic dimension is of the order 10^{-7} cm.
- At the microscopic scale (10^{-4} cm) as Tiller and Rutter[11] and others[47,54] for alloys and euterics.
- At the visible scale (10^{-2} cm) following Hardy and Corriel[48] or the present text (see following chapter).

In the theory of asymptotic developments, this corresponds to the interior description of the problem.

The exterior problem is treated using the Fourier equations.

Generally, the intermediate description is left aside, or the interior problem treated by an *ad hoc* procedure: the introduction of mean values; integration over a wavelength, and so on, to get matching with the exterior solutions.[49,50] This type of approach, which in fact has been used in the preceding paragraph, arises because the solution of the interior problem is extremely tricky since, on the one hand, the exact temperature of the interface and its form are not known simultaneously, and on the other hand, the equation of conservation of heat flux at the interface makes the system of equations nonlinear.

A second questionable hypothesis, generally accepted in this type of work, is the equality of the densities of the liquid and solid phases ($\rho_S = \rho_L$).

The point has been discussed by Carslaw[4] and Horway[51] who take account of a convective term in the liquid phase, appearing when $\rho_S \neq \rho_L$. The hydrodynamic and thermal problems must then be treated simultaneously.

In particular, Horway[52] has determined the pressure field for the growth of a spherical nucleus and shown the important role of the diminished pressure appearing at the interface when the crystallization velocity is large.

In the case of water and ice, which concerns us here, one should recall two special points:

- Contrary to the majority of materials, the density of the liquid is greater than that of the solid.
- The difference $(\rho_L - \rho_S)/\rho_L$ is around 8%.

One can therefore deduce from this that a water–ice interface, moving at velocity V, is equivalent, from the hydrodynamic point of view, to a homogeneous distribution of sources of intensity $V(\rho_L - \rho_S)/\rho_L \cong V/10 = u$.

For a stationary-plane interface, taking account of this, and of the coupling with the thermal problem, it is very simple to show that the solutions given previously [(4) to (8)] are modified according to:

$$X_i(V) < X_i(V+u)$$

$$\left(\frac{\partial T_S}{\partial x}\right)_{X_i} = G_S(V) > G_S(V+u)$$

$$\left(\frac{\partial T_L}{\partial x}\right)_{X_i} = G_L(V) < G_L(V+u)$$

For a perturbed interface, the complete solution is very difficult to establish; a detailed study of these phenomena will be presented in a further article.

A third hypothesis that had to be made in order to solve the problem concerns the isotropy and the homogeneity of the solid phase, and it is probably here that the greatest error is introduced, an error whose importance cannot be readily estimated.

The stability of plane or perturbed interfaces has been examined here by evaluating the entropy production in a well-defined thermodynamic system. One should note, however, that the only parameters used to describe the solid phase (as indeed for the liquid phase) are the thermal conductivity λ_S, the coefficient of diffusion α_S, and the surface energy γ_{eg}. In fact, the introduction of interfacial perturbations, whether it be in an arbitrary manner as adopted here, or from small perturbations or from normal mode analysis, remains nevertheless very artificial. In reality, the wavelength of a perturbation which can grow on an initially plane interface is predetermined by the crystalline structure of the solid phase. Apart from monocrystalline faces of low index which, under certain very strict growth conditions[53] (V very small), can be perfectly plane even on the molecular scale, a solid–liquid interface is formed by blocks whose size and number depend on the thermal history which has brought the interface to the state observed. In particular, it is well known that if no precautions are taken, especially during crystalline growth, the solid region will have numerous faults (dislocations, substitutions, vacancies, inclusions) and will be formed of an agglomeration of monocrystals of varied orientation and size.

Under these conditions, a macroscopically planar interface will present numerous local faults of planarity at the grain grooves (where the monocrystals constituting the solid phase join,[28,55] at the emergence of screw dislocations,[54] etc.). The validity of the conclusions based on theoretical developments which neglect the existence of these phenomena can be questioned. We are therefore brought back to the formulation of the interior problem as indicated earlier.

Now, at the level of a solid–liquid interface the data which we have are only fragmentary and very approximate, as are the estimations of: surface energy γ_{eg}, which according to authors, varies from 10×10^{-3} to 50×10^{-3} J/m^2[57,63]; the structure and the energy of the grain boundaries, which are very difficult to estimate[58,59] and the values of parameters as ordinary as the thermal conductivity and diffusivity which can be subjects for caution.[60] All this proves, therefore, that the determination and prediction of the structure of a solidification interface remains a very open problem.

In this chapter, the study of the problem of the stability of form of a solidification interface in an open system has been based on the irreversible character of the phenomena of heat transfer. It has been shown that, for a pure material, the possibility of existence of stationary interfaces of varied form is controlled by the thermal and mechanical constraints (T_F, T_C, V) imposed on the system. At the interface the effects of crystallization kinetics as well as curvatures have been taken into account.

A discussion of the diverse hypotheses adopted for this analysis has permitted us to point out the insufficiencies of the model studied; in the following chapter we uncover a certain number of phenomena and qualitatively estimate their effects on the fine structure of the interface.

IV. EXPERIMENTAL STUDY

Numerous experimental studies of the microstructure of solidification interfaces are reported in the literature. Many of these concern the solidification of metals, alloys, and eutectics[1,10,47,61,62]; others concern the freezing of water or aqueous solutions[62–68,82] under varying conditions.

For the metals, the determination of the exact interfacial structure is very difficult because the materials are opaque. The interface is therefore studied by taking a replica after a rapid decanting of the sample. Despite all the precautions taken, it is likely that the structure observed after this treatment is not the initial one. In particular, the surfaces with large curvature and the cusp zones must be somewhat eroded and smoothed.[78]

With transparent materials such as ice and numerous organic bodies, it is, on the contrary, possible to observe the interface directly during its movement and to follow its evolution and the modifications of its microstructure due to exterior contraints. The experimental setups for this are numerous[65–70] and furnish much information about interfacial phenomenon. Among other things one can cite:

- Preferential orientation
- Grains encroachment in polycristalline solid[54,68,70]
- Formation of air bubbles and impurities segregation[71]
- Repulsion of insoluble particles[72,73,77]
- Freezing of dispersed media and so on...[72,75]

A discussion of each of these phenomena would be out of context here and we will examine two points which illustrate the results of the preceding chapter:

- The existence of interfaces of a perturbed stationary form.
- The influence of the direction of crystalline growth on the form of the perturbations.

A. The Structure of a Moving Water–Ice Interface

Using an experimental setup described elsewhere[43,70] the moving water–ice interface was rendered visible. The ice front was observed in a film of water $8\times14\times0.1$ mm. Its evolution was controlled by two independently controlled modules working on the Peltier effect. Several qualities of water were tested (tap-water, triply distilled water, degassed or not) and although the purity exerts a quantitative influence on the evolution, the qualitative aspects of the results proved to be not closely dependent of it. Figure 12 shows a plane interface moving at 2.3×10^{-3} cm/sec in a liquid phase where the thermal gradient is 15°C/cm. One notices the polycrystalline structure of the ice, shown up by polarized light, and the competition between different grains of ice,[54,68] which grow or disappear. Figure 13 shows a grain groove[74] (triple interface) where two ice grains and the liquid phase come together. Here the front is motionless and one notices the local fault of planarity whose exact form depends on the surface energy balance at the triple point.[55] When a front having such grooves is forced to advance, these triple points remain behind the front and form the base of perturbations that may develop.

Figure 14 shows the very random aspect of crystalline growth in a supercooled liquid (negative thermal gradient in the liquid) where no

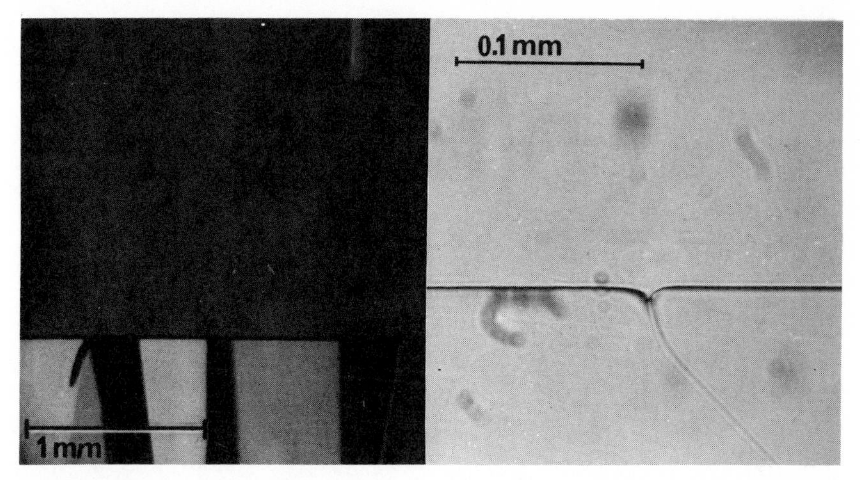

Fig. 12. Stationary plane interface. Polarized light shows the polycrystalline structure of the ice.

Fig. 13. Grain boundary on an immobile interface. One notices clearly the groove between the two grains.

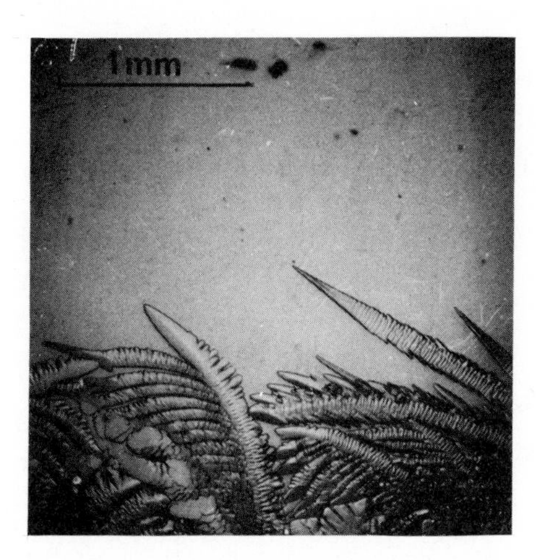

Fig. 14. Dendrites growth in supercooled water at $-5°C$.

stationary state appears. On the contrary, the two series, Figs. 15 (*a* to *c*) and 16 (*a* to *c*) show stationary interfaces with perturbed structures.

Figure 15 corresponds to an interface advancing at 4×10^{-3} cm/sec in water whose temperature gradient is 1°C/cm. The front is characterized by small almost equilateral sawtooth perturbations with a base of about 0.05 mm. One notices the stationary aspect of the interface during its displacement.

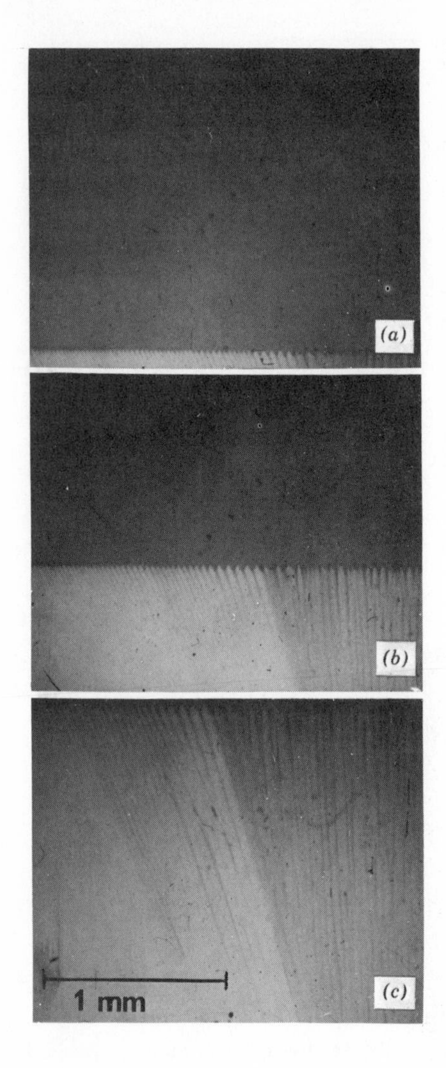

Fig. 15. Interface showing stationary perturbations of an eroded sawtooth type.

Figure 16 shows the fine structure of an interface moving, more rapidly than in the preceeding case, at 5.4×10^{-3} cm/sec. The thermal gradient in the solid varies from 10 to 8°C/cm, while the gradient in the liquid phase is zero. Again very well developed stationary perturbations can be noticed. The perturbations are quasi-isoceles sawtooth type with a base of 0.15 to 0.2 mm and 0.3 mm in height. The real surface is 3.5 times great as the plane one.

For both series of shots, the grains of ice crystallize in a direction normal to the basal plane (optical axis of each grain in the plane of the photo,

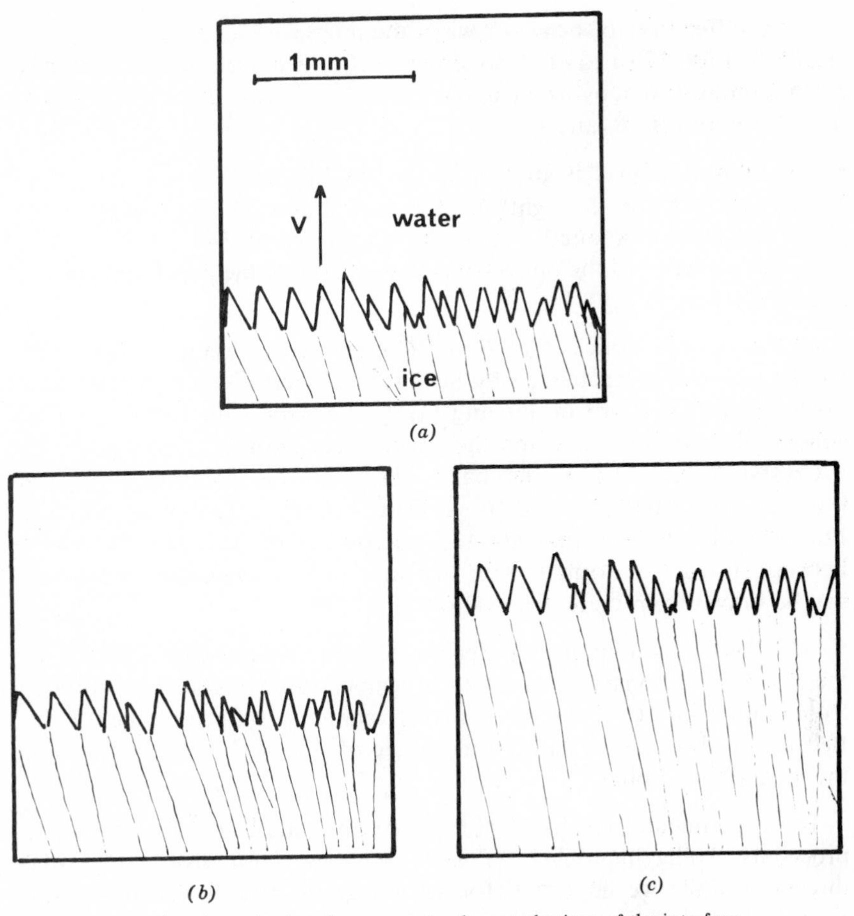

Fig. 16. Large, sharply wedged stationary sawtooth perturbations of the interface.

basal plane normal to the plane of the photo). Now it is known that at atmospheric pressure, water crystallizes into a hexagonal system.[76] Two types of crystals are therefore particularly interesting:

- Those crystallizing in the basal plane, direction [*a*], [1120] or [1100]; optical axis [*c*] of the crystal is then perpendicular to the plane of the paper and the crystals are easily identifiable in polarized light because they are always opaque.
- Those crystallizing in a plane determined by the velocity vector V and the direction [1000] such that the intensity of transmitted light depends on the angle between the polarizer/analyzer and the optical axis.

These differences of behavior when the interface moves are shown very clearly in Figs. 17 (*a* to *c*). A solidification front is seen having stationary perturbations and moving at about 4×10^{-3} cm/sec. It is composed of three ice grains *A*, *B*, and *C*.

- The central grain *A* is growing in the basal plane.
- The grain *B* (on the right) has the optical axis [*c*] in the plane of the paper and tilted at about 30° from the velocity vector *V*.
- *C* (on the left) has the optical axis in the plane of the paper and tilted at about 60° from *V*.

First of all one notices that *B* and *C* growing in a plane normal to the basal plane are characterized by very marked perturbations whose exact sawtooth form depends on the angle ([*c*], *V*) and on planes of small index compatible with this angle and the imposed constraints.

Crystal *A*, growing in the basal plane, on the contrary, has small rounded perturbations and is a little behind neighboring crystals. Furthermore one sees that grain *A* encroaches on *B*, while the groove between *A* and *C* is normal to the interface. Several important conclusions can be drawn from these observations:

- The thermal conditions imposed on the three grains being identical and the interface stationary, then the local temperature for grain *A* is different from grain *B* and *C* (it can be estimated very approximately to be 0.02°C). Because of this, the crystallization kinetic of *A* is different from those of the neighboring grains.[83]

The experimental results of Hillig[40] are thus illustrated by an inverted procedure. Hillig imposes a ΔT and measures *V* for diverse crystalline directions while we impose *V* for several grains and observe the spatial shift of the interfaces depending on ΔT.

These observations, however, cannot be interpreted from the theories of preferential orientation of certain crystalline directions[55,65,79,80] established for metallic materials. In particular, it is clearly seen that it is not the grains that protrude most into the liquid that encroach on their neighbors.

On the contrary, the theory proposed by Ketcham and Hobbs,[54] established from results concerning ice, is easily see to be true for grains growing in the plane normal to the basal plane, but the preferential direction between these grains (*B*, *C* in Figs. 17*a* to 17*c*) and those crystallizing in the base plane, (*A*) is more difficult to determine because the boundary of the grains *CA* remains normal to the interface while the orientation of the grain boundary *AB* is such that *B* disappears.

One should mention here the fundamental role played by the interfacial

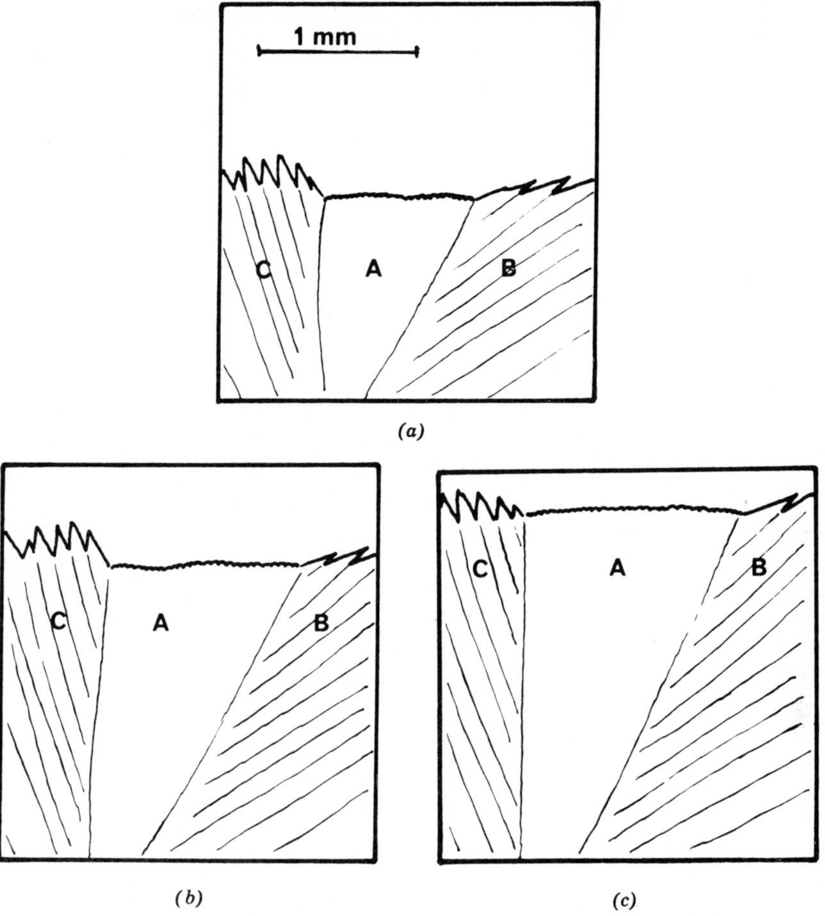

Fig. 17. Interface showing the dependence of the stationary perturbations of form, according to the direction of crystallization. Grain A, axis [*C*] normal to the paper; grains B and C, axis [*C*] is in the plane of the paper.

parameters:

- The anisotropy of the liquid–solid interfacial energy.
- The energy balance of the grain boundaries.[58,81]
- Crystallization kinetics related to the direction of crystalline growth.

We can then conclude that it is the above factors which control effectively the structure of the interface and that thermal conditions and other external constraints modulate only the intensity of this control.

B. Interaction between the Direction of Thermal Flux and Perturbations One

The effect of the direction of the thermal flux on the structure of the interface can be established by other, more direct methods. Figure 18 shows the modifications of an ice grain according to the direction of the thermal flux.

For the three interfaces (*1*), (*2*), and (*3*) the thermal conditions are identical. Interface (*1*) is observed during its first freezing cycle φ_1, from the bottom to the top, and presents quasi-isoceles sawtooth perturbations. After the complete freezing, the sample is turned through 88° and partially thawed. A second cycle of freezing φ_2, identical with the first, is then imposed from right to left and the perturbations shown on the interface (*2*) are obtained. The sample is then frozen completely and thawed partially after the heat source and sink are inverted, in order to submit the sample to a new freezing cycle φ_3 from left to right. The interface is as shown in (*3*). The diverse angular values indicated in Figure 18 are too approximate to determine the index corresponding to these facets but qualitatively they are sufficient to show the effect of the direction of thermal flux on the microstructure of the interface. It shows clearly that exact geometrical knowledge of the solidification interfaces cannot be obtained from purely

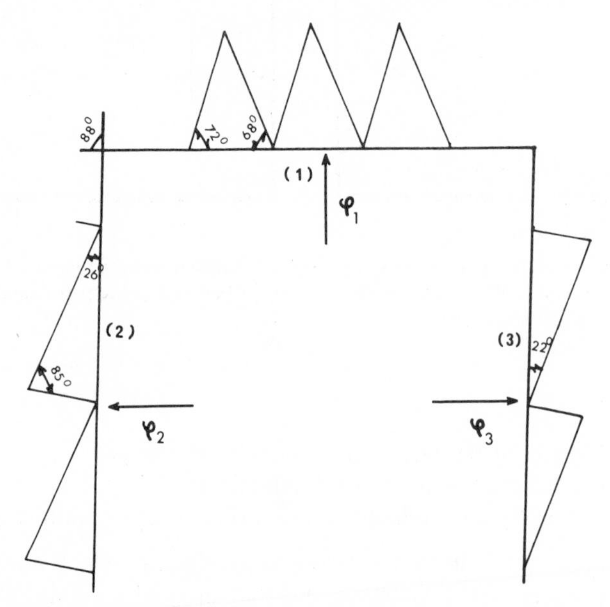

Fig. 18. Angle of facettes and form of the perturbations with respect to the direction of heat flux provoking crystalline growth.

thermal considerations, however complete. Here we find the justification of our conclusion of the previous chapter where we pointed out the grave danger introduced by the hypothesis of the isotropy of interfacial energy.

V. CONCLUSION

The study presented here permits us to clarify the role of several interfacial phenomena which occur along a solidification front in movement, together with their interactions with the external constraints.

Because the processes occurring in these systems are irreversible, a simple stability criterion has been developed which permits us to predict the existence of stationary interfaces having a variety of geometrical forms.

For this we have chosen a simple mathematical model, permitting complete analytical solutions, consisting of a two-phase system containing only a pure substance and defining a process of phase change such that the thermal and dynamical variables are independent.

A discussion of certain hypotheses, generally adopted for this type of problem, has shown the inadequacy of theoretical-schemes aimed at determining the exact morphology of the liquid–solid interface. We have suggested how this difficulty can be surmonted, at least partially, in looking for a solution as an asymptotic development and in properly choosing the scale of the analysis. We have also mentioned the lack of precise data for the parameters characterizing the interface, and the solid phase, and have shown that without this knowledge, it is not possible to propose a rigorous solution to the problem.

With the help of an experimental setup of great versatility permitting the imposition of very varied thermal conditions and the possibility of observing directly the solidification front during its motion, we have shown the existence of stationary states for interfaces of different forms.

The coupling between the external conditions and the observed behavior of the interface has been examined qualitatively. It seems that the real causes of phenomena are linked to the actual nature of the solid–liquid interfaces and that only the magnitude of their manifestations depend on exterior conditions. This analysis shows clearly that the solution of this problem is extremely difficult and numerous studies seem necessary.

Acknowledgments

This work has been financed partially by the DGRST.

References

1. J. J. Gilman, Ed., The Art and Science of Growing Crystals, Wiley, New York, 1963.
2. R. L. Parker, in *Solid State Phys.*, **25**, 151 (1970).
3. J. Stefan, *Ann. Phys. Chem.*, **42**, 269 (1891).

4. H. S. Carslaw and J. C. Jaeger, *Conduction of Heat in Solids*, Oxford Press, New York, 1959.
5. D. Turnbull, in *Solid State Phys.*, **3**, 225 (1956).
6. J. W. Gibbs, *Collected Works*, Longman, New York, 1931.
7. G. Herring, in *Structure and Properties of Solid Surfaces*, Univ. of Chicago Press, 1953.
8. I. Prigogine, Introduction à la Thermodynamique des Processus Irréversibles, Dunod Paris, 1968.
9. S. R. de Groot and P. Mazur, Non-Equilibrium Thermodynamics, North Holland, Amsterdam, 1962.
10. B. Chalmers, *Trans. Am. Inst. Mining Met. Eng.*, **200**, 519 (1954).
11. W. A. Tiller and J. W. Rutter, *Can. J. Phys.*, **34**, 96 (1956).
12. J. W. Rutter and B. Chalmers, *Can. J. Phys.*, **31**, 15 (1953).
13. W. A. Tiller, J. W. Rutter, K. A. Jakson and B. Chalmers, *Acta Met.*, **1**, 428 (1953).
14. G. P. Ivantsov, *Dokl. Akad. Nauk. SSSR*, **58**, 567 (1947).
15. C. Zener, *J. Appl. Phys.*, **20**, 950 (1949).
16. F. C. Franck, *Proc. Roy. Soc., (London) Ser., A*, **201**, 586 (1950).
17. D. E. Temkin, *J. Eng. Phys.*, **5**(4), 89 (1962).
18. F. S. Ham, *J. Appl. Phys.*, **30**, 1518 (1959).
19. F. S. Ham, *Quart. Appl. Math.*, **17**, 137 (1959).
20. G. Horvay and J. W. Cahn, *Acta Met.*, **9**, 695 (1961).
21. A. Seeger, *Phil. Mag.*, **44**, 1 (1953).
22. G. P. Ivantsov, in *Growth of Crystals*, A. V. Shubnikov and N. N. Sheftal, Eds., Vol. I, Consultants Bureau New York, 1960.
23. B. Chalmers, *Principle of Solidification*, Wiley, New York, 1964.
24. W. W. Mullins and R. F. Sekerka, *J. Appl. Phys.*, **34**, 323 (1963).
25. W. W. Mullins and R. F. Sekerka, *J. Appl. Phys.*, **35**, 444 (1964).
26. R. F. Sekerka, *J. Cryst. Growth*, **3**, 74 (1968).
27. R. F. Sekerka, *J. Cryst. Growth*, **10**, 239 (1971).
28. S. R. Coriel et R. F. Sekerka, *J. Cryst. Growth*, **19**, 90 (1973).
29. D. E. Temkin, *Soviet Phys. Dok.*, **5**, 609 (1960).
30. P. Glansdorff and I. Prigogine, *Structure, Stabilité, Fluctuation*, Masson, Paris, 1971.
31. W. G. Pfann, *Zone Melting*, Wiley, New York, 1957.
32. P. Viaud, *Int. Conf. Cryst. Growth*, *4th*, Tokyo, 1974 Col. Abstr. p. 526.
33. A. C. Zettlemoyer, *Nucleation*, Dekker, New York, 1968.
34. K. A. Jakson, in *Growth and Perfection of Crystals*, Wiley, New York, 1958, p. 319.
35. K. A. Jakson, in *Liquids Metals and Solidification*, Am. Soc. Metals, Cleveland, Ohio, 1958, 174.
36. J. W. Cahn, *Acta Metal.*, **8**, 554 (1960).
37. J. W. Cahn, W. B. Hillig and G. S. Sears, *Acta Met.*, **12**, 1421 (1964).
38. L. A. Tarshis and S. O'Hara, *J. Appl. Phys.*, **38**, 2086 (1967).
39. D. Turnbull, *J. Appl. Phys.*, **21**, 1022 (1960).
40. W. G. Hillig, in *Growth and Perfection of Crystals*, Wiley, New York, 1958, 350.
41. J. Frenkel, *J. Phys. Soviet.*, **9**, 392 (1945).
42. J. Frenkel, *Kinetic Theory of Liquids*, Oxford University Press, New York 1946.
43. P. Viaud, Thèse Doctorat, Univ. Paris VI, 1973.
44. P. Viaud, *J. Cryobiol.*, **9**, 231 (1972).
45. D. S. Choi, M. S. John and H. Eyring, *J. Chem. Phys.*, **53**, 2608 (1970).
46. E. Hesstvet, Norges Geotekniske Institutt, Pub. No. 66, 1964.
47. B. Daigne and F. Girard, *La Rech. Aérosp.*, **4**, 209 (1971).
48. S. C. Hardy and S. R. Coriell, *J. Cryst. Growth*, **5**, 329 (1969).

49. J. S. Darrozes, Personal Communication.
50. M. Van Dick, *Perturbation Methods in Fluid Mechanics*, Vol. 8, Acad. Press, New York, 1964.
51. G. Horway, *Proc. Natl. Congr. Appl. Mech. ASME, 4th*, **1962**, 1315.
52. G. Horway, *Int. J. Heat Mass Transfer*, **9**, 195 (1965).
53. J. K. Landauer, *S.I.P.R.E. Res. Rep.*, No. 48 (1958). (U.S. Snow Ice and Permafrost Res. Estab.)
54. W. M. Ketcham and P. V. Hobbs, *Phil. Mag.*, **18**, 659 (1968).
55. G. F. Bolling and W. A. Tiller, *J. Appl. Phys.*, **38**, 8, 1345 (1960).
56. R. J. Schaefer and M. E. Glicksman, *Acta Metal.*, **16**, 1009 (1968).
57. D. R. H. Jones and G. A. Chadwick, *J. Cryst. Growth*, **11**, 260 (1971).
58. P. Chaudhari and J. W. Matthews, Eds. *Grain Boundaries and Interfaces, Surface Science*, **31** (1972).
59. D. Nason and W. A. Tiller, *Surface Sci.*, **40**, 109 (1973).
60. R. W. Powel and Y. S. Touloukian, *Science*, **181**, 999 (1973).
61. K. H. Chien and T. Z. Kattamis, *Z. Metalk.*, **61**, 475 (1970).
62. T. Z. Kattamis, J. C. Coughlin and M. C. Flemings, *Trans. Met. Soc. AIME*, **239**, 1504 (1967).
63. R. J. Good and E. Elbing, *Ind. Eng. Chem.*, **62**, No. 3, 54 (1970).
64. L. J. Thomas and J. W. Westwater, *Chem. Eng. Progr.*, **69**, 41 (1963).
65. J. D. Harrison and W. A. Tiller, *J. Appl. Phys.*, **34**, 3349 (1963).
66. H. R. Pruppacher, *J. Chem. Phys.*, **47**, No. 5, 1807 (1967).
67. T. C. Bannister and B. E. Richard, *AIAA* paper 69–95 (1969).
68. C. A. Knight, *J. Appl. Phys.*, **32**, No. 2, 568 (1966).
69. D. R. H. Jones, *Rev. Sci. Instr.*, **41**, 1509 (1970).
70. P. Viaud, in *Proc. All Union Heat and Mass Transfer Conf. 4th Minsk*, **9**, 218 (1972) in Russian)
71. A. E. Carte, *Proc. Phys. Soc.*, **77**, 751 (1961).
72. J. Aguirre-Puente, M. Vignes and P. Viaud, *Proc. Int. Conf. Permafrost*, **1972**, 161, Vol. 4. (in Russian)
73. J. Cisse and G. F. Bolling, *J. Cryst. Growth*, **10**, 67 (1971).
74. D. R. H. Jones and G. A. Chadwick, *Phil. Mag.*, **22**, 291 (1970).
75. M. Azouni, *C. R. Acad. Sci. Paris*, **3**, 155 (1972), Table 275.
76. N. H. Fletcher, in *The Chemical Physics of Ice*, Cambridge Univ. Press, London, 1970.
77. C. Jaccard et L. Levi, *Z. Angew. Math. Phys.*, **12**, 70 (1961).
78. G. A. Chadwick, *Acta Met.*, **10**, 1 (1962).
79. W. A. Tiller, *Trans. AIME*, **209**, 847 (1957).
80. G. F. Bolling, J. J. Kramer and W. A. Tiller, *Trans. AIME*, **227**, 1453 (1963).
80. N. H. Fletcher, *J. Appl. Phys.*, **35**, 234 (1964).
82. W. D. Kingery, *Ice and Snow-Properties, Processes and Applications*, Tech. Press., Cambridge, Mass., 1963.
83. A. A. Chernov, *An. Rev. Mat. Sci.* **3** 397 (1973).

STELLAR EVOLUTIONARY STABILITY IN RELATION TO SPECTRAL THEORY

J. PERDANG*

Institut d'Astrophysique,
Cointe-Ougrée, Belgique

CONTENTS

Abstract

The concept of stability of a stellar model evolving over a finite time interval is analyzed in the framework of spectral theory of the linearized evolution equations, allowance being made for chemical abundance fluctuations. We introduce two distinct definitions of stability, termed strict and weak evolutionary stability, which prove to be particularly convenient for the purposes of stellar evolution. Weak stability is found to be in direct relation to the spectral properties of the linearized evolution equations. Strict stability is tackled by means of an auxiliary self-adjoint eigenvalue problem. A detailed physical classification of the spectra

*Chargé de Recherches du Fonds National de la Recherche Scientifique (Belgique).

of eigenvalues of the linearized evolution equations in the presence of abundance perturbations is developed. Attention is focused on continuous spectra, which are shown to arise as a consequence of the chemical abundance fluctuations, and, under particular circumstances, as an outcome of entropy perturbations. Furthermore we provide arguments in favor of continuous modes belonging to the nonradial dynamical spectrum.

I. INTRODUCTION

Statistical mechanics of a system of classical self-gravitating particles meets with difficulties connected with the peculiar nature of the gravitation interaction potential: (*a*) this potential is not (weakly) tempered; (*b*) it has no lower bound.[31] With the Liouville equation as the starting point, it is found[102,105,114] that the weak-coupling asymptotic collision operator exhibits a long-range divergence, and the distribution function displays a nonergodic behavior. Recent numerical experiments by Miller[86] clearly illustrate the particular status of the statistics of gravitationally interacting particles, and lend some support to the radical views expressed by Kurth[58,59] who denies a statistical mechanics of self-gravitating bodies.

While statistical studies of classical particles interacting with Newtonian forces are usually concerned with stellar systems, we consider here systems of n identical microscopic particles (nuclei) which make up a star. In that case gravitational interactions fail to be significant at small interparticle distances where repulsive forces come into play. The above-mentioned difficulty (*b*) therefore disappears. Moreover we assume that the configuration is enclosed in a spherical vessel whose volume is adjustable by means of an external pressure P_e which is equal to the internal pressure P at the surface of the vessel under equilibrium conditions. This model bears some analogy with those investigated by Ebert,[30] Bonnor,[13] McCrea,[81] Yabushita,[136] Penston,[96] and Horedt,[50] as well as with one model discussed by Lynden-Bell and Wood.[78] We mention also that it is not directly comparable with the configurations discussed by Antonov,[7] Thirring,[124,125] Hertel and Thirring,[49] and Aronson and Hansen,[8] nor with the numerical experiments by Miller.[86] By postulating a finite radius of the vessel, we ensure that difficulty (*a*) is eliminated, so that statistical mechanics, and hence thermodynamics, apply to our configuration. Regarded as a one-phase thermodynamic system, our configuration is divariant according to the Gibbs phase rule. We can choose the (uniform) temperature T and the pressure P arbitrarily. However, the existence of repulsive forces at short interparticle distances means that the system can live in a condensed phase for a sufficiently low kinetic energy of the particles. For a certain range of T and P the system can display two phases, a condensed phase and a gas phase, coexistence of both phases in equilibrium being allowed on the line segment of the (T,P) plane of vanishing affinity for the phase transition

$$A(T,P)=0 \tag{I.1}$$

Suppose then we start with a sufficiently high temperature T_1 and a given pressure P_1, and consider an isochoric transformation in which temperature is sufficiently decreased. In general the representative point (T,P) of the transformation will intersect the phase-line (I.1) at (T_t,P_t). Beyond this point the system undergoes a first-order phase transition toward the condensed phase. The edge of the phase-line defines a state of marginal stability, the critical point (T_c,P_c). As is well known from Tisza's theory of phase transitions,[126,127] the elements of the compliance matrix become singular at the critical point. In our one-component system, the fact that the critical behavior may be regarded as arising through entropy and volume fluctuations implies that the specific heat at constant pressure, the isothermal compressibility, and the thermal expansion coefficient tend to infinity.

This simple picture changes if we introduce a further degree of freedom to allow for one-dimensional critical sets in the (T,P) plane. An obvious additional variable of nonthermal origin which plays an important role in the present problem is a macroscopic velocity field $\mathbf{v}$. For a state of thermodynamic equilibrium this variable appears to be a quasithermodynamic variable in the sense defined by Tisza.* Energy density being an even function of $\mathbf{v}$, this variable can be regarded as the analog of the order parameter responsible for λ-points. However, since beyond the λ-point a nonzero velocity field is induced, classical thermodynamics fails to describe the phase subsequent to the transition point. This qualitative argument in favor of λ-points is not in contradiction with the simplified statistical analyses carried out by Ebert,[30] Bonnor,[13] and Lynden-Bell and Wood.[78] For example, in the case of an isothermal self-gravitating sphere subjected to an external pressure an instability is found at the point where the specific heat at constant pressure has a singularity for a given density contrast. For a higher density contrast the specific heat is found to be negative, which corresponds to the absence of stable thermodynamic equilibrium configurations.

It is tempting to resort to the previous ideas to understand the broad aspects of stellar evolution. As suggested by observation,[48,65] star formation originates in globules of masses of the order of 10^2 to $10^3 M_\odot$. Consider an initially stable globule in thermal equilibrium with an external heat bath, for example, a surrounding radiation field, exerting a pressure P_e

*If u is the energy density, $\partial u/\partial \mathbf{v}=\rho\mathbf{v}$, which vanishes under thermal equilibrium conditions.

upon the globule. As the temperature of the heat bath is decreased, the globule ultimately undergoes a phase transition. The globule contracts, and possibly breaks up into several fragments. Owing to the negative specific heat which characterizes the collapse stage, the globule is heated up while it is contracting and radiating. Once its temperature has become high enough to ignite thermonuclear reactions, contraction toward a condensed object is momentarily halted—the proper main sequence star is born.

The previous cursory qualitative considerations are intended to emphasize the peculiar status of a star: A star appears to be a transient configuration in the stage of jumping from one thermodynamic phase to another one. As such, a proper star is *thermodynamically unstable*. Conventional thermostatics enables us to deal with the stages before and after the phase transition. It is of no resource for a study of the star phase. Less trivial instabilities than the one mentioned are entirely inaccessible to classical thermostatics. However, nonequilibrium thermodynamics as developed by Glansdorff and Prigogine[37–40] can be adapted to the problem at hand by formally regarding the gravity field as an external force. In this approach the stellar medium is actually in stable local thermodynamic equilibrium, so that the Glansdorff-Prigogine formalism becomes meaningful. This method has been applied to general aspects of stellar stability theory in a somewhat modified form divised by Unno,[128] and has met with some success in particular problems.[101,128–130,132] We wish to point out that the original Glansdorff-Prigogine theory disregards fluctuations in the external force. If applied to magnetohydrodynamics an explicit contribution of the electromagnetic field to the quantity ψ is obtained.[66] In view of the formal analogy between magnetohydrodynamics and gravihydrodynamics (cf. also Appendix A), it remains to be seen whether similar contributions stemming from the gravity field fluctuations need to be included.

In the framework of nonequilibrium thermodynamics the process of star formation can be pictured as the spontaneous growing of gravity field fluctuations (Appendix A), which generate in turn a velocity field. After a transient nonstationary phase during which the medium breaks up into contracting cells, each cloudlet evolves toward a stationary nonequilibrium state, the actual star phase. The situation bears some analogy with Bénard convection. For a Rayleigh number larger than a critical value, velocity fluctuations are spontaneously amplified, and after a transient nonstationary stage, a steady state of cellular circulation is built up.

Due to radiative heat losses, evolution during the protostar phase is characterized essentially by a decreasing entropy. As the protostar approaches the main sequence radiation losses are progressively compensated by the onset of thermonuclear energy generation, and the rate of change of entropy tends to vanish. In later evolutionary phases (red giant stage) energy generation may overshoot the heat losses, and entropy may increase slightly with evolution. During most of its lifetime, the star appears to be in a state of *minimum entropy*. It is therefore a typical example of a *dissipative structure*.[104]

The present paper is devoted to a systematic stability analysis of the stellar phase, particular attention being paid to chemical abundance perturbations. In spite of our lengthy thermodynamic introduction, this analysis is not carried out along the lines of reasoning of irreversible thermodynamics, since our main interest is in the stability concept of evolutionary stellar sequences, rather than in steady-state models. As we endeavor to show, spectral analysis remains a powerful tool of investigation of stability of evolving systems.

II. EVOLUTION EQUATIONS. STABILITY

A. The Full Set of Evolution Equations

We denote by K compact sets, by R^n real n-spaces, and by T closed bounded time intervals. Let $\mathbf{r}$ be the space coordinate referred to a Galilean reference frame, and t the time variable

$$\mathbf{r}\in\mathsf{K}_r,\ t\in\mathsf{T},\ \mathsf{K}_r\subset\mathsf{R}^3,\ \mathsf{T}\subset\mathsf{R}^1 \tag{II.1}$$

We further introduce the following quantities

$$\begin{aligned} \underset{\sim}{p} &= (\mathbf{p}, \underset{\wedge}{p}) = (\mathbf{p}, \underline{\underline{p}}, p_{N+4}) \in \mathsf{K}_p \subset \mathsf{R}^{N+4} \\ \underset{\sim}{\sigma} &= (\boldsymbol{\sigma}, \underset{\wedge}{\sigma}) = (\boldsymbol{\sigma}, \underline{\underline{\sigma}}, \sigma_{N+4}) \in \mathsf{K}_\sigma \subset \mathsf{R}^{N+4} \\ \underset{\sim}{\mathbf{F}} &= (\mathcal{F}, \underset{\wedge}{\mathbf{F}}) = (\mathcal{F}, \underline{\underline{\mathbf{F}}}, \mathbf{F}_{N+4}) \in \mathsf{K}_F \subset \mathsf{R}^3 \times \mathsf{R}^{N+4} \end{aligned} \tag{II.2}$$

The vector valued field $\underset{\sim}{p}(\mathbf{r},t)$, conveniently referred to as the generalized momentum per unit volume, defines the state of the star at point $\mathbf{r}$ and time t. The law of evolution of $\underset{\sim}{p}$ is specified in terms of the sources $\underset{\sim}{\sigma}$ and the fluxes $\underset{\sim}{\mathbf{F}}$, by means of the following partial differential equation

$$\partial_t \underset{\sim}{p} + \partial_{\mathbf{r}}\cdot(\mathbf{v}\underset{\sim}{p} + \underset{\sim}{\mathbf{F}}) = \underset{\sim}{\sigma}, \qquad \mathbf{v} = \frac{\mathbf{p}}{\underline{\underline{1}}\cdot\underline{\underline{p}}} \tag{II.3}$$

where the notations are self-explanatory*. Equation (II.3) embodies the

*Scalar products in any $\mathbf{R}^n$ will be indicated by a dot $(\cdot)$. By $1\in\mathbf{R}^n$ we understand the vector $(1,1,\ldots,1)$. To guarantee a physically meaningful definition of the norm, or the scalar product in $\mathbf{R}^{N+4}$, it will be assumed that all components of $\underset{\sim}{p}$ have the same physical dimensions.

usual hydrodynamic balance equations. The first three components of $\underset{\sim}{p}$ correspond to the ordinary momentum density vector

$$\mathbf{p} = \mathbf{v}\rho \in \mathbf{R}^3 \tag{II.4}$$

ρ being the mass density, and $\mathbf{v}$ the velocity field. The next N components represent the partial mass densities $\underline{\underline{\rho}}$ of the N different chemical species

$$\underline{\underline{p}} \equiv \underline{\underline{\rho}} = \underline{\underline{X}}\rho \in \mathbf{R}^N \tag{II.5}$$

$\underline{\underline{X}}$ being the set of abundances by mass of the species.

It may prove convenient to resort to a mapping $\mathfrak{T} : \mathbf{R}^N \to \mathbf{R}^N$ ($\mathfrak{T}$ independent of $\mathbf{r}$ and t, and regular) endowed with the following properties:

a. Any $\underline{\underline{Q}} \in \mathbf{R}^N$ is transformed into

$$\underline{\underline{Q}}^T = \mathfrak{T} \cdot \underline{\underline{Q}} \tag{II.5$'$}$$

b. If $\underline{\underline{p}}$ is transformed by $\mathfrak{T}$, its thermodynamically conjugate variables are transformed by $\mathfrak{T}^{-1}$.

c. If $\underline{\underline{p}}$ is transformed by $\mathfrak{T}$, $\underline{\underline{1}}$ is transformed by $\mathfrak{T}^{-1}$.

With these conventions a $\mathfrak{T}$-transformation leaves all equations of this section invariant, and an explicit introduction of the adjoint space is avoided.

The last component of $\underset{\sim}{p}$ is the entropy density

$$p_{N+4} = S\rho \tag{II.6}$$

S being the entropy per unit mass.

In the absence of exterior disturbances and electromagnetic effects, the sources and fluxes relevant to stellar interiors are as follows:

1.
$$\boldsymbol{\sigma} = \mathbf{g}\rho \tag{II.7}$$

$\mathbf{g}$ being the gravity field (cf. Appendix A).

2. The rate of change of the mass densities, $\underline{\underline{\sigma}}$, is given as a function of the generalized momenta (but independent of the hydrodynamic momentum density $\mathbf{p}$)

$$\underline{\underline{\sigma}} = \underline{\underline{\sigma}}(\hat{p}) = (R_1, R_2, \ldots, R_N) \tag{II.8}$$

3. The rate of entropy generation is given by

$$\sigma_{N+4} = T\mathbf{F}_{N+4} \cdot \partial_{\mathbf{r}} \frac{1}{T} - \frac{\underline{\underline{\sigma}} \cdot \underline{\underline{\mu}}}{T} \tag{II.9}$$

where $\mathbf{F}_{N+4}$ is the entropy flux, $\underline{\underline{\mu}}$ the set of chemical potentials, and T the temperature.[28,83,103] In (II.9) we ignore contributions due to viscosity, as well as particle diffusion effects $(-\underline{\underline{\mathbf{F}}} \cdot \partial_{\mathbf{r}} \underline{\underline{\mu}}/T)$, which are inefficient in stellar interiors. In view of the important energy release in thermonuclear reactions, the integration constants $\underline{\underline{q}}$ entering into the chemical potentials (rest mass energies, cf. Ref. 64) lead to the most important contribution to $\underline{\underline{\mu}}$. It is therefore convenient to split up the chemical potentials as follows

$$\underline{\underline{\mu}} = \underline{\underline{q}} + \underline{\underline{\mu}}' \tag{II.10}$$

Then $(-\underline{\underline{\sigma}} \cdot \underline{\underline{q}})$ represents the rate of change of rest mass energy per unit volume, or equivalently the rate of nuclear energy generation per unit volume, $\epsilon\rho$. As first emphasized by Rosseland,[110] the change in rest mass should imply a modification of the usual continuity equation, which should become

$$\partial_t \rho + \partial_{\mathbf{r}} \cdot \rho \mathbf{v} = \underline{-\frac{\epsilon\rho}{c^2}} \tag{II.11}$$

c being the speed of light. However the underlined correction term is a relativistic effect, negligible in the frame of a Newtonian theory (cf. Appendix A). Therefore we have to postulate that

$$\underline{\underline{1}} \cdot \underline{\underline{\sigma}} = 0 \tag{II.12}$$

We shall separate the energy flux $\mathbf{W}$ into two components, $\mathbf{W}_R$, the flux due to radiation and electronic energy transport, and $\mathbf{W}_0$, the energy flow associated with other mechanisms. For $\mathbf{W}_R$ a Fourier type law is adopted

$$\mathbf{W}_R = -K\,\partial_{\mathbf{r}} T, \qquad K = K(\hat{p}\ ) \geqslant 0 \tag{II.13}$$

The entropy flux is then defined by[83]

$$\mathbf{F}_{N+4} = -KT^{-1}\,\partial_{\mathbf{r}} T + \mathbf{W}_0 T^{-1} \tag{II.14}$$

The entropy production becomes

$$\sigma_{N+4}=T^{-1}\Big[KT^{-1}(\partial_{\mathbf{r}}T)^2-\mathbf{W}_0T^{-1}\cdot\partial_{\mathbf{r}}T+\epsilon\rho-\underline{\underline{\sigma}}\cdot\underline{\underline{\mu}}'\Big] \qquad \text{(II.15)}$$

4. Viscosity effects being negligible under current stellar conditions, the flux $\mathcal{F}$ reduces to scalar pressure P

$$\mathcal{F}=P\,\mathcal{E}_3 \qquad \text{(II.16)}$$

$\mathcal{E}_n$ being the unit matrix in R^n.

5. Since we disregard particle diffusion,

$$\underline{\underline{F}}=0 \qquad \text{(II.17)}$$

To investigate system (II.3) the following complementary data are needed:

1. Two independent thermodynamic equations which characterize the material under investigation, for example, the thermal and caloric equations of state.
2. The set of reaction rates, which entails the rate of energy generation.
3. The energy transport coefficient K.
4. The flux $\mathbf{W}_0$. Under all currently encountered situations $\mathbf{W}_0$ is the energy transport due to neutrinos. Matter being usually transparent to neutrinos, this flux is given by

$$\partial_{\mathbf{r}}\cdot\mathbf{W}_0=\epsilon_\nu\rho \qquad \text{(II.17')}$$

ϵ_ν being the rate of energy production per unit mass in the form of neutrinos.

Equations (II.3) yield the balance equations in Eulerian form. This system is immediately transformed into

$$\rho D_t\underset{\sim}{p}^L+\partial_{\mathbf{r}}\cdot\underset{\sim}{\mathbf{F}}^L=\underset{\sim}{\sigma}^L \qquad \text{(II.18)}$$

where D_t is the convective derivative, and

$$\underset{\sim}{p}^L=(\mathbf{v},\underline{X},V,S)$$

$$\underset{\sim}{\mathbf{F}}^L=(P\,\mathcal{E}_3,\underline{\mathbf{0}},-\mathbf{v},-KT^{-1}\partial_{\mathbf{r}}T) \qquad \text{(II.19)}$$

$$\sigma^L=(\rho\mathbf{g},\underline{\sigma},0,\sigma_{N+4})$$

V is the specific volume $(1/\rho)$, and $\underline{X}$ and $\underline{\sigma} \in \mathrm{R}^{N-1}$ are $N-1$ *independent* chemical abundances and the corresponding $N-1$ rate functions.

The formal steps to generate (II.18) from (II.3) are as follows: We apply a $\mathfrak{T}$-transformation defined by

$$\mathfrak{T} = \begin{pmatrix} 1 & & & 0 \\ & 1 & & \\ 0 & & \ddots & \\ 1 & 1 & \vdots & 1 \end{pmatrix} \qquad \text{(II.19}'\text{)}$$

(all elements vanish except the principal diagonal and Nth row elements which are 1). The new system is then multiplied by the diagonal matrix $\mathrm{diag}\left[\mathcal{E}_3, \mathcal{E}_{N-1}, -\frac{1}{\rho}, 1\right]$, and finally we introduce the operator D_t. Equation (II.19) will be referred to as the (improper) Lagrangian form of the balance equations.

In connection with the entropy equation, several remarks are in order. For stars of full radial symmetry (K_h symmetry) the $N+4$th equation of (II.19) reduces to

$$TD_tS + D_mL = \epsilon - \underline{\underline{\sigma \cdot \mu'/\rho}} \qquad \text{(II.20)}$$

where $m(r)$ is the mass contained in a sphere of radius r, D_m is the (Lagrangian) derivative with respect to the mass variable, and $L(m)$ is the total luminosity that is, the surface integral of the total energy flux $\mathbf{W}$ over the surface of the sphere containing the mass m. On the basis of a purely thermostatic argument Kutter and Savedoff[60,61] on one hand, and Strittmatter et al.[121] on the other hand arrived at two incompatible entropy equations. Our equation (II.20) formulated in the correct frame of irreversible thermodynamics is seen to be equivalent to the entropy equation given by Strittmatter et al.[121] in the absence of convection [$\underline{\underline{\sigma}} = \rho D_t \underline{\underline{X}}$; the contribution of neutrinos to luminosity obeys $D_m L_0 = \epsilon_\nu$, by virtue of (17′)]. Although we do not make any attempt to discuss convection theory, we wish to emphasize that in the frame of simple pictures (cf. the quasiparticle scheme, Ref. 98), (II.26) remains essentially correct. In fact, convective regions being chemically homogeneous, the contribution due to convective particle diffusion, $-\underline{\underline{F}}_C \cdot \partial_r \underline{\underline{\mu}} T^{-1}$, $\underline{\underline{F}}_C$ being the particle flux, vanishes. However, in the underlined term of (II.20) we are not entitled to write $\underline{\underline{\sigma}} = \rho D_t \underline{\underline{X}}$ for a convective region, so that the form of this contribution as indicated by Strittmatter et al.[121] is not satisfactory in the presence of convection.

As regards the mathematical problem defined by the evolution equation (II.3) we formulate the following set of *assumptions*:

A.1. Consider the partial differential evolution equations (II.3) and (18) written in the form

$$\partial_t \underset{\sim}{p} = \mathfrak{N}^E(\underset{\sim}{p}) \qquad \text{or} \qquad D_t \underset{\sim}{p}^L = \mathfrak{N}^L(\underset{\sim}{p}^L) \tag{II.22}$$

The nonlinear operators $\mathfrak{N}(\underset{\sim}{p})$ admit of continuous Fréchet derivatives, denoted by $\mathfrak{L}^E(t)$ and $\mathfrak{L}^L(t)$ at any $t \in \mathsf{T}$.

A.2. Given a set of initial conditions $\underset{\sim}{p}(\mathbf{r},0) = \underset{\sim}{p}_0(\mathbf{r})$, and natural* boundary conditions, there exists a unique differentiable map $\mathfrak{M}: t \to \underset{\sim}{p}(\mathbf{r},t)$ of T into K_p such that $\underset{\sim}{p}$ satisfies (II.22) together with the initial and boundary conditions.

A.3. The class of allowed initial conditions $\underset{\sim}{p}_0(\mathbf{r})$ is continuous in K_r. The family $\underset{\sim}{p}(\mathbf{r},t)$ is continuous in $\mathsf{K}_r \times \mathsf{T}$. For fixed $t, \underset{\sim}{p}(\mathbf{r},t) \in C(\mathsf{K}_r) \times \mathsf{R}^{N+4} = \mathsf{B}$; B is assumed to be a Banach space (Ref. 54, 1). We represent $\underset{\sim}{p}(\mathbf{r},t)$ by

$$\underset{\sim}{p}(\mathbf{r},t) \to \sum_i \Pi_i(t) \underset{\sim}{\xi}_i(\mathbf{r}) \tag{II.23}$$

$\{\underset{\sim}{\xi}_i(\mathbf{r})\}$ being termed the basis of the Banach space B. Upon limiting the expansion (II.23) to a finite number M of terms, a norm can be defined in B such that

$$\|\underset{\sim}{p}(\mathbf{r},t) - \underset{\sim}{p}_M(\mathbf{r},t)\| < \epsilon(M), \qquad \forall t \in \mathsf{T} \tag{II.24}$$

where $\underset{\sim}{p}_M$ represents the truncated series, and where the positive number ϵ can be rendered arbitrarily small for M sufficiently large. The space $\Gamma = \Gamma^{N+4} \times \Gamma^M$ spanned by the M basis functions $\underset{\sim}{\xi}_i(\mathbf{r}), i = 1, 2, \ldots, M$, will be referred to as the *phase space of stellar evolution.*

Under assumption A.3 the evolution equations can be cast into the form

$$\partial_t \Pi = \mathfrak{N}(\Pi) \tag{II.25}$$

*We define the surface of the star by $F\{\underset{\sim}{p}(\mathbf{r},t)\} = 0$, where F is a convenient function or functional; in simple cases F is the surface pressure. The natural boundary conditions follow from the classical method of integration of (II.3) over an arbitrarily small cylindrical volume with bases parallel to the surface of the star and lateral surface normal to the surface of the star.

where Π and $\mathfrak{N}$ are $M\times(N+4)$ vectors in Γ. Assumptions A.1 and A.3 are implicit in all numerical stellar evolutionary codes. Attempts to define the precise conditions of applicability of assumption A.2 (uniqueness of the solution) have been made by Kähler and Weigert.[52] For any evolutionary sequence of stellar models whose stability is to be investigated, we postulate that assumptions A.1, A.2, and A.3 are fulfilled. However, to leave space for several interesting physical phenomena, the continuity assumptions A.3 and their implications will be relaxed at some stages of the stability discussion.

B. Definition of Stability. Connection with the Spectral Problem

The current definitions of stability introduced in the field of ordinary differential equations[89,108,111] are concerned with the asymptotic behavior of the solution $t\to\infty$, which is not available, in general, in the stellar case. We shall adopt in this paper two modified definitions of the stability concept encountered in mechanics,[95] which cover the situation of finite time intervals, and, furthermore, can be directly related to usual spectral theory as handled in stellar evolution.

Let $\Sigma(t)$ be a sequence of models specified by their state vector $\underset{\sim}{p}(\mathbf{r},t)\in C(\mathsf{K}_r)$ in the time interval T of length Δt, computed with initial conditions $\underset{\sim}{p}_0(\mathbf{r})$. Let $\Sigma'(t)$ be a perturbed sequence of models of state vector $\underset{\sim}{p}'(\mathbf{r},t)$ not necessarily continuous in $\mathsf{K}_r, t\in\mathsf{T}$, with initial conditions $\underset{\sim}{p}'_0(\mathbf{r})$ in a sufficiently small neighborhood N of $\underset{\sim}{p}_0(\mathbf{r})$.

D.1. If for *any* initial condition $\underset{\sim}{p}'_0(\mathbf{r})\in\mathsf{N}$ the sequence $\Sigma'(t)$ fulfills

$$\|\underset{\sim}{p}'(\mathbf{r},t)-\underset{\sim}{p}(\mathbf{r},t)\|\leqslant\exp(-\alpha t)\|\underset{\sim}{p}'_0(\mathbf{r})-\underset{\sim}{p}_0(\mathbf{r})\| \tag{II.26}$$

α, real positive, $t\in\mathsf{T}$, and $\|\ \|$ denoting a convenient norm, then $\Sigma(t)$ is said to be strictly stable in T in the sense of the norm considered.

D.2. If for any initial condition $\underset{\sim}{p}'_0(\mathbf{r})\in\mathsf{N}, \exists t_*: t_*>0$ and $t>t_*, t, t_*\in\mathsf{T}$ and the sequence $\Sigma'(t)$ fulfills

$$\|\underset{\sim}{p}'(\mathbf{r},t)-\underset{\sim}{p}(\mathbf{r},t)\|\leqslant\exp[-\alpha(t-t_*)]\|\underset{\sim}{p}'(\mathbf{r},t_*)-\underset{\sim}{p}(\mathbf{r},t_*)\| \tag{II.27}$$

then $\Sigma(t)$ is weakly stable in T in the sense of the norm considered.

It is implicitly understood that t_* is small as compared to Δt.

D.3. If definition D.1 (D.2) is violated, then the sequence $\Sigma(t)$ is termed not strictly (weakly) stable.

D.4. If for any *fixed* $t \in \mathsf{T}$ we can find at least one initial condition $\underset{\sim}{p}\,'_{0t}(\mathbf{r})$ such that the corresponding evolution $\underset{\sim}{p}\,'(\mathbf{r},t)$ obeys

$$\|\underset{\sim}{p}\,'(\mathbf{r},t)-\underset{\sim}{p}\,(\mathbf{r},t)\| \geqslant \exp(\beta t)\|\underset{\sim}{p}\,'_{0t}(\mathbf{r})-\underset{\sim}{p}\,_0(\mathbf{r})\| \qquad \text{(II.28)}$$

where β is real positive, then the sequence $\Sigma(t)$ is unstable in the sense of the norm considered.

If the time interval T becomes unbounded, definition D.1 coincides with the concept of exponentially asymptotic stability.[75,79]

In the sequel we shall be concerned exclusively with stability considerations in the first approximation.[111]

In the particular case of an equilibrium model $[\underset{\sim}{p}\,(\mathbf{r},t)=\underset{\sim}{p}\,_0(\mathbf{r}), t\in(-\infty,+\infty)]$, weak stability is directly investigated in terms of the associated spectral problem:

$$\mathsf{S}=\{s \mid s\Delta\underset{\sim}{p} = \mathfrak{L}^{\mathrm{E}}\Delta\underset{\sim}{p} \qquad \text{or}$$

$$s\delta\underset{\sim}{p}\,^L = \mathfrak{L}^L\delta\underset{\sim}{p}\,^L, \delta\underset{\sim}{p}, \delta\underset{\sim}{p}\,^L \in \mathsf{C}^{N+4}\mathfrak{L}^E, \mathfrak{L}^L \in \mathsf{R}^{N+4}\times\mathsf{R}^{N+4}, s\in\mathsf{C}^1\} \qquad \text{(II.29)}$$

The Fréchet derivatives $\mathfrak{L}$ are computed for the equilibrium model, Δ and δ are the Eulerian and Lagrangian perturbation operators, and C^n denotes a complex n-space. If the spectrum of eigenvalues S is such that

$$\mathrm{Re}\,\mathsf{S}<-\alpha \qquad \alpha \text{ arbitrary real positive} \qquad \text{(II.30)}$$

the equilibrium model is (at least) weakly stable, the time interval T being chosen sufficiently large. On the other hand if

$$\exists s_k\in\mathsf{S}:\mathrm{Re}\,s_k>\beta \qquad \beta \text{ arbitrary real positive} \qquad \text{(II.31)}$$

the equilibrium is unstable. In the case of a spectrum whose real parts are nonpositive, with at least one zero real part, our definition D.3 applies. In conventional stellar stability theory this case is sometimes called stable.[70]

For nonequilibrium models a spectral problem of type (II.29) still remains meaningful. In fact let T' be a time interval of sufficiently short length $\Delta\tau$, such that the Fréchet derivatives $\mathfrak{L}$ computed for a model at time t are invariant under time translations

$$t\to t+a, \qquad |a|<\frac{\Delta\tau}{2} \qquad \text{(II.32)}$$

Since the irreducible representations of the translation group have characters e^{st}, a spectral problem (II.29) can be defined for $t \in \mathsf{T}'$. However while a knowledge of S provides us with full information on (weak) stability of equilibrium models, a knowledge of the spectra $\mathsf{S}(\tau)$, for any $\tau \in \mathsf{T}$, defined by (II.29) does not necessarily yield this information. Fortunately many situations arise, where simple practical criteria are obtained in terms of $\mathsf{S}(\tau)$ alone.

C.1. Let T be a time interval of length Δt, and let M be a sufficiently large number. If the spectrum obeys the inequality

$$\Delta t |\mathrm{Re}\, \mathsf{S}(\tau)| > M, \qquad \forall \tau \in \mathsf{T} \tag{II.33}$$

together with condition (II.30) (for any $\tau \in \mathsf{T}$), then $\Sigma(t)$ is weakly stable in T.

C.2. If

$$\exists s_k(\tau) \in \mathsf{S}(\tau): \Delta t\, \mathrm{Re}\, s_k(\tau) > M, \qquad \forall \tau \in \mathsf{T} \tag{II.34}$$

then the sequence is unstable in T.

C.3. Let T be a time interval of length ΔT, and let M be a sufficiently large number. A convenient choice of ΔT can be made such that the spectrum $\mathsf{S}(\tau)$ can be partitioned as follows:

$$\mathsf{S}(\tau) = \mathsf{S}^+(\tau) + \mathsf{S}^-(\tau) + \mathsf{S}^0(\tau) \tag{II.35}$$

All eigenvalues $s(\tau)$ which obey

$$\Delta T |\mathrm{Re}\, s(\tau)| > M, \qquad \forall \tau \in \mathsf{T} \tag{II.36}$$

belong to $\mathsf{S}^+(\tau) + \mathsf{S}^-(\tau)$, with $s(\tau) \in \mathsf{S}^+(\tau) [\mathsf{S}^-(\tau)]$ if $\mathrm{Re}\, s(\tau) > 0 (< 0)$; all remaining eigenvalues obey

$$\Delta T |\mathrm{Re}\, s(\tau)| < 1$$

and belong to $\mathsf{S}^0(\tau)$. If the conditions of criteria C.1 and C.2 are not satisfied, the stability properties of the sequence $\Sigma(t)$ in T depend on the subspectrum $\mathsf{S}^0(\tau)$, and possibly on the corresponding eigenfunctions.

Statements C.1 and C.2 are intuitively obvious. A formal verification of C.1, C.2, and C.3 is provided in Appendix B.

C.4. If the trace of the operator $\mathfrak{L}(\tau)$ is positive for any $\tau \in \mathsf{T}$, then the evolutionary sequence $\Sigma(t)$ is unstable over T.

We show this property by means of a semiintuitive argument. A rigorous proof is obtained along the lines of reasoning given in Appendix B (part a). According to assumption A.3 we can investigate stability in the frame of the discrete representation (II.23). Consider then a cluster of ν initial conditions in the neighborhood of the initial condition of the sequence whose stability is investigated. Assume ν large enough so that a continuous density w of representative points in the phase space Γ becomes meaningful. Since according to assumption A.2 any initial contition gives rise to one and only one path in Γ, the number of phase points ν is conserved over T, so that we can write a continuity equation in Γ

$$\partial_t w + \partial_\Pi \cdot [\mathfrak{N}(\Pi) w] = 0 \qquad \text{or} \qquad D_t w = -(\operatorname{tr} \partial_\Pi \mathfrak{N}(\Pi)) w \quad \text{(II.37)}$$

D_t being the hydrodynamic derivative in Γ, and $\partial_\Pi \mathfrak{N}$ being the discrete representation of $\mathfrak{L}(\tau)$. Thus if $\operatorname{tr}\mathfrak{L}(\tau)>0, \forall \tau \in \mathsf{T}$, the volume occupied by the cluster of points in Γ increases with time. Evolutionary paths generated by nearby initial conditions therefore tend to diverge.

C.5. If the sequence $\Sigma(t)$ is invariant under a group $\mathcal{G}$, and if the sum of all eigenvalues of $\mathfrak{L}(\tau)$ belonging to the same irreducible representation Γ^{f_ν} of $\mathcal{G}$ is positive for any $\tau \in \mathsf{T}$, then the evolutionary sequence is unstable.

$\Sigma(t)$ being invariant under $\mathcal{G}$, the linearized evolution equation is invariant under $\mathcal{G}$ for any $t \in \mathsf{T}$. Consider the time-dependent solution of the latter equation expanded in the form (II.23) (infinite series). In the (infinite dimensional) phase space Γ, let $D(G_i)$ be the matrix representation of the group element G_i. Let $D^\nu(G_i)$ be the irreducible matrix representation of dimension f_ν of the same group element. Then

$$D(G_i) = \sum_\nu a_\nu D^\nu(G_i) \quad \text{(II.38)}$$

a_ν being a set of positive integers and the summation going over all irreducible representations.[45] The space Γ is decomposed into the direct sum of subspaces Γ^{f_ν} of dimension f_ν,

$$\Gamma = \underbrace{\Gamma^{f_1} \oplus \Gamma^{f_1} \oplus \cdots \oplus \Gamma^{f_1}}_{a_1} \oplus \underbrace{\Gamma^{f_2} \oplus \Gamma^{f_2} \oplus \cdots \oplus \Gamma^{f_2}}_{a_2} \oplus \cdots \oplus \underbrace{\Gamma^{f_\nu} \oplus \Gamma^{f_\nu} \oplus \cdots \oplus \Gamma^{f_\nu}}_{a_\nu} \oplus \cdots \quad \text{(II.39)}$$

that is, we can select a set of basis functions $\{\xi\}$ such that the functions of the subset $\{\xi_1^\nu, \xi_2^\nu, \ldots, \xi_{f_\nu}^\nu\}$ which span the space Γ^{f_ν} transform among themselves under $\mathcal{G}$. Statement C.5 then follows from an application of criterion C.4 to the space Γ^{f_k}. We note that group-adapted eigenfunctions for most symmetry groups of practical interest can be found in Melvin.[84]

Let $\mathfrak{L}^A$ be the adjoint operator of the linear operator $\mathfrak{L}$. We define a symmetrized eigenvalue problem as follows

$$\mathsf{S}'=\left\{s' \mid s'\underset{\sim}{\varphi}=\mathfrak{L}'\underset{\sim}{\varphi},\underset{\sim}{\varphi}\in\mathsf{R}^{N+4},\mathfrak{L}'=\tfrac{1}{2}(\mathfrak{L}+\mathfrak{L}^A),s'\in\mathsf{R}^1\right\} \tag{II.40}$$

$\mathfrak{L}$ being either the Lagrangian or the Eulerian linear operator.

C.6. If the spectrum of the symmetrized eigenvalue problem (II.40) is negative for any time $\tau\in\mathsf{T}$, then the sequence $\Sigma(\tau)$ is strictly stable over T in the sense of the Euclidean norm in Γ.

C.7. If the spectrum S' of the symmetrized eigenvalue problem is positive at any time $\tau\in\mathsf{T}$, the sequence $\Sigma(\tau)$ is unstable over T in the sense of the Euclidean norm in Γ.

C.8. If the spectrum S' of the symmetrized eigenvalue problem has at least one positive eigenvalue $s_+(0)$ at the initial time $\tau=0$, then $\Sigma(\tau)$ is unstable over an interval T' of sufficiently short duration $\Delta\tau$, in the sense of the Euclidean norm in Γ.

Criteria C.6 and C.7 follow from Ważewski's inequalities applied to the linearized finite dimensional equations (II.25). Let $s_i(\tau)$ be the set of eigenvalues of the symmetrized matrix $\partial_\Pi\mathfrak{N}$, that is

$$\tfrac{1}{2}(\partial_\Pi\mathfrak{N}+\widetilde{\partial_\Pi\mathfrak{N}})\psi=s_i(\tau)\psi \tag{II.41}$$

where the tilde denotes the adjoint matrix. Then

$$\exp\left[\int_0^t d\theta\,\lambda(\theta)\right]|\Delta\Pi_0|\leqslant|\Delta\Pi(t)|\leqslant\exp\left[\int_0^t d\theta\,\Lambda(\theta)\right]|\Delta\Pi_0| \tag{II.42}$$

with

$$\Lambda(\tau)=\max_{i=1,2,\dots}s_i(\tau),\qquad \lambda(\tau)=\min_{i=1,2,\dots}s_i(\tau) \tag{II.43}$$

$|\ \ |$ denotes the Euclidean norm in the finite dimensional space Γ.[134] Thus

$$\exp\left[\inf_{\tau\in\mathsf{T}}\lambda(\tau)t\right]|\Delta\Pi_0|\leqslant|\Delta\Pi(t)|\leqslant\exp\left[\sup_{\tau\in\mathsf{T}}\Lambda(\tau)t\right]|\Delta\Pi_0| \tag{II.44}$$

which entails C.6 and C.7. To prove criterion C.8 select as an initial condition for the perturbation the eigenvector $\Delta\Pi_{0+}$ associated with the positive eigenvalue $s_{+}(0)$ of the symmetrized matrix of (II.41). Direct integration of the time-dependent linearized evolution equation yields

$$|\Delta\Pi(t)| = \exp[s_{+}(0)t]|\Delta\Pi_{0+}| + O(\epsilon^2), t \in \mathsf{T}' \qquad \text{(II.45)}$$

provided $\Delta\tau s_{+}(0) = O(\epsilon)$, ϵ being a sufficiently small number. This relation obeys the instability definition D.4 to lowest order in ϵ.

Since criterion C.6 is also a sufficient condition for weak stability, the symmetrized eigenvalue problem, which is much easier to investigate numerically than the starting non self-adjoint problem, yields in fact valuable information on stability. It should be emphasized however that the instabilities as derived from criterion C.8 do not always meet our intuitive feeling. Examples of such pathological cases are displayed in Appendix C.

III. GENERAL PROPERTIES OF THE SPECTRUM

Apart from the trivial properties that S is not empty (Ref. 54, I.5)* and that it is symmetric in the complex plane with respect to the real axis, no mathematical property valid for arbitrary fluxes and sources has been proved so far. Depending on the particular structure of $\underset{\sim}{\mathbf{F}}$ and $\underset{\sim}{\sigma}$ we can expect therefore that the spectrum S will be made up of a discrete, a continuous, and a residual part.[42] However under special circumstances of interest for stellar conditions several useful results can be established.

P.1. If the balance equations obey a local conservation law [$\mathbf{F}_k = 0, \sigma_k = 0$ for component k of (II.3)], and if the velocity field vanishes in the nonperturbed model, then S contains an infinitely degenerate zero spectrum S_0^k.

The kth component in the Eulerian eigenvalue problem obeys

$$s[\Delta p_k + \partial_{\mathbf{r}} \cdot (\delta \mathbf{r} p_k)] = 0, \qquad \text{with} \qquad s\delta\mathbf{r} = \Delta\mathbf{v} \qquad \text{(III.1)}$$

If we set $s = 0$, this equation can be satisfied by taking $\Delta p_k + \partial_{\mathbf{r}} \cdot (\delta \mathbf{r} p_k)$ equal to an arbitrary function of $\mathbf{r}$ (compatible with the boundary conditions). Expand this function in terms of a complete basis $\{\xi_i(\mathbf{r})\}$ (obeying the relevant boundary conditions), and eliminate Δp_k expressed as

$$\Delta p_k = -\partial_{\mathbf{r}} \cdot (\delta \mathbf{r} p_k) + c\xi_i(\mathbf{r}), \qquad c, \text{ arbitrary constant} \qquad \text{(III.2)}$$

for any of the basis functions, from the remaining $N+3$ spectral equations

* The results quoted in Kato apply in the frame of the restrictive assumptions A.3.

in which $s=0$. The latter is transformed into a nonhomogeneous set of $N+3$ equations in $N+3$ unknowns. Thus, provided its homogeneous part is nonsingular, this system admits of a unique solution for any basis function ξ_i, the corresponding eigenvalue being $s=0$. If the homogeneous part is singular, nonunique solutions exist for any ξ_i as is seen in an elementary way by representing the spectral operator $\mathfrak{L}^E$ by its finite dimensional approximation $\partial_\Pi \mathfrak{N}$. The nullspace of $\mathfrak{L}^E$ connected with the conservation law k is denoted by Γ_0^k.

In the Lagrangian formulation, proposition P.1 takes a more general form.

P.1′. If the components $\mathbf{F}_k^L$ and σ_k^L vanish identically in (II.18), then S contains an infinitely degenerate zero spectrum S_0^k. Since the kth component of the spectral equation reduces now to $s\delta p_k^L=0$, the proof is obvious.

P.2. If for a certain class of perturbations the balance equations obey a local conservation law in perturbed form ($\partial_\mathbf{r}\cdot\Delta\mathbf{F}_k=\Delta\sigma_k$ for component k in the Eulerian spectral equations), and if the velocity field vanishes in the nonperturbed model, then S contains a zero spectrum.

The linearized form of the kth balance equation is identical with (III.1). The proof is similar to that of proposition P.1.

P.3. We define invariance of a quantity $Q(\underset{\sim}{p})\in\mathsf{R}^n$, $\underset{\sim}{p}\in\mathsf{R}^n$, with respect to permutations of a set of coordinates $(k_1, k_2, \ldots, k_\nu)$ in R^n as

$$\mathcal{P}_{ij}\underset{\sim}{Q}(\underset{\sim}{p}) = \mathcal{P}_{ij}[Q_1(p_1\cdot\cdot p_i\cdot\cdot p_j\cdot\cdot p_n),\ldots,Q_i(p_1\cdot\cdot p_i\cdot\cdot p_j\cdot\cdot p_n),\ldots,$$

$$Q_j(p_1\cdot\cdot p_i\cdot\cdot p_j\cdot\cdot p_n),\ldots,Q_n(p_1\cdot\cdot p_i\cdot\cdot p_j\cdot\cdot p_n)]$$

$$=[Q_1(p_1\cdot\cdot p_j\cdot\cdot p_i\cdot\cdot p_n),\ldots,Q_j(p_1\cdot\cdot p_j\cdot\cdot p_i\cdot\cdot p_n),\ldots,$$

$$Q_i(p_1\cdot\cdot p_j\cdot\cdot p_i\cdot\cdot p_n),\ldots,Q_n(p_1\cdot\cdot p_j\cdot\cdot p_i\cdot\cdot p_n)] \qquad \text{(III.3)}$$

ij being any pair of the $(k_1 k_2, \ldots, k_\nu)$.

If the fluxes and sources are invariant under permutations of the N chemical coordinates $(\underset{=}{p})$, then the spectrum S contains two classes of modes: nondegenerate modes, and $(N-1)$-fold degenerate modes in R^N.

Invariance of $\underset{\sim}{\mathbf{F}}$ and $\underset{\sim}{\sigma}$ under the group of permutations $S(N)$ of the N chemical coordinates $\underset{=}{p}$ implies invariance of the spectral equations (II.29) under the same group. The linear operators $\mathfrak{L}$ have the following aspect

$$\mathfrak{L} = \sum_k A_{ij}^{(k)} D^{(k)} \tag{III.4}$$

where the $A_{ij}^{(k)}$ $(i,j=1,2,\ldots,N+4)$ are matrices operating in $\mathbf{R}^{N+4}$ and $D^{(k)}$ are operators in the configuration space $\mathbf{R}^3$. The invariance property then implies that:

a. In the matrices $A_{ij}^{(k)}$ the chemical variables are involved in a form invariant under $S(N)$, that is, in terms of combinations

$$I_l(\underline{\underline{p}}), \qquad l=1,2,\ldots,r \qquad \mathcal{P} I_l(\underline{\underline{p}}) = I_l(\underline{\underline{p}}) \tag{III.5}$$

for any element $\mathcal{P}$ of the group.

b. In $\mathbf{R}^{N+4}$ the group elements of $S(N)$ are represented by matrices of the form

$$\mathcal{P}^{(N+4)} = \left(\begin{array}{ccc|c|c} 1 & & & & \\ & 1 & & & \\ & & 1 & & \\ \hline & & & \mathcal{P}^{(N)} & \\ \hline & & & & 1 \end{array} \right) \tag{III.6}$$

where $\mathcal{P}^{(N)}$ is a permutation matrix in $\mathbf{R}^N$, and where the matrix elements not indicated in (III.6) vanish.

It follows that the eigenvectors $\Delta \underset{\sim}{p}$ parametrically depend on the invariants (III.5):

$$\Delta \underset{\sim}{p} = \Big[\mathbf{f}\big(I_1(\underline{\underline{p}}),\ldots,I_r(\underline{\underline{p}}),\ldots\big), \\ \underline{\underline{f}}\big(I_1(\underline{\underline{p}}),\ldots,I_r(\underline{\underline{p}}),\ldots\big), f_{N+4}\big(I_1(\underline{\underline{p}}),\ldots,I_r(\underline{\underline{p}})\big),\ldots \Big] \tag{III.7}$$

Therefore the only components of $\Delta \underset{\sim}{p}$ which are not trivially affected by the permutation group are the perturbed chemical variables $\Delta \underline{\underline{p}}$. It is readily shown that the N dimensional representation Γ^N of the symmetric group $S(N)$ realized by the matrices $\mathcal{P}^{(N)}$ of (III.6) is fully reducible into a one-dimensional representation Γ^1 and an irreducible $(N-1)$-dimensional representation Γ^{N-1}.[100] Hence the eigenfunctions of the linear operator $\mathfrak{L}$ classified with respect to their behavior under $S(N)$ belong either to the nondegenerate Γ^1 representation or to the $(N-1)$-times degenerate representation Γ^{N-1}.[63]

An evolutionary sequence $\Sigma(t)$ computed over the time interval T always displays a simple geometric symmetry. Since the degeneracies connected with geometrical groups have been discussed in the frame of adiabatic oscillations of stars,[97] and since the extension to arbitrary perturbations is trivial, we merely note that in the presence of both a geometrical group G and a permutation group of chemical species $S(N)$ the full symmetry group of the sequence is the direct product $G \times S(N)$.

P.4. If the perturbations δp^L are allowed to be generalized functions, and if among the fluxes $\mathbf{F}_k^{L\sim}(k=4,5,\ldots,N+4)$ a subset $\mathbf{F}_{\nu_i}(i=1,2,\ldots,m)$ vanishes while the corresponding sources $\sigma_{\nu_i}^L$ are nonvanishing functions of $\underset{\sim}{p}^L$, then the spectrum S contains m continuous branches.

The explicit form of the m relevant equations among the linearized system (II.18) reduces to

$$s\delta p_{\nu_i}^L = \sum_{j=1}^{m} \partial_{p_{\nu_j}^L} \sigma'^L_{\nu_i} \delta p_{\nu_j}^L + \sum_{\substack{l=4\\ l\neq \nu_i}}^{N+4} \partial_{p_l^L} \sigma'^L_{\nu_i} \delta p_l^L \quad \underset{\sim}{\sigma}'^L = \frac{\underset{\sim}{\sigma}_L}{\rho} \tag{III.8}$$

If $\mathfrak{M}$ denotes the $m \times m$ matrix of elements $\partial_{p_{\nu_j}^L} \sigma'^L_{\nu_i} = \mathfrak{M}_{ij}$, and $\mathfrak{N}$ the $m \times (N+1-m)$ matrix of components $\partial_{p_l^L} \sigma'^L_{\nu_i} = \mathfrak{N}_{il}$, (III.8) becomes

$$[s\mathcal{E}_m - \mathfrak{M}(\mathbf{r})] \cdot \mathbf{q} = \mathfrak{N}(\mathbf{r}) \cdot \boldsymbol{\varphi} \tag{III.9}$$

where $\mathbf{q}$ is the set of m perturbed momenta $\delta p_{\nu_i}^L$, and $\boldsymbol{\varphi}$ the $(N+1-m)$ remaining perturbed variables δp_l^L. We recall now that the solution $D(x)$ of the equation

$$(x - x_0) D(x) = 1 \tag{III.10}$$

in the frame of generalized functions is given by

$$D(x) = C\delta(x - x_0) + PP(x - x_0)^{-1} \tag{III.11}$$

(cf. Ref. 85), where C is an arbitrary constant and δ is the Dirac distribution, and where the symbol PP denotes the Cauchy principal part. Let $\mathcal{S}(\mathbf{r})$ be a nonsingular matrix that diagonalizes $\mathfrak{M}(\mathbf{r})$, that is,

$$\mathcal{S}^{-1}(\mathbf{r}) \cdot \mathfrak{M}(\mathbf{r}) \cdot \mathcal{S}(\mathbf{r}) = \mathrm{diag}[\mu_1(\mathbf{r}), \mu_2(\mathbf{r}), \ldots, \mu_m(\mathbf{r})] \tag{III.12}$$

Upon applying $\mathcal{S}^{-1}$ to (III.9) we transform the latter into a set of m decoupled equations of the form (III.10). It remains to solve these equations according to (III.11), and to apply $\mathcal{S}(\mathbf{r})$ upon the result:

$$\mathbf{q}(\mathbf{r}) = PP\left\{[s\mathcal{E}_m - \mathfrak{M}(\mathbf{r})]^{-1}\cdot \mathfrak{N}(\mathbf{r})\cdot\boldsymbol{\varphi}(\mathbf{r})\right\} + \delta[s\mathcal{E}_m - \mathfrak{M}(\mathbf{r})]\cdot\mathbf{C} \quad \text{(III.13)}$$

$\mathbf{C}$ being a set of m arbitrary constants. We finally eliminate the perturbations $\mathbf{q}(\mathbf{r})$ among the remaining $(N+4-m)$ equations, which yields a spectral equation of the form

$$s\phi = \Lambda_s\phi + f_s \quad \text{(III.14)}$$

where the linear operator Λ_s involves the parameter s through a Cauchy principal part, and where the inhomogeneous part f_s arises through the δ distribution. In simple cases so far studied[5] (III.14) can be cast into an inhomogeneous integral equation which belongs to a class discussed in the literature.[88, 120] For these cases the nonhomogeneous problem has a unique solution for any s which is not an eigenvalue of the homogeneous problem $f_s = 0$. Although detailed theorems devoted to equations of type (III.14) with coefficients involving generalized functions seem to be lacking,[44] we expect this result to hold under all practical situations of stellar interest. If $\mathbf{r}_*$ is a point lying within the volume V of the star, the inhomogeneous term f_s vanishes unless $s = \mu_i(\mathbf{r}_*)$ $(i = 1, 2, \ldots, m)$. The spectral equation is therefore inhomogeneous only for those values of s which coincide with an eigenvalue of $\mathfrak{M}(\mathbf{r})$ at a point $\mathbf{r}_* \in \mathsf{V}$. Thus S contains m continuous branches

$$\mathsf{S}_{c_i} = \{s \mid s = \mu_i(\mathbf{r}_*) \qquad i = 1, 2, \ldots, m,\ \mathbf{r}_* \in \mathsf{V};\ s\phi \neq \Lambda_s\phi\} \quad \text{(III.15)}$$

IV. APPLICATIONS TO EVOLVING STARS OF SPHERICAL SYMMETRY

A. Null Spectra

Propositions P.1 and P.2 provide us with a simple tool for an analysis of the zero spectra which arise in a sequence of evolutionary stellar models $\Sigma(t)$. In conformity with most detailed computations, we assume that all models of the sequence have K_h symmetry. We recall that the spectrum S at time t is the spectrum of the operator $\mathcal{L}(t)$, which is the linearized form of $\mathfrak{N}[\underset{\sim}{p}(t)]$, a given model of the sequence $\Sigma(t)$ being regarded as "frozen in" at time t *after* the linearization has been performed. For any model of

$\Sigma(t)$ the following null spectra exist:

Z.1. In the frame of Newtonian gravitation theory mass is rigorously conserved. Thus any model has a zero spectrum S_0^M of infinite multiplicity, associated with eigenvectors, which can be assumed to be orthonormalized (in a convenient sense) and which span the null space Γ_0^M.

Z.2. A vector field in $\mathbf{R}^3$ can be expanded in terms of a spheroidal and a toroidal field.[11,18,87] A trivial calculation shows that the toroidal component of the momentum density field obeys a local conservation law provided the velocity field in the nonperturbed model vanishes. This entails a second zero spectrum S_0^T whose orthonormalized eigenvectors span the null space Γ_0^T, orthogonal to Γ_0^M. Since in (III.1) s^2 appears as a factor for the components of the momentum density field, this spectrum is in fact double. We note that S_0^T is related to the specific structure of the momentum equations of a self-gravitating ideal fluid in the absence of internal motions: If elasticity (case of planetary bodies) or rotation were to be taken into account, the infinite degeneracy would be lifted at least partially, and nonzero modes would appear.[19,62,77] As a particular example of eigenfunctions of S_0^T we mention the class of arbitrary infinitesimal rotations around a polar axis (the ${}_n\mathrm{T}_1$ modes in the geophysical notation), with ${}_0\mathrm{T}_1$ corresponding to rigid-body rotation.[36] The null spectrum S_0^T is practically always guaranteed in nonrotating stars and in the absence of internal motions (convection). In the elementary convection theory used in stars the equations of mean motion remain essentially identical with (II.3), so that even in that case this zero spectrum survives.

Further conservation laws apply under particular evolutionary phases.

Z.3. For sufficiently low temperatures and densities the thermonuclear reaction rates vanish. The N partial densities $\underline{\underline{p}}$ then obey local conservation laws, which imply $(N-1)$ zero spectra $\mathsf{S}_0^{X_i}(\mathrm{i}=1,2,\ldots,N-1)$ connected with the conservation of $N-1$ independent chemical abundances, in addition to S_0^M. The corresponding eigenvectors span $(N-1)$ mutually orthogonal null spaces $\Gamma_0^{X_i}(i=1,2,\ldots,N-1)$, each $\Gamma_0^{X_i}$ being orthogonal to both Γ_0^M and Γ_0^T. Conditions of validity of Z.3 apply to all configurations of masses $M \lesssim 0.1 M_\odot$ at any stage of their evolution.[43] For higher masses they are significant during the pre-main sequence phase, as well as in latest evolutionary stages.

Z.4. If temperature can be regarded as uniform throughout a model, and if conditions Z.3 remain fulfilled, entropy conservation is warranted, which implies a further zero spectrum S_0^S. The associated eigenfunctions span the null space Γ_0^S which is orthogonal to $\Gamma_0^M \oplus \Gamma_0^T \oplus \Gamma_0^{X_1} \oplus \cdots \oplus \Gamma_0^{X_{N-1}}$. These requirements are roughly met in later phases of white dwarf evolution.[92, 135]

Z.5. Finally if both temperature and density tend to vanish (Δp remaining finite), conservation of all generalized momentum density components holds, so that we obtain two additional (double) zero spectra $\mathsf{S}_0^{\Sigma_1}$ and $\mathsf{S}_0^{\Sigma_2}$ connected with the two spheroidal components of the velocity field. The corresponding orthogonal spaces are $\Gamma_0^{\Sigma_1}$ and $\Gamma_0^{\Sigma_2}$. The three zero spectra due to momentum density conservation belong to the irreducible representations Γ^1 and Γ^2 of the group $S(3)$ of permutations of coordinates of the configuration space R^3 (argument similar to P.3). In the limit of the unphysical conditions Z.5 the whole spectrum collapses toward the origin of the complex plane.

B. Adiabatic Approximation

If we successively relax restrictions Z.5, Z.4, and Z.3, we obtain a simple physical classification of the spectrum S. We start with superimposing a finite density upon a configuration obeying Z.5. The conservation laws for the three Cartesian components of the momentum field cease to hold. Moreover the corresponding source term ceases to be invariant under permutations of the coordinates in R^3. The twofold degeneracy (representation Γ^2) is therefore lifted, and we obtain *two distinct* (double, nonzero) *spectra*. The same conclusion holds true if we release the particular assumptions Z.4 and Z.3, but we discard entropy and abundance fluctuations ($\delta S=0, \delta X_i=0$). In the frame of this adiabatic approximation the two discrete nonzero spectra S^{Σ_1} and S^{Σ_2} inform us about the *dynamical stability* of the star. In Cowling's classification[23] these spectra correspond to the p- and g-modes. For a review of the properties of these spectra we refer to Ledoux and Walraven.[74] We merely note here that the time scales of the adiabatic eigenvalues are typically many orders of magnitude shorter than the evolution time scales, so that the usual stability definition for equilibrium models is entirely acceptable for this class of perturbations.

If the function space of the allowed perturbations is enlarged to include generalized functions, an argument similar to P.4 suggests the possibility of continuous branches of eigenvalues given by

$$s^2 = -\Gamma_1(r_*)P(r_*)\frac{l(l+1)}{r_*^2\rho(r_*)} \tag{IV.1a}$$

and

$$s^2 = A(r_*)g(r_*), \qquad r\in(0,R) \tag{IV.1b}$$

(cf. Appendix D for details of the algebra and the definition of the symbols). While the set of eigenvalues (IV.1a) corresponds to localized pure oscillations, eigenvalues (IV.1b) are real (positive and negative) if the quantity $A(r_*)$ is positive. Since the inequality $A(r_*)>0$ embodies the Schwarzschild convection criterion, these continuous modes appear to be directly related to the conditions for occurrence of convection at the layer r_* in a star. Further aspects of this question are discussed in Appendix D.

C. Nonadiabatic Case

We now relax requirements Z.4 but suppose that conditions Z.3 remain satisfied. The nonadiabatic effects thus introduced first alter the dynamical spectra ($\mathsf{S}^{\Sigma_1,\Sigma_2} \to \mathsf{S}^{\Sigma_1',\Sigma_2'}$). Furthermore, as a direct consequence of nonconservation of entropy a discrete, nonzero spectrum $\mathsf{S}^{S'}$ emerges from the previous zero spectrum S_0^S. This conclusion applies also in the general case (no restrictive condition Z.3), provided we assume that the chemical abundances are not allowed to fluctuate. This new spectrum is known in the literature as the *secular spectrum* . In spite of recent detailed numerical investigations of this spectrum carried out for almost all mass ranges and evolutionary phases, its mathematical properties remain unresolved. A simplified model which mimics the structure of $\mathsf{S}^{S'}$ suggests however that this spectrum always has an upper bound on the real axis and an accumulation point at $(-\infty)$.[1,3]

The concept of secular stability deserves a few comments under conditions of an evolutionary time scale τ_E of the order of, or shorter than, the inverse of the smallest real part (in absolute value) of the secular modes. A typical example of such a situation is provided by the pre-main sequence contraction phase. As demonstrated by Gabriel[34] in the case of a $1.1 M_\odot$ star, all eigenvalues of S^S are real and negative on the Hayashi track, while during the Henyey phase the lowest order eigenvalue becomes positive. For a $0.8 M_\odot$ carbon star a similar result holds.[90] Criterion C.3 is readily applied to the analysis carried out by Noels.[90] We choose a time interval T of duration $\Delta T = 1.2\times 10^7$ years, which covers models 1 to 7. Denoting by $s_1(\tau), s_2(\tau), \ldots, s_i(\tau), \ldots,$ the set of eigenvalues ordered as follows: $\mathrm{Re}\, s_1 \geqslant \mathrm{Re}\, s_2 \geqslant \ldots \geqslant \mathrm{Re}\, s_i \geqslant \ldots,$ we have for the models under consideration $s_1 \Delta T \cong 0.6$, $|s_2|\Delta T \cong 2.5$, $|s_3|\Delta T \cong 6 \ldots$. If we take $M=2.5$, the set S^0 contains only one eigenvalue, $s_1(\tau)$, $\mathsf{S}^+ = \varnothing$, and $\mathsf{S}^- = \{s_2(\tau), s_3(\tau), \ldots\}$. As follows from criterion C.3, the stability properties of the sequence of models depend on set S^0 alone (in the sense that initially excited modes of S^- lead to a perturbation which is less than $O(e^{-M}) = 0.082$ times the initial perturbation, at $t = \Delta T$). Application of criterion C.4 to the projection of the spectral operator into Γ^0 reveals that the sequence is unstable over the interval T in the sense of definition D.4. A less precise conclusion is reached with the help of criterion C.8. As is well known from the algebra of matrices, the eigenvalues of the symmetrized operator obey

$$\max_{s' \in \mathsf{S}'} s' \geqslant \max_{s \in \mathsf{S}} \operatorname{Re} s \tag{IV.2}$$

Therefore S' has at least one positive eigenvalue at any time $t \in \mathsf{T}$. Criterion C.8 then warrants that the sequence is unstable during any sufficiently short interval T'. It is physically obvious that the length of this interval should fulfill $\Delta\tau s_+(0) \ll 1, s_+$ being the positive eigenvalue. The conclusion drawn from criterion C.8 is seen to be similar to Schwarzschild and Härm's[112] interpretation of roots with positive real parts in the secular spectrum of evolving models.

Current secular stability codes compute the secular spectrum $\mathsf{S}^{S'}$ by means of an algebraic system of the form

$$(sA + B)\chi = 0 \tag{IV.3}$$

where A is a singular matrix, while the determinant dtm $B \neq 0$ (except at isolated evolutionary stages), and where χ is the vector of the perturbed physical parameters. We can recast this equation in the form

$$C\chi = s^{-1}\chi, \qquad C = -B^{-1}A \tag{IV.4}$$

The eigenvalue problem (IV.4) is immediately symmetrized $[C' = \frac{1}{2}(C + \tilde{C})]$. So far numerical investigations of the symmetrized secular spectrum are lacking. It is of interest to point out that (IV.4) immediately exhibits an eigenvalue $s = (\pm)\infty$, dtm C being zero.

As a final question connected with the secular spectrum, we inquire into the possibility of continuous modes. From proposition P.4 we infer that such modes occur at least under special conditions of vanishing heat flux. As was first argued by Reeves,[107] neutrino luminosity may substantially exceed photon luminosity in carbon white dwarfs, so that to a reasonable approximation the photon luminosity becomes negligible. On the basis of this property, Ledoux[71] obtained a simple analytic stability criterion for the spectrum of radial vibrational modes and showed that those stars are vibrationally unstable. Ledoux's result was subsequently confirmed by detailed numerical integrations which included the slight contributions of the photon flux.[91] Recent models computed by Mariska and Hansen,[80] and Paczyński and Kozlowski[94] justify the approximation of negligible photon luminosity, which is found to hold for a sufficiently small mass, central density, and temperature*. According to data by Hansen and Spangenberg,[46] similar results prevail for helium models on the white dwarf branch.

*$M < 1.4099\, M_{\odot}$, $\log \rho_c < 9.93 \mathrm{g\ cm}^{-3}$, $\log T_c < 8.053\mathrm{K}$

If we now ignore the photon heat flux, the thermal equation becomes algebraic in the entropy. From proposition P.4 we infer that the secular spectrum contains a continuous branch. The analytic expression of the continuum of eigenvalues is obtained from the linearized equation (II.20), with the underlined part neglected. The heat flux reduces to the neutrino flux given by (II.17′). From the discussion of proposition P.4 we obtain

$$s = \frac{\epsilon_N(\mathbf{r}_*)}{c_v(\mathbf{r}_*)T(\mathbf{r}_*)}[\nu_N(\mathbf{r}_*) - \nu_\nu(\mathbf{r}_*)] \tag{IV.5}$$

where $c_v(\mathbf{r}_*)$ is the local specific heat at constant volume and per unit mass at position $\mathbf{r}_*$ in the star, and where the coefficients ν_N, ν_ν are defined by

$$\nu_N = \left(\frac{\partial \ln \epsilon_N}{\partial \ln T}\right)_\rho, \qquad \nu_\nu = \left(\frac{\partial \ln \epsilon_\nu}{\partial \ln T}\right)_\rho \tag{IV.6}$$

The entropy fluctuation $\delta S(\mathbf{r})$ associated with an eigenvalue (IV.5) is singular at $\mathbf{r} = \mathbf{r}_*$. For $\nu_\nu > \nu_N$, all eigenvalues of the continuum are negative. This condition is actually fulfilled in carbon white dwarfs, where $\nu_\nu = 6$ and $\nu_N = 0$. Regarded as equilibrium models, carbon white dwarfs are secularly stable.

As already mentioned above, the dynamical spectra are modified by the nonadiabatic contributions. The new spectra $\mathsf{S}^{\Sigma_1', \Sigma_2'}$ inform us about the *vibrational stability* of a star. Under conditions of thermal imbalance, the analysis of vibrational stability has met so far with serious obstacles.[24,25,29,115,116,117] We believe that the paradoxical situation encountered in the frame of this problem is essentially due to a lack of a precise definition of the concept of stability. Our definitions D.1 to D.4 eliminate this semantic difficulty, and furthermore dictate the choice of variables to use in a stability test. The criteria of Section II enable us to discuss vibrational stability of evolving stars in terms of the spectral properties of a sequence of frozen in models. To avoid the mathematical complications connected with perturbation theory, which is straightforward only in equilibrium models,[74] the vibrational spectrum can be readily computed with a program such as that of Castor[17] with the thermal imbalance contributions included. Criterion C.8, together with inequality (IV.2) and criterion C.3 are expected to yield unambiguous answers on vibrational instability in most situations of practical interest.

D. The Role of Chemical Kinetics

In the general case, the absence of local conservation laws for the chemical abundances implies that the $(N-1)$-fold degeneracy of the spectra $\mathsf{S}_0^{X_i}$ (in R^N), which prevailed as a consequence of proposition P.3 in the frame of the approximations discussed above, is lifted, and generates $(N-1)$ distinct nonzero spectra $\mathsf{S}^{X_i''}$, referred to as the *chemical or nuclear spectra*.* The vibrational and secular modes belong to the Γ^1 (nondegenerate) representation of the permutation group under conservation of the chemical abundances. As a consequence, the inclusion of the details of chemical kinetics will merely displace those modes ($\mathsf{S}^{\Sigma_1',\Sigma_2'} \rightarrow \mathsf{S}^{\Sigma_1'',\Sigma_2''}$ and $\mathsf{S}^{S'} \rightarrow \mathsf{S}^{S''}$). Note also that the presence of abundance fluctuations cannot lift any degeneracies connected with geometrical symmetries.

1. Continuous Branches

Among the N chemical balance equations, $N-1$ involve independent chemical abundances which can be isolated from equations (II.3) by means of the $\mathfrak{T}$-transformation (II.19′). The requirements of proposition P.4 being satisfied, we conclude that the nuclear spectra display $N-1$ continuous branches defined by

$$s = \mathrm{diag}\, \mathfrak{R}(\mathbf{r}_*); \qquad \mathfrak{R}(\mathbf{r}_*) = \partial_{\underline{X}}\, \underline{\sigma}' \tag{IV.7}$$

The corresponding abundance perturbations are singular at $\mathbf{r} = \mathbf{r}_*$. Numerical evidence in favor of these continuous branches has been found in the case of one independent abundance variable[2,27] and two independent abundance variables,[6] by a direct computation of the spectrum (for purely radial perturbations). In the framework of a simplified model a detailed investigation of the associated integral formulation of the spectral equations (cf. P.4) has been carried out both analytically and numerically.[5]

The existence of continuous nuclear modes was discussed some time ago[99] in the particular case of negligible coupling between the chemistry and the thermodynamics of the star. The physical interpretation of such modes adduced in that case also holds true if the coupling is taken into account: In the absence of particle diffusion, a given point $\mathbf{r}_*$ of the star essentially behaves as if it was chemically isolated from the rest of the star. If we add some material i at point $\mathbf{r}_*$, this material reacts with the remaining species at point $\mathbf{r}_*$, the time scale of this process being entirely fixed by the physical conditons at point $\mathbf{r}_*$. The presence of disintegrations among the thermonuclear reaction network may imply that one or several eigenvalues of the reaction matrix $\mathfrak{R}$ reduce in absolute value to disintegration constants λ_i. The corresponding continuum of stellar eigenvalues then collapses to a single point $s = -\lambda_i$ of infinite multiplicity. As an example, consider a phase A of duration τ_A during which an unstable species U is synthesized, and let λ_U be its disintegration constant, with

*We disregard accidental degeneracies, or further invariances under a subgroup of the permutation group in $\mathbf{R}^n$.

$1/\lambda_U \gg \tau_A$. If the particles U do not undergo reactions with other species, $-\lambda_U$ will be a rigorous eigenvalue of the reaction matrix. Phase A can be regarded as the trigger of the nuclear mode λ_U of the star. In the subsequent phase B the properties of the star change with the characteristic time scale $1/\lambda_U$, provided the excitation was small enough, so that a linearized formulation remains significant. Although this latter requirement is not fulfilled in supernovae explosions, where the luminosity is multiplied by a factor of 10^6 to 10^9 during the ignition phase, it is a remarkable fact that the luminosity in the postexplosion phase of type I supernovae decays almost strictly exponentially, with a period of 35 to 65 days. This value is close to the periods of several unstable nuclei formed in the explosion.[10,12,14,32] Detailed nonlinear numerical investigations by Colgate and McKee[21,22] lend support to the idea that the disintegration relevant to type I supernovae is $^{54}Co \rightarrow {}^{56}Fe$ (period: 77 days).

Since the eigenvalues of the reaction matrix $\mathfrak{R}(\mathbf{r})$ are eigenvalues of the star, it is of considerable relevance to inquire into the possibility of unstable steady states in the stellar reaction chains. It is well known from the theory of chemical instabilities in homogeneous reaction systems, that beyond the onset of instability a reaction system evolves either toward a new stable steady state, a limit cycle[41] or toward a multidimensional closed limit set.[99] It is manifest that if an unstable steady state of a stellar reaction occurs, it stimulates evolution toward a new steady state, a periodic, or a multiply periodic oscillation of the star. Instabilities of this type could provide simple explanations for the large number of nonstationary and recurrent phenomena observed in stars which still remain a riddle.

In spite of the huge number of reactants usually involved in a thermonuclear reaction chain, especially in the network of nucleosynthesis,[9] one notices that the stoichiometric schemes have an essentially simple structure. Catalytic steps, necessary for the existence of unstable steady states[104] are very rare among stellar reactions. Nonetheless it is possible to single out a set of reactions from the classic PP I and PP III chains, the carbon cycle, the triple alpha process, and the first steps of carbon burning, which does exhibit an unstable steady state. If we assume that the abundances of all species involved in this reaction network are held constant, except the ^4He and ^{12}C abundances, denoted by X and Y respectively, the stoichiometric scheme of relevance can be simplified as follows

$$\begin{aligned} A + X &\rightarrow 2X, & K_{AX} \\ 3X &\rightarrow Y, & K_X \\ B + Y &\rightarrow X + Y, & K_{BY} \\ 2Y &\nearrow X + C, & K_{Y1} \\ 2Y &\searrow D, & K_{Y2} \end{aligned} \tag{IV.8}$$

Species A, B, C, and D are not allowed to fluctuate. The reaction parameters K depend on the local physical conditions of the star. An inspection of the corresponding kinetic equations reveals that the system admits of two steady-state solutions in the positive quadrant of the (X, Y) plane, given by

$$\left.\begin{aligned} &(1)\quad X_1 = Y_1 = 0,\\ &(2)\quad X_2, Y_2 \neq 0, \qquad \text{solutions of}\\ &Y_2 = \left(\frac{K_X}{2K_{Y1}+2K_{Y2}}\right)^{1/2} X_2^{3/2}\\ &\left(3 - \frac{K_{Y1}}{2K_{Y1}+2K_{Y2}}\right) X_2^2 = A\frac{K_{AX}}{K_X} + B\frac{K_{BY}}{[2K_X(K_{Y1}+K_{Y2})]^{1/2} X_2^{1/2}} \end{aligned}\right\} \tag{IV.9}$$

The reaction matrix evaluated at the second steady state has negative eigenvalues, while the eigenvalues for state (1) are given by

$$s = 0 \qquad \text{and} \qquad s = +K_{AX}A$$

$$\text{(inverse of the time scale of the proton–proton chain)} \tag{IV.10}$$

showing that this state is unstable. The phase portrait of the reaction is exhibited in Figs. 1 and 2. It stresses that under the influence of any perturbation of the steady state (1), the reaction evolves toward the second steady state (2). Comparison with stellar model calculations by Boury,[15] Van der Borght,[131] and Kovetz and Shaviv[57] illustrates that the approximation of (IV.8) to the real reaction system is not unreasonable for massive pure hydrogen stars, except that the PP I steps produce a nonnegligible amount of ^{4}He. If we formally include this mechanism by setting

$$E \to X \tag{IV.11}$$

where the abundance E is not allowed to fluctuate, it is easily seen that the singular point (1) of the previous set of kinetic equations is shifted toward the unphysical region of the phase plane. This circumstance does not alter our previous qualitative conclusions. The origin of the phase plane can be viewed as a perturbation of this unstable equilibrium point. At the very onset of nuclear burning a spontaneous transition of the representative point in the (X, Y) plane from the origin toward the separatrix $s-s$ (Figs. 1 and 2) takes place, followed by a slow evolution up the separatrix. For

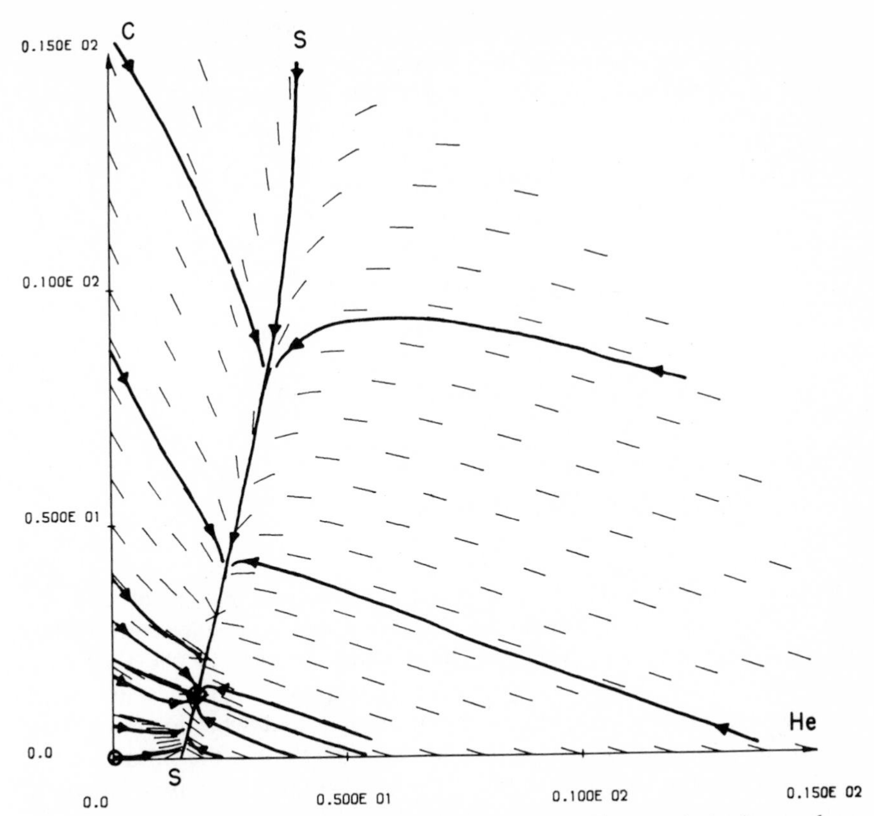

Fig. 1. Phase portrait of the idealized reaction scheme (IV.8.). The numerical values used are: $K_{AX}\,A = K_{BY}\,B = 1$; $K_X = 0.2$; $K_{Y1} = K_{Y2} = 0.1$. The equilibrium states are marked by circles.

massive pure hydrogen stars we can maintain that it is an instability within the reaction scheme that is responsible for the initial evolution of these stars.

So far any attempts to isolate further unstable states in stellar reaction networks have proved unsuccessful. It seems to us that nontrivial unstable states, inducing phenomena such as hystereses or oscillations, encountered in the framework of biochemical reactions, demand substantially more complex stoichiometric schemes than those presently included in nucleosynthesis.

2. Discrete Branches

The $N-1$ spectra $\mathsf{S}^{X_i''}$ induced by the zero spectra $\mathsf{S}_0^{X_i}$ when abundance fluctuations are included are not expected to coincide with the continuous branches mentioned in Section IV.D.1. It can be shown explicitly for radial perturbations that $N-1$ further nuclear branches will show up under general circumstances. To do so we expand the fluctuations in terms of Fourier series

$$\left.\begin{aligned}
\delta r(r) &= \sum_{n=1}^{\infty} \sin[(2n-1)kr]\, r_n, \qquad k=\frac{\pi}{2R} \\
\delta\rho(r) &= \sum_{n=1}^{\infty} \cos[(2n-1)kr]\, \rho_n \\
\delta X_i(r) &= \sum_{n=1}^{\infty} \cos[(2n-1)kr]\, X_{in}, \qquad i=1,2,\ldots,N-1 \\
\delta S(r) &= \sum_{n=1}^{\infty} \cos[(2n-1)kr]\, S_n
\end{aligned}\right\} \tag{IV.12}$$

(Quarter range Fourier series,[76]). If all perturbations belong to class C, these expansions are known to be absolutely and uniformly convergent (Ref. 20, Sect. 38). In (IV.12) we have taken account of the regularity conditions at the center of the star.* The expansion satisfies the usual boundary conditions adopted in radial secular and vibrational stability ($\delta r=0$ and $\delta L=0$ at the center, and $\delta P=0$ and $\delta T=0$ at the surface of the star). In principle, expansions (IV.12) hold also for generalized functions (Ref. 76, Chapt. 5), provided the expansion coefficients r_n, ρ_n, X_{in}, S_n are computed with respect to a unitary function U, integrations over the finite radius of the star being replaced by integrations weighted by U and going over an infinite range. The classical frame of Fourier series is inappropriate to represent distributions (cf. example 40 in Ref. 76, Chapter 5). Therefore, if we restrict ourselves to the classical theory of Fourier expansions, we are sure that the continuous spectra discussed above are discarded from our formulation. We introduce then expansion (IV.12) into the spectral equations (II.29), multiply by $\sin(2n'-1)kr$ and $\cos(2n'-1)kr$, respectively, and integrate over the finite volume of the star. It is easily seen that all integrals converge under usual stellar conditions. If the series (IV.12) is cut off after M terms, the spectral problem is transformed into an algebraic eigenvalue problem, which has $[2+(N-1)+1]\times M$ roots. $2M$ eigenvalues

*It is shown by group theoretic arguments that in the neighborhood **N** of the origin $\mathbf{r}=0$ a vector field is odd and vanishes at the origin, and a scalar field is even in $\mathbf{r}$, and has a vanishing gradient at the origin, provided the origin is an inversion center, and the fields are analytic in $\mathbf{r}$.

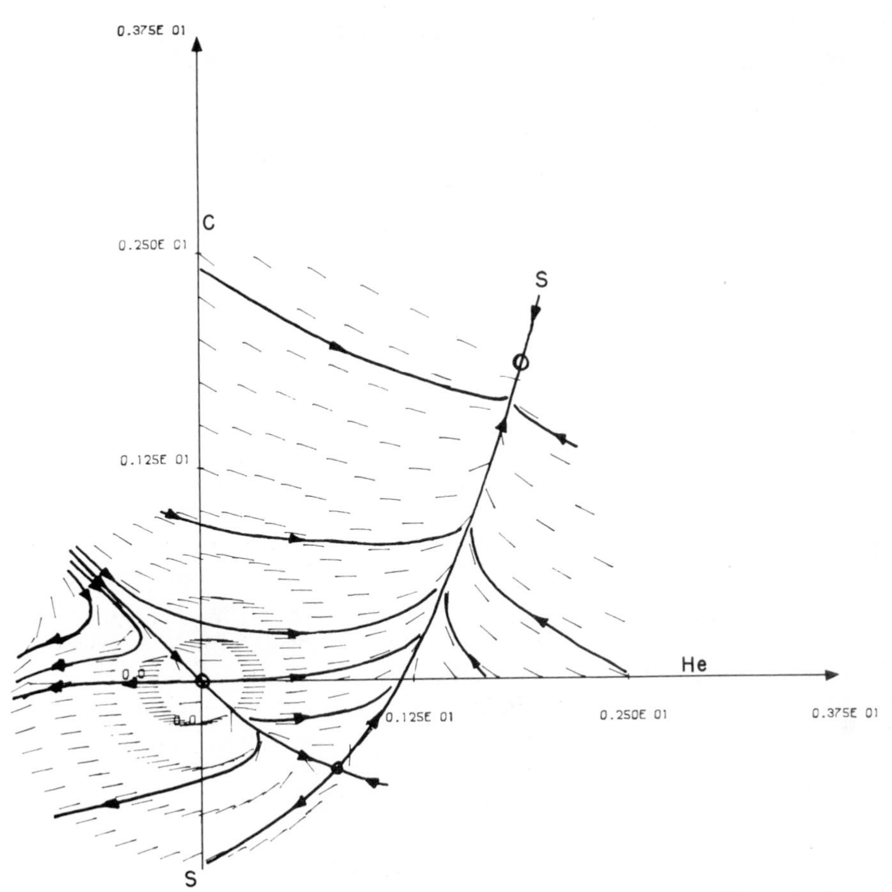

Fig. 2. Details of the nature of the unstable singular point at the origin. This point is a nonelementary singular point of the type of a node-saddle point. The corresponding Poincaré index is zero. The second singular point in the physical quadrant is of the nature of a parabolic node.

correspond to the discrete (double) vibrational spectrum, M eigenvalues represent the discrete secular spectrum, and the remaining $(N-1)\times M$ discrete modes are connected with the perturbations of the chemical abundances. As $M\rightarrow\infty$ the new chemical spectra may become continuous, due to special singularities in the operator $\mathfrak{L}$. This situation would be similar to the continuous modes found by Kopal[55] in the radial dynamical spectrum, where a continuum of eigenvalues is generated from particular density distributions. In any case the corresponding eigenfunctions are ordinary functions, so that the new modes are necessarily distinct from the continuous modes referred to above.

A particular feature of these additional nuclear branches is exhibited if the linearized chemical balance equations

$$s\,\delta \underline{X} = \mathfrak{R}(\mathbf{r})\cdot\delta \underline{X} + (\partial_S \underline{\sigma}')_{\rho,\underline{X}}\delta S + (\partial_\rho \underline{\sigma}')_{S,\underline{X}}\,\delta\rho \qquad \text{(IV.13)}$$

are divided by s, and if $\delta \underline{X}$ is eliminated from the right-hand side of these equations. For any s different from an eigenvalue of the reaction matrix $\mathfrak{R}(\mathbf{r})$, we have

$$\delta \underline{X} = s^{-1}\mathfrak{R}(\mathbf{r})\cdot\left\{\left[s\mathcal{E}_{N-1} - \mathfrak{R}(\mathbf{r})\right]^{-1}\cdot\left[(\partial_S \underline{\sigma}')_{\rho,\underline{X}}\delta S + (\partial_\rho \underline{\sigma}')_{S,\underline{X}}\,\delta\rho\right]\right\}$$

$$\underline{+ s^{-1}(\partial_S \underline{\sigma}')_{\rho,\underline{X}}\delta S + s^{-1}(\partial_\rho \underline{\sigma}')_{S,\underline{X}}\,\delta\rho} \qquad \text{(IV.14)}$$

The reaction rates are temperature and density dependent, of the form $\rho^\mu T^\nu(\mu,\nu \gtrsim 1)$. If the behavior of the particles can be described by an ideal gas law, we notice that the matrix elements $\mathfrak{R}_{ij}$ behave as $\rho^\mu T^\nu$, while for example $(\partial_\rho \underline{\sigma}')_{S,\underline{X}}$ behaves as $\rho^{\mu-1}T^\nu$. Therefore, as long as s differs sufficiently from the eigenvalues of the matrix $\mathfrak{R}(\mathbf{r})$, the major contribution to $\delta \underline{X}$ stems from the underlined part of (IV.14), the remaining contributions being important only in those regions of the star where energy generation is most efficient. In current secular stability investigations, the radial eigenvalue problem is reformulated in terms of the variables $\chi = (\delta r/r, \delta P/P, \delta T/T, \delta L/L)$. If abundance fluctuations are neglected, the relevant spectral equation becomes

$$D_m \chi = [A(m) + sB(m)]\chi \qquad \text{(IV.15)}$$

A and B being 4×4 matrices depending on the mass variable m. With abundance perturbations included, and eliminated by means of (IV.14), the new spectral problem becomes

$$D_m \chi = [A(m) + sB(m) + s^{-1}C(m) + s^{-1}F(m,s)]\chi \qquad \text{(IV.16)}$$

The additional matrix C depends on m, and involves those contributions due to the $\underline{X}$ dependence of the equations of state and opacity law, which arise through the underlined factors in (IV.14). The matrix $F(m,s)$ is a function of both the mass and the parameter s, and it contains contributions of the $\underline{X}$ dependence of the energy generation induced by the full expression (IV.14), as well as contributions from the equations of state and opacity due to the nonunderlined quantities in (IV.14). Essentially the matrix elements of F display a higher temperature or density sensitivity than those of C, so that the latter are dominant in those parts of the star where energy generation is negligible. As a reasonable approximation we may then disregard F in a large region of the star. If we assume furthermore that the eigenvalue $s=(-\infty)$ is an accumulation point of the ordinary secular spectrum, as suggested by model calculations, it is intuitively clear, and it can be shown in detail, that in general (IV.14) admits of a discrete branch of eigenvalues with an accumulation point at the origin of the complex plane. This argument fails if we neglect fluctuations of the chemical abundances everywhere except in the energy generation, since in that case the matrix C vanishes.

Numerical calculations of the spectral problem defined by (IV.15) have been performed for a variety of stars evolving from the main sequence to the red giant branch, a single independent abundance perturbation being investigated (hydrogen).[4] It is found that the discrete nuclear branch S^X does present an accumulation point at $s=0$, and, furthermore, *that all its eigenvalues lie in the positive complex half plane* at any time τ of the evolution considered. An extension of this work is now under investigation in which H, ^{3}He and ^{4}He are allowed to fluctuate.[6] Defouw, Hansen, and Siquig[27] discussed the problem of ^{3}He perturbations in the energy generation (all other abundances being held constant). Under these circumstances no discrete nuclear spectrum shows up. In a simple analytical model in which the matrix C vanishes, it is found again that this discrete branch is missing.[5] It appears that this branch collapses toward the origin as the matrix C tends to zero.

The numerical data on the discrete nuclear branch of eigenvalues available at the present time allow us to make the following statements:

1. Any stellar sequence $\Sigma(t)$ covering the time interval T from the main sequence to the red giant branch is unstable in any sufficiently short interval $\mathsf{T}'\subset\mathsf{T}$. Since at any time τ a model of the sequence has at least one eigenvalue with a positive real part ($\max \mathrm{Re}\, \mathsf{S}^X > 0$), instability follows from relation (IV.2) and criterion C.8.
2. Any stellar sequence $\Sigma(t)$ covering the time interval from the main-sequence to the red giant branch is unstable in any subinterval over which its evolutionary time scale is of the order of the nuclear burning time.

The eigenvalues of the nuclear discrete branch correspond to time scales longer than, and of the order of the nuclear burning time[4] while all other discrete spectra (vibrational and secular) are connected with substantially shorter times.* Therefore a partition can always be found (II.35) such that

*The orders of magnitude of the eigenvalues for the different spectra can be estimated by means of a systematic application of dimensional analysis.[99]

S^0 contains the nuclear discrete branch, and the vibrational and secular spectra fall into $\mathsf{S}^+ + \mathsf{S}^-$. If the secular or vibrational spectra have at least one root s_k such that $\mathrm{Re}\, s_k > 0$, the sequence is unstable according to C.2. If $\mathsf{S}^+ = \varnothing$, stability entirely depends on the projection of the linear operator $\mathfrak{L}$ into the space Γ^0 (criterion C.3) The projection of $\mathfrak{L}$ into Γ^0 has a positive trace, all eigenvalues of the nuclear discrete branch having positive real parts in T. Instability follows from criterion C.4 (or C.5).

The physical meaning of this instability can be illustrated as follows. Consider a model A with a given initial continuous and almost homogeneous hydrogen profile $X_A(m)$, and a model B of the same physical characteristics, but with a slightly different hydrogen profile $X_B(m)$. Then the sequences $\Sigma_A(t)$ and $\Sigma_B(t)$ generated by models A and B tend to diverge in the phase space Γ, with a time scale of the order of the evolution time. In other words, if the initial model is imperfectly known, so that its representative point in the phase space may be found anywhere in a small volume of measure ϵ, then our uncertainty about the model at subsequent times increases:

The measure of the most probable volume in Γ^0 is amplified with an e-folding time equal to the inverse of the trace of the nuclear discrete branch. *Our information about a star is degraded as evolution proceeds.*

3. *Effects on the Vibrational and Secular Spectra*

As a final point we mention that the inclusion of abundance perturbations alters the eigenvalues of both the vibrational ($\mathsf{S}^{\Sigma_1', \Sigma_2'} \to \mathsf{S}^{\Sigma_1'', \Sigma_2''}$) and secular spectra ($\mathsf{S}^{S'} \to \mathsf{S}^{S''}$), the effect being as a rule destabilizing, in the sense that the largest real parts of these spectra are displaced toward the right in the complex plane. As an example we mention that Robe et al.[109] computed the nonradial vibrational spectrum of a $0.5 M_\odot$ star, and found that it is stable if abundance fluctuations are neglected. If abundance fluctuations are taken into account in the energy generation, the spectrum exhibits vibrationally unstable g-modes.[16] The ordinary secular eigenvalues of a main sequence star have all negative real parts. In the presence of hydrogen perturbations the eigenvalues are slightly displaced, by less than 0.1%,[4] toward the origin of the complex plane. If ^{3}He perturbations alone are taken into consideration in the energy generation,[26,27,113] a shift of 20 to 30% of the eigenvalues toward the origin of the complex plane is observed.

Intuitively these results may seem fairly obvious: If we test a configuration with respect to a set K_0 of perturbations, and afterwards with respect to a larger class $K_1 \supset K_0$, the likelihood of finding an instability in the second trial is manifestly greater than in the first case. This statement is actually rigorous in the frame of our strict stability definition (for a trivial proof, use the finite representation of the linear operator $\mathfrak{L}$). It does not apply, however, to the spectrum of $\mathfrak{L}$, in the sense that $\max \operatorname{Re} \mathsf{S}(K_1) \geqslant \max \operatorname{Re} \mathsf{S}(K_0)$, where the spectrum $\mathsf{S}(K_i)$ is computed under the constraint that the only variables allowed to fluctuate belong to K_i, $i=0,1$. For example, a stellar model "stable" with respect to adiabatic perturbations (all eigenvalues of $\mathsf{S}^{\Sigma_1,\Sigma_2}$ are purely imaginary) may become unstable (at least one eigenvalue has a positive real part) or stable (all eigenvalues have negative real parts) as entropy perturbations are included.

E. Final Remarks

The physical classification of the full spectrum S is summarized in Table I. The list of subspectra recorded is thought to be exhaustive, in the sense that within the frame of the approximations stated at the beginning of Section IV and the physics specified in Section II, any eigenvalue computed for any model of an evolutionary sequence belongs to one of the spectral branches mentioned. This does not preclude finer classifications (cf. Ledoux,[72] who classifies the dynamical spectra of p- and g-modes in respect to their model dependence), nor does it exclude the cases where some branches may appear as multiple branches (cf. Tassoul[122], and Tassoul and Tassoul,[123] who show that in models in which the quantity A defined by D.5 changes sign n times, one has to expect $n+1$ g-modes). Moreover some branches listed may "disappear," in particular unphysical models; such branches either collapse toward the origin of the complex plane, or they are sent to infinity (example: the adiabatic p-modes in an incompressible star).

If a less restrictive set of physical assumptions is adopted, the classification of Table I becomes incomplete, particularly under the following circumstances:

1. In the presence of magnetic fields, or rotation, the toroidal momentum component is not conserved, and a new class of modes (S^T) is exhibited.
2. If the details of the transfer equation for photons are fully taken into account, that is, if the photon gas is described in terms of an additional balance equation,[118] a further spectrum connected with nonconservation of photon density arises. A similar remark holds for neutrinos which induce an additional spectrum if their behavior is described more rigorously in terms of a complete transfer equation.

In this paper no attempt was made to provide a consistent stability treatment for convective regions of a star. The numerical calculations

TABLE I

No fluxes and sources	Adiabatic approximation	Nonadiabatic approximation	Chemical kinetics included
Conservation of mass: zero spectrum S_0^M			
Conservation of toroidal momentum component: zero spectrum S_0^T			
	Nonconservation of spheroidal momentum components		
Conservation of spheroidal momentum components: zero spectra $S_0^{\Sigma_1,\Sigma_2}$	S^{Σ_1,Σ_2}2: 2 discrete dynamical spectra ("g and p modes") $S_c^{\Sigma_1,\Sigma_2}$(?): 2 continuous dynamical spectra[a]	$S^{\Sigma_1',\Sigma_2'}$: 2 discrete vibrational spectra ("g and p modes") $S_c^{\Sigma_1',\Sigma_2'}$(?): 2 continuous vibrational spectra	$S^{\Sigma_1'',\Sigma_2''}$: full discrete vibrational spectra ("g and p modes") $S_c^{\Sigma_1'',\Sigma_2''}$(?): full continuous vibrational spectra
		Nonconservation of entropy	
Conservation of entropy: zero spectrum S_0^s		$S^{s'}$: discrete secular spectrum $S_c^{s'}$(??): continuous secular spectrum[b]	$S^{s''}$: full discrete secular spectrum $S_c^{s''}$(??): full continuous secular spectrum

Conservation of independent chemical species: zero spectrum $\mathbf{S}_0^{X_i}, i=1,2,\ldots,N-1$	Nonconservation of chemical species $\mathbf{S}^{X_i''}$: $(N-1)$ discrete nuclear branches $\mathbf{S}_c^{X_i''}$: $(N-1)$ continuous nuclear branches

[a]The absence of a rigorous mathematical proof of existence is indicated by a question mark (?).
[b]The lack of an investigation of the possibility of a general continuous secular spectrum is indicated by a double question mark (??).

quoted disregard in particular the role of convective particle diffusion fluxes within the perturbed equations. However a direct extension of the discussion presented here can be obtained in the framework of the quasiparticle picture of convection.[98] The most important modification of the structure of the hydrodynamic equations for the mean flow is the appearance of diffusion fluxes in the chemical balance equations, and of eddy viscosity effects in the momentum equation, as well as the occurrence of a new balance equation governing the evolution of the quasiparticle gas. It is not yet clear how these modifications will affect the previous continuous branches. The new balance equation generates a further spectrum of eigenvalues.

Acknowledgments

The author is indebted to Professor E. A. Spiegel for a most useful conversation on the subject of stability of evolving systems, which initiated the present work. He wishes to thank the Institute of Astronomy, Cambridge, for the hospitality shown him. He is indebted to the Fonds National de la Recherche Scientifique, Belgium, and the Leverhulme Trust, London, for financial support.

APPENDIX A

The gravitational field equations

$$\partial_{\mathbf{r}} \cdot \mathbf{g} = -4\pi G\rho; \qquad \partial_{\mathbf{r}} \wedge \mathbf{g} = 0 \tag{A.1}$$

where $\mathbf{g}$ is the gravity field, G the constant of gravitation, and ρ the mass density, enable us to write the gravity force density in terms of the divergence of a tensor $\mathcal{G}$:

$$\rho\mathbf{g} = -\partial_{\mathbf{r}} \cdot \mathcal{G} \tag{A.2}$$

$$\mathcal{G} = (4\pi G)^{-1}(\mathbf{gg} - \tfrac{1}{2}|\mathbf{g}|^2 \mathcal{E}_3) \tag{A.3}$$

The tensor $\mathcal{G}$ (gravity pressure tensor) plays a role similar to the Maxwell tensor in magnetohydrodynamics. If we regard equations (A.1) as the nonrelativistic limit of a vector-field of same structure as the Maxwell field, they imply the existence of a continuity equation, in the sense that they entail a mass flux $\mathbf{p}$ such that

$$\partial_t \rho + \partial_{\mathbf{r}} \cdot \mathbf{p} = 0 \tag{A.4}$$

This relation is trivially deduced from a differential form version of the field equations (cf. Ref. 33 for the similar problem of charge conservation in electromagnetism). In this sense quasiclassical amendments of the continuity equation, taking mass loss due to thermonuclear reactions into account[110] are inconsistent in the frame of Newtonian gravitation theory.

The following simple problem of gravitational stability is easily worked out in the framework of the Glansdorff-Prigogine stability theory.[39] We consider a homogeneous and isotropic part of a volume V immersed in a large self-gravitating mass devoid of internal motions, and of volume W. Upon restricting perturbations in V to purely adiabatic fluctuations, the contributions to the stability criteria

$$\text{(a)} \quad \delta^2 Z<0; \qquad \text{(b)} \quad \partial_t \delta^2 Z \geqslant 0 \tag{A.5}$$

reduce to the kinetic energy contained in the fluctuations. (The boundary conditions on which conditions (A.5) rely are clearly fulfilled for the problem under investigation). We obtain

$$\delta^2 Z=-\int_{\mathsf{V}} d\mathsf{V}\left[\tfrac{1}{2} T^{-1} \rho|\Delta \mathbf{v}|^2\right]<0 \tag{A.6}$$

showing that condition (a) is permanently satisfied for nonzero Eulerian velocity field fluctuations. Condition (b) is easily rewritten, upon scalar multiplication of the perturbed momentum density equation [cf. (II.3)], by $\Delta\mathbf{v}^*/T$, and upon adding the complex conjugate part of the result:

$$\begin{aligned} \partial_t \tfrac{1}{2} \rho|\Delta \mathbf{v}|^2= & -\partial_{\mathbf{r}} \cdot\left[\gamma P \rho^{-1} T^{-1} \tfrac{1}{2}\left(\Delta \mathbf{v}^* \Delta \rho+\Delta \mathbf{v} \Delta \rho^*\right)\right] \\ & -\tfrac{1}{2} \gamma P \rho^{-2} \partial_t|\Delta \rho|^2+(8 \pi G)^{-1} \partial_t|\Delta \mathbf{g}|^2 \end{aligned} \tag{A.7}$$

and

$$\begin{aligned} \partial_t \delta^2 Z= & \partial_t\left\{\int_{\mathsf{V}} d\mathsf{V}(8 \pi G T)^{-1}\left[(\gamma P / \rho)(4 \pi G \rho)^{-1}|\operatorname{div} \Delta \mathbf{g}|^2-|\Delta \mathbf{g}|^2\right]\right\} \\ & +\int_{\partial \mathsf{V}} d \partial \mathsf{V}\, \mathbf{n} \cdot\left[\left(\gamma \frac{P}{\rho}\right) T^{-1} \tfrac{1}{2}\left(\Delta \mathbf{v}^* \Delta \rho+\Delta \mathbf{v} \Delta \rho^*\right)\right] \end{aligned} \tag{A.8}$$

In this equation $\mathbf{n}$ is the outer normal to the surface, γ is the ratio of specific heats, and T and P are temperature and pressure. If we assume that no perturbations occur at the boundary, the surface integral vanishes. We then obtain

$$\begin{aligned} & \partial_t \delta^2 Z \equiv \partial_t \phi \\ & \phi(t)=\int_{\mathsf{V}} d\mathsf{V}(8 \pi G T)^{-1}\left(k_J^{-2}|\operatorname{div} \Delta \mathbf{g}|^2-|\Delta \mathbf{g}|^2\right) \\ & k_J^2=\frac{4 \pi G \rho}{(\gamma P / \rho)} \end{aligned} \tag{A.9}$$

Stability properties of the configuration are entirely ruled by the time behavior of the function $\phi(t)$. If we consider in particular normal modes ($\Delta\mathbf{g}\,\alpha\,e^{st}$), we have

$$\partial_t\phi = 2(\mathrm{Re}\,s)\phi$$

so that unstable modes correspond to

$$\phi < 0 \tag{A.10}$$

From definition (A.9) we see that the length scale of unstable normal modes

$$L = \left(\frac{\overline{|\Delta\mathbf{g}|^2}}{\overline{|\mathrm{div}\,\Delta\mathbf{g}|^2}} \right)^{\frac{1}{2}} \tag{A.11}$$

must be larger than the Jeans length $\lambda_J = 2\pi/k_J$. (The overbar represents a volume integral.) For example, if V is spherical and of radius R, and if we consider perturbations of space dependence

$$\frac{\sin kr}{r}\mathbf{r}^0 \tag{A.12}$$

$\mathbf{r}^0$ being the unit radial vector, and r the radial spherical coordinate, the function ϕ becomes

$$\frac{\phi}{R} = C[k^2 - k_J^2 + O(R^{-1})] \tag{A.13}$$

where C is a positive quantity. The surface contribution in (A.8) divided by R being of order R^{-1}, the assumption of unperturbed boundary conditions in the perturbations is valid to order R^{-1}. With R large enough we recover the classical Jeans criterion for gravitational instability.

APPENDIX B

The time interval T of duration ΔT is partitioned into a finite number n of subintervals of same length $\Delta t_i = \Delta T/n = \epsilon\Delta T$

$$\mathsf{T} = \sum_{i=1}^{n} \mathsf{T}_i \tag{B.1}$$

n is chosen large enough, such that in any T_i the spectral operator $\mathfrak{L}(\tau)$ is time invariant. In line with assumption (A.3), we postulate that $\mathfrak{L}(\tau)$ can be represented by a matrix in the finite dimensional phase space Γ. Let $\mathfrak{L}_i = \partial_\Pi \mathfrak{N}$ be this discrete representation in the interval T_i.*

Consider furthermore the following partition of the spectrum

$$\mathsf{S} = \mathsf{S}^+ + \mathsf{S}^- + \mathsf{S}^0 \tag{B.2}$$

$$\begin{aligned} \mathsf{S}^+ &= \{s | \Delta T \operatorname{Re} s(\tau) > \ \ M, \quad \tau \in \mathsf{T}\} \\ \mathsf{S}^- &= \{s | \Delta T \operatorname{Re} s(\tau) < -M, \quad \tau \in \mathsf{T}\} \\ \mathsf{S}^0 &= \{s | \Delta T |\operatorname{Re} s(\tau)| \leqslant \ \ 1, \quad \tau \in \mathsf{T}\} \end{aligned} \tag{B.3}$$

M is a sufficiently large number, and both ΔT and M are chosen in such a way that if $s^k(0) \in \mathsf{S}^+(\mathsf{S}^-, \mathsf{S}^0)$, then $s^k(\tau) \in \mathsf{S}^+(\mathsf{S}^-, \mathsf{S}^0)$ at any time $\tau \in \mathsf{T}$. Define then $\bar{\chi}_i^+, \bar{\chi}_i^-, \bar{\chi}_i^0$ as the sets of conveniently normalized eigenvectors of $\mathfrak{L}_i$ whose eigenvalues are the sets $\bar{s}_i^+ \in \mathsf{S}^+, \bar{s}_i^- \in \mathsf{S}^-, \bar{s}_i^0 \in \mathsf{S}^0$. Since these eigenvectors form a basis in Γ, any initial condition $\Delta\Pi_0$ can be written

$$\Delta\Pi_0 = \bar{a}_i^+ \cdot \bar{\chi}_i^+ + \bar{a}_i^- \cdot \bar{\chi}_i^- + \bar{a}_i^0 \cdot \bar{\chi}_i^0 \tag{B.4}$$

$\bar{a}$ being a set of expansion coefficients. The solution of the time-dependent linearized evolution equation can now be written

$$\Delta\Pi(t) = e^{\mathfrak{L}_\nu \Delta t} \cdot e^{\mathfrak{L}_{\nu-1} \Delta t_{\nu-1}} \cdots e^{\mathfrak{L}_2 \Delta t_2} \cdot e^{\mathfrak{L}_1 \Delta t_1} \cdot \Delta\Pi_0 \tag{B.5}$$

with

$$t = \sum_{i=1}^{\nu-1} \Delta t_i + \Delta t, \qquad \Delta t \in \mathsf{T}_\nu$$

We first note that

$$\begin{aligned} e^{\mathfrak{L}_1 \Delta t_1} \cdot \Delta\Pi_0 = {} & \bar{a}_1^+ \circ e^{(\bar{s}_1^+ \Delta T)\epsilon} \cdot \bar{\chi} k_1^+ + \bar{a}_1^- \circ e^{(\bar{s}_1^- \Delta T)\epsilon} \cdot \bar{\chi}_1^- \\ & + \bar{a}_1^0 \circ e^{(\bar{s}_1^0 \Delta T)\epsilon} \cdot \bar{\chi}_1^0 \,{}^{**} \end{aligned} \tag{B.6}$$

*This assumption discards the possibility of continuous branches of the spectrum, which are either replaced by discrete eigenvalues, or entirely eliminated. Continuous branches of the type discussed in Section III (proposition P.4) can always be excluded by postulating that the eigenfunctions are of class C.

**The following notations are used: $\bar{c} = \bar{a} \circ \bar{b} \Leftrightarrow c_i = a_i b_i$ (no summation); $\bar{a} \circ \bar{b} \cdot \bar{d} = \bar{a} \cdot \bar{b} \circ \bar{d} = \sum_i a_i b_i d_i$; $(e^{\bar{s}})_i = e^{s_i}$.

The eigenfunctions of $\mathfrak{L}_i$ and $\mathfrak{L}_{i-1}$ are interrelated by

$$\begin{pmatrix} \bar{\chi}_{i-1}^{+} \\ \bar{\chi}_{i-1}^{-} \\ \bar{\chi}_{i-1}^{0} \end{pmatrix} = \begin{pmatrix} \bar{\chi}_{i}^{+} \\ \bar{\chi}_{i}^{-} \\ \bar{\chi}_{i}^{0} \end{pmatrix} + \begin{pmatrix} C_i^{++} & C_i^{+-} & C_i^{+0} \\ C_i^{-+} & C_i^{--} & C_i^{-0} \\ C_i^{0+} & C_i^{0-} & C_i^{00} \end{pmatrix} \begin{pmatrix} \bar{\chi}_{i}^{+} \\ \bar{\chi}_{2}^{-} \\ \bar{\chi}_{i}^{0} \end{pmatrix} \tag{B.7}$$

Due to the continuity of $\mathfrak{L}(\tau)$ (assumption A.1), the blockmatrices C_i are of order ϵ. More precisely, as is obtained from an elementary perturbation calculation, if $\chi_i^{(r)}$ is an eigenfunction belonging to the eigenvalue $s_i^{(r)}$ of $\mathfrak{L}_i$, and $\chi_{i-1}^{(r)}$ is the corresponding eigenfunction associated with $s_{i-1}^{(r)}$ of $\mathfrak{L}_{i-1}$, then the correction factor $\Delta\chi_i^{(rr')} = C_i^{(rr')}\chi_i^{(r')}$ (no summation) in B.7 is of the form $C_i^{(rr')} = A_i^{(rr')}/[s_i^{(r)} - s_i^{(r')}], r \neq r'$. $A_i^{(rr')}$ is independent of r and r' in order of magnitude. For r, r' belonging to the same subset $\mathsf{S}^+, \mathsf{S}^-$, or S^0, the coupling $C_i^{(r,r')}$ is of order ϵ. If $s_i^{(r)}$ and $s_i^{(r')}$ belong to different subsets, this coupling is at most of order ϵ/M (definition B.3). Successive expansions of the eigenfunctions of $\mathfrak{L}_{i-1}$ in terms of those of $\mathfrak{L}_i$, and applications of $e^{\mathfrak{L}_2\Delta t_2}, e^{\mathfrak{L}_3\Delta t_3}, \ldots, e^{\mathfrak{L}_\nu\Delta t}$ upon (B.6) yield

$$\Delta\Pi(t) = F^+(t) + F^-(t) + F^0(t) + O(M^{-1})\Delta\Pi_0$$

with

$$F^k(t) = \bar{a}_1^k \circ \left[e^{(\bar{s}_1^k \Delta T)\varepsilon} \cdot e^{C_2^{kk}} \circ e^{(\bar{s}_2^k \Delta T)\varepsilon} \ldots e^{C_{\nu-1}^{kk}} \circ e^{(\bar{s}_{\nu-1}^k \Delta T)\varepsilon} \cdot e^{C_\nu^{kk}} \circ e^{\bar{s}_\nu^k \Delta t} \cdot \bar{\chi}_\nu^k \right]$$

$$k = +, -, 0 \tag{B.8}$$

The factor $O(M^{-1})$ takes care of the contributions of the nondiagonal blocks in the matrix in (B.7). If we neglect contributions in M^{-1}, the separation (B.3) of the spectrum valid at any $\tau \in \mathsf{T}$ implies that we have a decomposition of the operator according to the decomposition $\Gamma = \Gamma^+ \oplus \Gamma^- \oplus \Gamma^0$ of the phase space,

$$\mathfrak{L}(\tau) = \mathfrak{L}^+(\tau) + \mathfrak{L}^-(\tau) + \mathfrak{L}^0(\tau) \tag{B.9}$$

the parts $\mathfrak{L}^+(\tau)$, $\mathfrak{L}^-(\tau)$, and $\mathfrak{L}^0(\tau)$ having spectra S^+, S^-, and S^0, respectively.[54]

a. $F^+(t)$ is the solution of

$$\partial_t F^+ = \mathfrak{L}^+(t) F^+ \tag{B.10}$$

with $F^+(0) = \Delta\Pi_0^+$, projection of $\Delta\Pi_0$ into Γ^+. Let $\mathcal{W}(t)$ be the matrix of fundamental solutions of (B.10), with $\mathcal{W}(0) = \mathcal{E}_{n^+}$, n^+ being the dimension of Γ^+. We have

$$dtm\,\mathcal{W}(t) \geqslant \exp\left[t \inf_{\tau \subset \mathsf{T}} \operatorname{tr} \mathcal{L}^+(\tau)\right] \tag{B.11}$$

as follows from elementary properties of the Wronskian.[51] With the definitions of the norms

$$\|V\| = \max_{i=1,2\ldots n^+} |V_i| \qquad \|\mathcal{M}\| = \max_{i=1,2,\ldots,n^+} \sum_j |\mathcal{M}_{ji}| \tag{B.12}$$

for vectors and matrices in Γ^+,[134] we have[93]

$$\|\mathcal{W}(t)\| \geqslant [dtm\,\mathcal{W}(t)]^{1/n^+} \tag{B.13}$$

For any fixed time $t \in \mathsf{T}$ at least one initial condition $\Delta\Pi_{ot}^+$ exists such that

$$\|F^+(t)\| = \|\mathcal{W}(t)\Delta\Pi_{0t}^+\| = \|\mathcal{W}(t)\|\,\|\Delta\Pi_{0t}^+\| \tag{B.14}$$

so that

$$\|F^+(t)\| \geqslant \exp\left(t \inf_{t \in \mathsf{T}} \frac{\operatorname{tr} \mathcal{L}^+(\tau)}{n^+}\right)\|\Delta\Pi_{0t}^+\| \tag{B.15}$$

From (B.3) we note that the argument of the exponential function is greater than $Mt/\Delta T$.

b. Definition B.8 implies that

$$\|F^-(t)\| \leqslant \exp\left[t \max_{\substack{i=1,\ldots,n \\ j=1,2,\ldots,n^-}} \operatorname{Re} s_i^{-(j)}\right]\left(\max_{i=1,2,\ldots,n} \|\exp C_i^{--}\|\right)^n \|\bar{a}_1^- \cdot \bar{\chi}_\nu^-\| \tag{B.16}$$

$s_i^{-(j)}$ being the jth eigenvalue of set S^- of $\mathcal{L}_i$. From definitions (B.3) we have

$$t \max_{\substack{i=1,2,\ldots,n \\ j=1,2,\ldots,n^-}} \operatorname{Re} s_i^{-(j)} \leqslant -M\frac{t}{\Delta T}, \qquad n^- = \dim(\Gamma^-) \tag{B.17}$$

$(\max \|\exp C_i^{--}\|)^n$ is of order 1, and $\bar{a}_1^- \cdot \bar{\chi}_\nu^-$ is of the same order as $\Delta\Pi_0^-$ the projection of the initial condition $\Delta\Pi_0$ into Γ^-. Thus

$$\|F^-(t)\| \leqslant O\left(\exp - \frac{Mt}{\Delta T}\right)\|\Delta\Pi_0^-\| \tag{B.18}$$

Provided M is large enough, it is always possible to select $t > t_*, t_* \in \mathsf{T}$, such that $O(\exp - Mt/\Delta T) \leqslant O(M^{-1})$. If we neglect quantities of order M^{-1}, $F^{-}(t)$ leads to no contribution to $\Delta\Pi(t)$ for $t > t_{*}$.

From inequalities (B.15) and (B.18) we can infer the following conclusions on stability of the evolutionary sequence $\Sigma(t)$ in terms of the nonempty sets of the partition (B.2):

Nonempty spectral sets	Implication on stability
S°	Stability depends on $\mathfrak{L}^{\circ}(t)$
S^{-}	Weak stability (criterion C.1)
S^{+}	Instability (C.2)
S° and S^{-}	Stability depends on $\mathfrak{L}^{\circ}(t)$
S° and S^{+}	Unstable
S^{-} and S^{+}	Unstable
S^{-} and S^{+} and S°	Unstable

APPENDIX C

To display the special character of the definitions of stability used in this paper we briefly devote our attention to two elementary examples.

a. Consider a closed reaction system in a motionless, time-independent, homogeneous medium. Assume that the coupling between the reaction and the hydrodynamics and thermodynamics of the medium is negligible. If the initial conditions of the reaction are defined by a point $\underline{\underline{p}}_0$ in the space of chemical variables $\mathbf{R}^N$, the representative point $\underline{\underline{p}}(t)$ of the reaction state moves on a plane

$$\underline{\underline{p}}(t) \cdot \underline{\underline{1}} = \underline{\underline{p}}_0 \cdot \underline{\underline{1}} = \rho = \text{constant} \tag{C.1}$$

as a consequence of the conservation of mass. Now we select a second set of initial conditions $\underline{\underline{p}}'_0 \in \mathbf{R}^N$ with the following constraint

$$\underline{\underline{p}}_0 - \underline{\underline{p}}'_0 = a\,\underline{\underline{1}}, \qquad a \text{ arbitrary (real)} \tag{C.2}$$

that is, $\underline{\underline{p}}'_0$ lies on the normal to the plane specified by (C.1). The track generated by the initial condition $\underline{\underline{p}}'_0$ evolves on the plane

$$\underline{\underline{p}}'(t) \cdot \underline{\underline{1}} = \underline{\underline{p}}'_0 \cdot \underline{\underline{1}} = \rho' \tag{C.3}$$

parallel to the plane C.1. Adopting the Euclidean norm in $\mathbf{R}^N$, the following inequality holds

$$|\underline{\underline{p}}'(t)-\underline{\underline{p}}(t)|\geqslant|\underline{\underline{p}}'_0-\underline{\underline{p}}_0|, \qquad \forall t\in\mathsf{T} \tag{C.4}$$

T being an arbitrary time interval. Thus *any* closed reaction system violates both the strict and the weak stability definitions (within the frame of the Euclidean norm).

It is manifest that we can avoid such situations by postulating that the allowed perturbations have zero projections in the null spaces associated with conservation laws (cf. proposition P.1). But even within the range of this convention, the definitions of stability used here do not always meet the wants of our intuitive feeling of stability as is illustrated in the next example.

b. We consider again the medium specified under example a and examine the evolution of the independent abundances X and Y in the disintegration scheme

$$X\to Y \qquad Y\to Z \tag{C.5}$$

λ_X and λ_Y being the relevant disintegration constants. For a sufficiently short time interval T' of duration

$$\Delta t \ll \min\left(\frac{1}{\lambda_X},\frac{1}{\lambda_Y}\right) \tag{C.6}$$

the Euclidean distance $d(t)$ between two evolutionary paths defined by the initial conditions (X_0, Y_0) and (X'_0, Y'_0) obeys the relation

$$d(t)^2=d(0)^2+2t(X'_0-X_0, Y'_0-Y_0)\mathbb{S}\begin{pmatrix}X'_0-X_0\\ Y'_0-Y_0\end{pmatrix} \tag{C.7}$$

where $\mathbb{S}$ is the symmetrized reaction matrix

$$\mathbb{S}=\begin{pmatrix}-\lambda_X & +\dfrac{\lambda_X}{2}\\ +\dfrac{\lambda_X}{2} & -\lambda_Y\end{pmatrix} \tag{C.8}$$

If $\mathbb{S}$ is negative definite, which occurs for

$$\lambda_Y>\frac{\lambda_X}{4} \tag{C.9}$$

the system is strictly stable in the sense of the Euclidean norm. For

$\lambda_Y < \lambda_X/4$, it is always possible to choose (X_0', Y_0') in the neighborhood of (X_0, Y_0), such that the norm increases with time in T'. This result is easily understood in the framework of the physical argument given under example a: If λ_Y is sufficiently smaller than λ_X, an approximate conservation law for Y holds during a sufficiently short time interval.

Example b shows that under the combined effect of two circumstances our strict stability condition does not satisfy our natural stability feeling: (*1*) The matrix representation of the spectral operator is strongly nonsymmetric. (*2*) An approximate conservation law operating over a short time interval is at work.

Note that if Δt is large enough, our weak stability condition can be satisfied.

APPENDIX D

If we exclude the eigenvalues $s=0$ and $s=\infty$, the equations governing nonradial adiabatic oscillations in the orthogonal complement $\Gamma^{\perp}$ of $\Gamma_0^M \oplus \Gamma_0^T \oplus \Gamma_0^S \oplus \Gamma_0^{X_1} \oplus \cdots \oplus \Gamma_0^{X_{N-1}}$ can be written for equilibrium models

$$\frac{d}{dr}u - \frac{\rho}{P\Gamma_1}gu = \left[-l(l+1)s^{-2} - \frac{\rho r^2}{(\Gamma_1 P)} \right] y - l(l+1)s^{-2}\Delta\phi \tag{D.1}$$

$$\frac{d}{dr}y + Ay = r^{-2}(Ag - s^2)u - \frac{d}{dr}\Delta\phi \tag{D.2}$$

$$r^{-2}\frac{d}{dr}r^2\frac{d}{dr}\Delta\phi - r^{-2}l(l+1) = 4\pi G\rho\left(\frac{\rho}{(\Gamma_1 P)}y - Ar^{-2}u \right) \tag{D.3}$$

The symbols used have the following meanings: r is the radial space coordinate,

$$u = r^2\delta r \qquad y = \frac{\Delta P}{\rho} \tag{D.4}$$

$g(r)$ is the modulus of the gravitational field in the equilibrium model, ϕ is the gravity potential,

$$A = \frac{1}{\rho}\partial_r\rho - (\Gamma_1 P)^{-1}\partial_r P \tag{D.5}$$

and Γ_1 is the adiabatic coefficient

$$\Gamma_1 = \left(\frac{\partial \ln P}{\partial \ln \rho}\right)_S \tag{D.6}$$

In these equations all perturbations are expanded in terms of the spherical harmonics $Y_l^m(\theta, \varphi)$. The details on boundary conditions can be found in Ledoux and Walraven (Ref. 74, Sections 75 and 79).

If we now eliminate y in the frame of generalized functions, we obtain after a trivial rearrangement

$$u = PP(Ag - s^2)^{-1} r^2 \left[\frac{d}{dr}\Delta\phi + \left(\frac{d}{dr} + A\right) \left\{ \frac{\left(\frac{d}{dr} - \frac{\rho g}{\Gamma_1 P}\right) u + \frac{l(l+1)}{s^2}\Delta\phi}{\left(-\frac{\rho r^2}{s^2 \Gamma_1 P}\right)\left[s^2 + \left(\frac{\Gamma_1 P}{\rho r^2}\right) l(l+1)\right]} \right.\right.$$

$$\left.\left. + C_1 \delta\left(s^2 + \frac{\Gamma_1 P}{\rho r^2} l(l+1)\right) \right\} \right] + C_2 \delta(Ag - s^2) \tag{D.7}$$

where C_1 and C_2 are arbitrary constants. System (D.7) together with (D.3) from which y is immediately eliminated, defines a nonhomogeneous linear differential system, provided the parameters s are chosen such that the contribution of one of the delta distributions does not vanish, that is,

$$s^2 = A(r_*) g(r_*) \tag{D.8a}$$

or

$$s^2 = -\Gamma_1(r_*) P(r_*) r^{-2} \rho(r_*)^{-1} l(l+1) \tag{D.8b}$$

with $r_* \in (0, R)$.

The interpretation of modes (D.8a) has a special physical appeal. The corresponding displacement δr has a singularity at $r = r_*$. It represents a fluid particle (or layer) at r_* displaced out of its hydrostatic equilibrium position, the neighboring fluid particles remaining essentially at rest. The displaced layer oscillates around its equilibrium position with a period

$$P=2\pi[-A(r_*)g(r_*)]^{-1/2} \tag{D.9}$$

provided $A(r_*)<0$, the value of the period being entirely fixed by the local conditions prevailing at r_*. If $A(r_*)>0$, the equilibrium position of the layer at r_* is unstable. This interpretation is essentially equivalent to the semiintuitive derivation of Schwarzschild's convection criterion by means of the bubble method. More generally it suggests that the latter technique when applied to more involved situations (cf. Ref. 119) relies in fact on the existence of continuous spectra connected with generalized functions as eigenfunctions. According to this point of view, the stability criteria obtained by the bubble method are not necessarily the most general ones, since they particularize the manifold of allowed perturbations.

The convective stability criterion $A(r_*)<0$ is well known to be related to the dynamical discrete spectrum (Ref. 74, Section 78; Refs. 67, 68, and 69). Although continuous modes have not been contemplated so far, it is of interest to note that the rigorous proof of the Schwarzschild convection criterion provided by Kaniel and Kovetz[53] takes account of the possibility of continuous modes among the adiabatic spectrum.

A mathematical proof of existence of the continuous modes (D.8a) and (D.8b)* is still lacking (cf. proposition P.4). However the alluring physical interpretation offered above entitles us to feel confident about their mathematical reality.

References

1. M. L. Aizenman and J. Perdang, *Astron. Astrophys.*, **15**, 200 (1971).
2. M. L. Aizenman and J. Perdang, *Astron. Astrophys.*, **23**, 209 (1973).
3. M. L. Aizenman and J. Perdang, *Astron. Astrophys.*, **25**, 53 (1973).
4. M. L. Aizenman and J. Perdang, *Astron. Astrophys.*, **28**, 327 (1973).
5. M. L. Aizenman and J. Perdang, to be published (1975).
6. M. L. Aizenman and J. Perdang, in preparation, (1975).
7. V. A. Antonov, *Vestn. Leningr. Gos. Univ.* **7**, 135 (1962).
8. E. B. Aronson and C. J. Hansen, *Astrophys. J.*, **177**, 145 (1972).
9. W. D. Arnett, *Ann. Rev. Astron. Astrophys.*, **11**, 73 (1973).
10. W. Baade, G. B. Burbidge, F. Hoyle, E. M. Burbidge, R. F. Christy, and W. A. Fowler, *Publ. Astron. Soc. Pac.*, **68**, 296 (1956).
11. G. E. Backus *Ann. Phys.*, **4**, 372 (1958).

*These modes correspond to pure oscillations under any physical conditions prevailing at r_*. Since physically one necessarily excites a finite interval of the continuum of periods, the resulting oscillations are rapidly damped, with a time scale of the order of the interval ΔP of excited periods, due to phase mixing, or, more rigorously, as a consequence of the generalized Riemann-Lebesgue theorem.[76] However, it remains to be seen whether these modes might become vibrationally unstable.

12. C. Bertaud, in *Colloque International sur les Novae, Supernovae et Novoïdes, CNRS Paris*, 183 (1965).
13. W. B. Bonnor, *Mon. Not. Roy. Astron. Soc. Lond.*, **116**, 351 (1956).
14. L. B. Borst, *Phys. Rev.*, **78**, 807 (1950).
15. A. Boury, *Mém. Soc. Roy. Sci. Liege*, **8**, fasc. 6 (1964).
16. A. Boury, M. Gabriel, A. Noels, and R. Scuflaire, *Astron. Astrophys.*, **35**, 185 (1974).
17. J. I. Castor, *Astrophys. J.*, **166**, 109 (1971).
18. S. Chandrasekhar, *Hydrodynamic and Hydromagnetic Stability*, Clarendon Press, Oxford 1961, Appendix III.
19. S. Chandrasekhar, *Ellipsoidal Figures of Equilibrium*, Univ. of Yale, London, 1969.
20. R. V. Churchill, *Fourier Series and Boundary Values Problems*, McGraw-Hill, Cambridge, Mass., 1941, Sect. 38.
21. S. Colgate and C. McKee, *Astrophys. J.*, **157**, 623 (1969).
22. S. Colgate and C. McKee, in *Stellar Evolution*, H. Y. Chiu and A. Muriel, Eds., MIT Press, Cambridge, Mass., 1972, p. 307.
23. T. G. Cowling, *Mon. Not. Roy. Astron. Soc. Lond.*, **101**, 367 (1941).
24. J. P. Cox, C. J. Hansen and W. R. Davey, *Astrophys. J.*, **182**, 885 (1973).
25. W. R. Davey and J. P. Cox, *Astrophys. J.* **189**, 113 (1974).
26. R. J. Defouw, *Astrophys. J.*, **182**, 983 (1973).
27. R. J. Defouw, C. J. Hansen, and R. Siquig, *Astrophys. J.*, **184**, 581 (1973).
28. S. R. DeGroot, *Thermodynamik irreversibler Prozesse*, Bibl. Institut, Mannheim, 1960, Chapt. VII.
29. J. Demaret, *Astrophys. Space Sci.*, **31**, 305 (1974).
30. R. Ebert, *Z. Astrophys.*, **42**, 263 (1956).
31. M. E. Fisher, *Arch. Ration. Mech. Anal.*, **17**, 377 (1964).
32. P. R. Fields, M. H. Studier, H. Diamond, J. F. Mech, M. G. Inghram, G. M. Stevens, S. Field, W. M. Manning, A. Ghiorso, S. G. Thompson, G. H. Higgins, and G. T. Seaborg, *Phys. Rev.*, **102**, 180 (1956).
33. H. Flanders, *Differential Forms*, Academic Press, New York, 1963.
34. M. Gabriel, *Astron. Astrophys.*, **18**, 242 (1972).
35. F. R. Gantmacher, *Théorie des Matrices*, Tome II, Dunod, Paris, 1966 Chap. 14.
36. F. Gilbert and G. J. F. McDonald, *Geophys. Res.*, **65**, 675 (1960).
37. P. Glansdorff and I. Prigogine, *Physica*, **20**, 773 (1954).
38. P. Glansdorff and I. Prigogine, *Physica*, **30**, 351 (1964).
39. P. Glansdorff and I. Prigogine, *Physica*, **31**, 1242 (1965).
40. P. Glansdorff and I. Prigogine, *Physica*, **46**, 344 (1970).
41. P. Glansdorff and I. Prigogine, *Thermodynamic Theory of Structure, Stability and Fluctuations*, Wiley-Interscience, New York, 1971, Chap. XIV.
42. I. M. Glazman, *Direct Methods of Qualitative Analysis of Singular Differential Operators*, Oldbourne Press, London, 1965.
43. A. S. Grossman, *Astrophys. J.*, **161**, 619 (1970).
44. I. M. Guelfand and G. E. Chilov, *Les Distributions*, Dunod, Paris, 1962, Sect. 2.6.
45. M. Hamermesh, *Group Theory*, Addison-Wesley, Reading, Mass., 1964, Chapter III.
46. C. J. Hansen and W. H. Spangenberg, *Astrophys. J.*, **168**, 71 (1971).
47. C. J. Hansen, J. P. Cox and M. A. Hertz, *Astron. Astrophys.*, **19**, 144 (1972).
48. C. Heiles, *Ann. Rev. Astron. Astrophys.*, **9**, 293 (1971).
49. P. Hertel and W. Thirring, CERN preprint TH 1330 and 1338, 1971.
50. G. Horedt, *Mon. Not. Roy. Astron. Soc. Lond.*, **151**, 81 (1970).
51. E. L. Ince, *Ordinary Differential Equations*, New York, Dover Publications Inc., 1956, Chap. V.

52. H. Kähler and A. Weigert, *Astron. Astrophys.*, **30**, 431 (1974).
53. S. Kaniel and A. Kovetz, *Phys. Fluids*, **10**, 1186 (1967).
54. T. Kato, *Perturbation Theory for Linear Operators*, Springer-Verlag, Berlin, Heidelberg, New York, 1966.
55. Z. Kopal, *Proc. Natl. Acad. Sci. USA*, **34**, 377 (1948).
56. A. Kovetz, *Astrophys. Space Sci.*, **4**, 365 (1969).
57. A. Kovetz and G. Shaviv, *Astrophys. Space Sci.*, **14**, 378 (1971).
58. R. Kurth, *Zeitschrift für angew Math. Phys.*, **6**, 115 (1955).
59. R. Kurth, *Introduction to the Mechanics of Stellar Systems*, Pergamon Press, London, 1957.
60. G. S. Kutter and M. P. Savedoff, *Astron. J.*, **72**, 810 (1967).
61. G. S. Kutter and M. P. Savedoff, *Astrophys. J.*, **156**, 1021 (1969).
62. H. Lamb, *Hydrodynamics*, Cambridge University Press, London, 1932.
63. L. D. Landau and E. M. Lifschitz, *Quantenmechanik*, Akademie-Verlag, Berlin, 1965.
64. L. D. Landau and E. M. Lifschitz, *Statistische Physik*, Akademie-Verlag, Berlin, 1966.
65. R. B. Larson, *Ann. Rev. Astron. Astrophys.*, **11**, 219 (1973).
66. G. Lebon, *J. Eng. Math.*, **2**, 381 (1968).
67. N. R. Lebovitz, *Astrophys. J.*, **142**, 229 (1965).
68. N. R. Lebovitz, *Astrophys. J.*, **142**, 1257 (1965).
69. N. R. Lebovitz, *Astrophys. J.*, **146**, 946 (1966).
70. P. Ledoux, *Handbook of Physics*, **LI**, Springer-Verlag, Berlin, Göttingen, Heidelberg, 1958, p. 604, Sect. 2.
71. P. Ledoux, *Astrophys. Norv.*, **9**, 187 (1964).
72. P. Ledoux *Questions récentes d'astrophysique théorique*, Chaire Francqui, Belge, 1967–1968.
73. P. Ledoux, *Oscillations et stabilité stellaires in La Structure Interne des Etoiles*, Observatoire de Genève, Genève, 1964.
74. P. Ledoux and T. Walraven, *Handbook of Physics*, Vol. **LI**, Springer-Verlag, Berlin, p. 353.
75. S. Lefschetz, *Differential Equations: Geometric Theory*, Interscience, New York, 1957, Chap. IV.
76. M. J. Lighthill, *Introduction to Fourier Analysis and Generalised Functions*, Cambridge University Press, 1958.
77. A. E. H. Love, *A Treatise on the Mathematical Theory of Elasticity*, Dover Publications, New York, 1944.
78. D. Lynden-Bell and R. Wood, *Mon. Not. Roy. Astron. Soc. Lond.*, **138**, 495 (1968).
79. I. G. Malkin, *Sb. Nauch. Tr. Kaz. Aviats. Inst.*, **7** (1937).
80. J. T. Mariska and C. J. Hansen, *Astrophys. J.*, **171**, 317 (1972).
81. W. H. McCrea, *Mon. Not. Roy. Astron. Soc. Lond.*, **117**, 562 (1957).
82. S. Meggitt, Thesis, Univ. of Canberra, 1969.
83. J. Meixner and H. G. Reik, *Handbook of Physics*, Vol III/2, Springer-Verlag, Berlin, 1959, Sect. 4–5.
84. M. A. Melvin, *Rev. Mod. Phys.*, **28**, 18 (1956).
85. A. Messiah, *Mécanique Quantique*, Tome 1, Dunod, Paris, 1962, p. 396.
86. R. H. Miller, *Astrophys. J.*, **180**, 759 (1973).
87. R. M. Morse and H. Feshbach, *Methods of Theoretical Physics*, Vol **1**, McGraw-Hill, Cambridge, Mass., 1953.
88. N. I. Muskhelishvili, *Singular Integral Equations*, P. Noordhoff N. V., Groningen, 1953.
89. V. V. Nemytskii and V. V. Stepanov, *Qualitative Theory of Differential Equations*, Princeton Univ. Press, 1960, Chap. IV, V.

90. A. Noels-Grötsch, *Astron. Astrophys.*, **18**, 350 (1972).
91. A. Noels-Grötsch, A. Boury, and M. Gabriel, *Ann. d'Astrophys.*, **30**, 13 (1967).
92. J. P. Ostriker and L. Axel, in *Low Luminosity Stars*, S. S. Kumar, Ed., Gordon and Breach, New York, 1969, p. 357.
93. A. M. Ostrowski, *Solutions of Equations and Systems of Equations*, Pergamon Press, New York, 1966, Chapt. 19.
94. B. Paczyński and M. Kozłowski, *Acta Astron.*, **22**, 315 (1972).
95. L. A. Pars, *A Treatise on Analytical Dynamics*, Heinemann, London, 1965, Sect. 23.7.
96. M. V. Penston, *Mon. Not. Roy. Astron. Soc. Lond.*, **145**, 457 (1969).
97. J. Perdang, *Astrophys. Space Science*, **1**, 355 (1968).
98. J. Perdang, in Notes on the Summer Study Program in Geophysical Fluid Dynamics, WHOI, Vol. II, ref. n° 69–41, 1969, p. 103
99. J. Perdang, Thesis, Université de Liège, 1969.
100. J. Perdang, 1973, unpublished.
101. S. Pinault and S. P. S. Anand, *Astrophys. Space Sci.*, **24**, 105 (1973).
102. I. Prigogine, *Nature*, **209**, 602 (1966).
103. I. Prigogine, *Introduction to Thermodynamics of Irreversible Processes*, Wiley, New York, 1967.
104. I. Prigogine, in *Theoretical Physics and Biology*, M. Marois, Ed., North-Holland, Amsterdam, 1969, p. 23.
105. I. Prigogine and G. Severne, *Physica*, **32**, 1376 (1966).
106. G. Rakavy and G. Shaviv, *Astrophys. Space Sci.*, **1**, 429 (1968).
107. H. Reeves, *Astrophys. J.*, **138**, 79 (1962).
108. G. R. Reissig, G. Sansone and R. Conti, *Qualitative Theorie Nichlinearer Differentialgleichungen*, ed. Cremonese, Roma, 1963, Chapt. I.
109. H. Robe, P. Ledoux and A. Noels, *Astron. Astrophys.*, **18**, 424 (1972).
110. S. Rosseland, *Astrophysik auf atomtheoretischer Grundlage*, Springer, Berlin, 1931, Chapt. 3.
111. G. Sansone and R. Conti, *Equazioni differenziali non lineari*, ed. Cremonese, Roma, 1956, Chapt. I, IV, IX.
112. M. Schwarzschild and R. Härm, *Astrophys. J.*, **142**, 855 (1965).
113. M. Schwarzschild and R. Härm, *Astrophys. J.*, **184**, 5 (1973).
114. G. Severne, *Physica*, **61**, 307 (1972).
115. N. R. Simon, *Astrophys. J.*, **159**, 859 (1970).
116. N. R. Simon, *Astrophys. J.*, **164**, 331 (1971).
117. N. R. Simon and V. R. Sastri, *Astron. Astrophys.*, **21**, 39 (1972).
118. R. Simon, *J. Quant. Spectr. Rad. Transfer*, **3**, 1 (1963).
119. R. Simon, 19e Colloque International d'Astrophysique de Liège, 1974, to appear in *Mém. Soc. Roy. Sci. Liège*.
120. V. I. Smirnov, *A Course of Higher Mathematics*, Vol IV, Pergamon Press, New York, 1964, Chap. I.
121. P. A. Strittmatter, J. Faulkner, J. W. Robertson, and D. J. Faulkner, *Astrophys. J.*, **161**, 369 (1970).
122. J. L. Tassoul, *Ann. Astrophys.*, **30**, 363 (1967).
123. J. L. Tassoul and M. Tassoul, *Ann. Astrophys.*, **31**, 251 (1968).
124. W. Thirring, *Z. Physik*, **235**, 339 (1970).
125. W. Thirring, *Essays Phys.*, **4**, 125 (1972).
126. L. Tisza, in *Phase Transformations in Solids*, R. Smoluchowski, J. E. Mayer, and W. A. Weyl, Eds., Wiley, New York, 1951. Reprinted in L. Tisza, *Generalized Thermodynamics*, MIT Press, Cambridge, Mass., 1966, p. 194.

127. L. Tisza, *Ann. Phys.*, **13**, 1 (1961). Reprinted in L. Tisza, *Generalized Thermodynamics*, MIT Press, Cambridge, Mass., 1966, p. 102.
128. W. Unno, *Publ. Astron. Soc. Japan*, **20**, 356 (1968).
129. W. Unno, *Publ. Astron. Soc. Japan*, **21**, 240 (1969).
130. W. Unno, *Publ. Astron. Soc. Japan*, **22**, 299 (1970).
131. R. Van der Borght, *Aust. J. Phys.*, **17**, 165 (1964).
132. R. Van der Borght, *Publ. Astron. Soc. Japan*, **23**, 539 (1971).
133. T. Ważewski, *Studia Math.*, **10**, 48 (1948).
134. W. Wasow, *Asymptotic Expansions for Ordinary Differential Equations*, Interscience, New York, 1965, Chapt. 1.
135. V. Weidemann, *Ann. Rev. Astron. Astrophys.*, **6**, 351 (1968).
136. S. Yabushita, *Mon. Not. Roy. Astron. Soc. Lond.*, **140**, 109 (1968).

STELLAR ATMOSPHERES, NONEQUILIBRIUM THERMODYNAMICS, AND IRREVERSIBILITY

RICHARD N. THOMAS

Institut d'Astrophysique
Paris, France

Abstract

This article is a summary of current investigations of the stellar atmosphere considered as a transition zone between stellar interior and interstellar medium, thus between regions described by linear and by nonlinear nonequilibrium thermodynamics; the properties of the transition zone are fixed by the particular storage modes for the star considered as a concentration of matter and energy in a nonequilibrium, unstable-against-concentration medium; irreversibility is a consequence of the evolution of such concentrations in such media, and thus *toward* greater structural organization.

CONTENTS

I. INTRODUCTION

This conference treats various aspects of instability and dissipative structures in hydrodynamics. I was asked to contribute something from astronomy, possibly connected with our studies on the non-LTE (nonlocal-thermodynamic-equilibrium) states of radiation and matter in stellar atmospheres. Logically, then, my focus should lie on the nonequilibrium thermodynamics of the coupling between radiation fields and instability and/or dissipation in those nonthermal velocity fields occuring in stellar atmospheres.

Such a presentation—a vignette of astronomy in the midst of more conventional hydrodynamics—must, I think, be descriptive and concise, outlining the current broad frontiers of such astronomical studies, especially in their relation to other things discussed here. Anything very narrow and detailed would be lost in the jargon of unfamiliar application. (Cf. re this difficulty, the proceedings of the several Symposia on Cosmical Gas Dynamics, sponsored jointly by the IAU and IUTAM.[15]) So, toward this objective, let me summarize how we at the Institut d'Astrophysique in Paris view: (*1*) the stellar atmosphere as exhibiting the structure of one kind of transition zone between the domains of linear and nonlinear nonequilibrium thermodynamics; (*2*) the properties of the transition zone as depending upon the storage modes for matter and energy in the subatmosphere and in the nonequilibrium environment; (*3*) any nonthermal velocity fields in the star as reflecting subatmospheric nonthermal storage modes, rather than any instability of atmosphere or star, with dissipative structures in such velocity fields exhibiting the amplitude-fixing process for such storage modes, either through their coupling with thermal modes or through fluxes of matter and energy to the environment; (*4*) the star itself as exhibiting one kind of concentration of matter and energy that arises in one species of the hierarchy of nonequilibrium, unstable-against-concentration media which characterize the Universe; and, finally, (*5*) possible extensions of this type of thinking on instability of medium, concentrations, and structural evolution to applications other than the stellar atmosphere. Concerning (*5*), I consider that most irreversible processes relating to structure originate in just this instability against concentration of nonequilibria, but steady-state and quasihomogeneous media, rather than in an extension of the entropy description of the change in state of equilibrium ensembles upon interaction. The evolution is then *toward organization* (i.e., a concentration with structure), rather than the entropy-described evolution *toward disorganization* (i.e., thermalized homogeneity).

II. Stellar Atmospheres: Empirical

By far the greatest amount* of work on stellar atmospheres, either diagnostic or theoretical-modeling, rests on the assumption that the atmosphere is a kind of boundary for LTE stellar material, distinguishing it from non-LTE interstellar material. The "boundary" itself is optical,

*For example, 80% of the papers in the bibliography of the 1973 IAU report on the Theory + Analysis of Stellar Atmospheres fall into this "classical" category.[21]

representing the very small region where the "opacity" of the atmosphere —either to radiation from another, eclipsed star or to a photon which escapes without further absorption from the star—becomes insignificant. The effect of such opacity varies exponentially with the opacity, and the opacity itself varies exponentially with geometrical height in the atmosphere, so the boundary should be abrupt. Furthermore, because the "temperature" of this LTE atmosphere varies as the fourth-root of the radiative energy density, the change in thermal excitation across this kind of atmosphere is relatively small. And finally, this stellar atmosphere model is assumed to be static, to ensure that the radiation field alone fixes the thermodynamic (LTE) state.

Unfortunately, when compared with observations, this simple "classical" model is characterized by four general kinds of "anomalies," each embracing many kinds of observational facts, which strongly imply its lack of internal physical consistency and inapplicability to the real world. These are summarized below.

A. Stellar Atmospheric "Turbulence"

This term was almost simultaneously introduced to describe: (*1*) an anomalous "broadening" of stellar spectral lines *attributed* to *random* atmospheric motions, which were occasionally supersonic; and (*2*) a "turbulent pressure" which was thought necessary to "geometrically extend" stellar atmospheres beyond the predictions of the classical model (cf. Section I.B). Some of these anomalies have been later understood as arising from height-gradients of systematic, rather than random, velocity fields; others, as the result of bad spectral diagnostics (cf. Ref. 13). But in any case the phenomena covered by this type of anomaly discredit the idea that the atmosphere can be treated as having only thermal motions; and the existence of either random or systematic nonthermal motions is equally damaging to the classical model. This implies nonthermal, instead of only thermal, kinetic energy storage modes in the subatmosphere.

B. Extended Atmospheres

This term was originally introduced to imply that the actual stellar atmosphere extended, in many cases, far beyond the limits predicted by the above classical atmosphere. Various explanations were offered, usually based on some kind of momentum transfer in the atmosphere: "turbulent" pressure, radiation pressure, and so on. And so, this anomaly links to the preceeding one. But now it has been recognized—especially from detailed solar studies (cf. Ref. 20)—that such atmospheric extension often represents an "apparent" extension arising from a confusion of height-gradient of excitation state with that of atmospheric density. The confusion arose from imposing LTE, rather than non-LTE diagnostics. In addition, there are the effects arising from the above-mentioned presence of systematic velocity fields and their coupling to thermal motions. That is, sometimes atmospheric extension has been inferred from a presumed necessity to explain emission lines as reflecting a greater emitting area of the star for lines than that for the continuous spectrum. But now we recognize that chromospheric effects—mechanical dissipation effects associated with non-thermal kinetic energy storage—can often explain such emission lines simply in terms of a rise in electron kinetic temperature associated with such dissipation, and "normal" atmospheric extension. A comparison of presence, and size, of such an effect in lines formed by photoionization and those formed by collisional excitation gives a measure of the extent of the chromosphere for the star.[3]

C. Stellar Symbiosis

This term was originally introduced to denote the simultaneous presence, in the spectra of a few, apparently single stars, of features characterizing both a "hot" and a "cool" star, as they would appear on the basis of the classical model. Hence it was thought either that symbiotic stars were double (cf. Ref. 23) or that they were single with an extremely peculiar geometry.[8] Since then it has been increasingly recognized that *most* stars exhibit such "multispectral" characteristics, relative to the single-region classical model, when broad spectral regions are observed (the far UV, the far IR, radio, etc.). For example, see the interpretation of the evolution of thinking on the solar atmosphere as "symbiotic evolution" by Gebbie and Thomas.[6] What, in fact, this "symbiotic" phenomenon implies, is that a stellar atmosphere consists of many regions, rather than the single one of the classical model, whose "states" vary widely. What the anomalies in Sections II.A and II.B imply, are that properties of these regions must include both non-LTE effects, *and* the effects of nonthermal velocities, as well as other, as-yet-unspecified effects explicitely set aside in the single-region "classical" model, and, indeed, even in the neoclassical model, where only non-LTE effects are included (as in Ref. 11).

D. Flux of Matter

Customarily, in computing stellar atmospheric models one ignores matter-fluxes. The exception lies in those considerations inspired by Parker's ideas on the solar wind,[12] which, however, focus on the existence of a stellar chromosphere-corona as the "cause" of the wind. One must of course then ask "why," and in how many stars, such coronae occur. It is readily shown that if one considers thermal storage modes only, in the subatmosphere, no coronae occur and only a negligible matter-flux arises from thermal evaporation of the outer atmosphere. However, observations of matter-flux across the whole range of stellar types give matter-fluxes exceeding these thermal-evaporation predictions by a factor 10 in the logarithm. The only way to remedy the situation for the wide variety of stellar types in which such nonthermal mass-fluxes are observed is to consider a variety of nonthermal kinetic energy storage modes in the subatmosphere. Such modes produce both chromospheres-coronae and nonthermal matter-fluxes as parallel consequences of the nonlinear transition from these nonthermal storage modes.[19]

As a consequence of the existence of these four categories of anomalies, we are led to the following "conceptual" formulation of the role, structure, and function of a stellar atmosphere, as contrasted to its "boundary" role in the "classical" model.

III. STELLAR ATMOSPHERES: CONCEPTUAL

Geometrically, we recognize that the stellar atmosphere is the region between stellar interior and interstellar medium. *Structurally*, we recognize that the stellar interior is quasiequilibrium, while the inter-stellar medium is strongly nonequilibrium. In terms of the language of nonequilibrium thermodynamics, quasiequilibrium implies *linear* nonequilibrium thermodynamics; that is, all microscopic distribution functions are those of thermodynamic equilibrium, expressed in terms of *local* thermodynamic-equilibrium-state parameters, and all fluxes propagate by *diffusion*, viz, they are the gradients of some potential. And in the same language, strongly nonequilibrium simply means that all phenomena are nonlinear; the usual TE state parameters of pressure, density, temperature do not suffice to describe the situation. Thus, without further sophistication we recognize the need for a transition zone between quasiequilibrium and nonequilibrium, linear and nonlinear, regions; and, geometrically, the stellar atmosphere must be that transition zone. It only remains, then, to decide what kind of properties it must have to fulfill this role. Section I provides an empirical guide to what the theory must produce.

Were stellar material indeed different in some way from inter-stellar—in the most extreme case a solid like a planetary surface as contrasted to the gaseous interplanetary medium–we might envision a boundary. But both star and interstellar medium are gaseous; so indeed the transition zone must be continuous and any distinction between stellar and interstellar material is vague. Rather, the distinction must lie in the "state" of the material, and it is in this that the transition must occur. Then it remains to specify the "transition" parameters. Taken in this sense, we recognize that the classical model implies that the transition occurs only in the properties of photon distribution functions: from the "opaque" regions of the subatmosphere, where the radiation field is *quasi*isotropic—the nonisotropy simply reflecting the presence of a "linear" flux in the above sense—to the "transparent" regions of the interstellar medium, where the energy density of the radiation corresponds in no way to its spectral distribution.

The simplest step beyond this first-approximation would be to grant the same kind of transition to internal energy states of atoms as to the energy states of photons. Then the "transition" would lie in radiative excitation—with the above nonequilibrium radiation field—replacing the condition of detailed balance in collisional excitations. So, the transition would lie in collisional rate, which implies, in particle density (the temperature dependence is, roughly, second order, so long as we admit only thermal storage modes; cf. the following). Consequently, many of the developments in stellar atmospheric structure of recent years consist simply in investigating this second kind of transition: collisional versus radiative excitation processes, the so-called "non-LTE" model of stellar atmospheres.[7,11,18]

But such considerations alone are of little interest in our discussion, they do not involve nonthermal velocity fields. Indeed, we can say that such models describe a star which is simply a *thermal* storage pot of matter and energy. They are the "real world" models of the original idea of a star as an isolated black-body; the "real world" results from recognizing the star has no enclosing walls.

But then we pose the question of why the star should be restricted to being a *thermal* storage pot for matter and energy; why not other, nonthermal, storage modes? And with this question comes the recognition of what must be the real direction of development of stellar atmospheric research: specifying these nonthermal modes of storage of matter and energy in the stellar interior, then specifying the various aspects of transition between this essentially linear storage and the nonlinear storage plus propagation in the environment. Each aspect of transition will, in a rough way, define a region of the atmosphere. In the classical model, the photon transition defined the LTE photosphere. In the "neoclassical" atmosphere, the collisional transition defined the non-LTE photosphere; see the work by Pecker, Praderie, and Thomas[14] for an outline of the general scheme of transitions—the photosphere and beyond. So, the primary problem, before we can discuss theoretical models of stellar atmospheres to any degree of completeness, is to specify the nonthermal storage modes in the interior, especially the subatmosphere. We expect the set of atmospheric regions to depend upon the set of storage modes.

IV. STELLAR INTERIOR: STORAGE MODES

We recognize two classical examples: convection and radial pulsation. The very first stellar models, constructed long before much attention was given radiative storage and transport, centered on quasiadiabatic storage associated with convection.[5] The problem was linear; indeed, the first models were wholly storage, neglecting energy production in the interior and energy fluxes at the exterior, forcing boundary temperatures and atmospheric motions to be zero. The situation is curious, in that then, as well as in subsequent discussions of the effect of including radiation, one spoke of "convective *transport*" of energy, or "radiative versus convective *transport*" of energy; whereas indeed the question of fluxes, in this linear situation, is secondary to that of storage. What *is* important is the question whether the storage is thermal only—radiative and kinetic— or also has nonthermal storage modes. The strictly adiabatic case is the best example: one has storage, but no fluxes and there is only that change in internal energy corresponding to the adiabatic change in the LTE state with position. One should not speak of "stability against convection," but, rather, ask what set of modes stores the most energy, this being the "reason" the star came into existence (cf. below).

In the same way, the first treatments of stellar pulsation were adiabatic, which again relates to the change of quasiequilibrium state, under this LTE approximation, but with time at the same position, as well as with change in position. Again, it is not a question of stability, but rather of energy storage. No energy was "dissipated" in these adiabatic changes of state; there was only energy exchange between thermal and nonthermal modes. Departure from adiabaticity, thus dissipation, came only at the "surface" (in the sense of the classical atmospheric model, the atmosphere simply rising and falling as a whole with the pulsation) via thermal radiation from the star.

The "amplitudes" of the storage did not enter in each of the above examples because of the adiabatic character. The polytropic exponent characterized the convective modes; the "linear" period, the pulsational. But in neither case do we know how much energy is stored. To determine this, we need to consider the nonadiabatic aspects: balance the fluxes into and out from the nonthermal storages modes, the fluxes being spatial, temporal, and between various energetic degrees of freedom. The situation thus differs from the thermal storage modes in that the latter have nonzero amplitude when the configuration "degenerates" to thermodynamic equilibrium, whereas in this degenerate equilibrium case the nonthermal storage modes vanish.

Contemporary models of pulsation allow nonadiabatic effects throughout the star.[9] Nonlinear effects enter in trying to predict the amplitudes of the pulsation, by considering interchange between nonthermal and thermal modes and fluxes of each from star to environment.[2] However, at present such treatments essentially ignore the transition zone character of the atmosphere. For example, the effects of matter fluxes from the star on amplitudes and atmospheric structure have been essentially ignored. Such matter (and associated nonthermal energy) fluxes may be small, but must be included in order to produce a complete set of atmospheric regions,[19] especially chromosphere and corona.

Those treatments of stellar chromospheres—the atmospheric region corresponding to the presence of nonthermal kinetic energy storage[14,19]—which exist, largely depend on the production of acoustic modes from convective modes.[1,16] More recent considerations also introduce the possibility of chromospheres reflecting what are basically pulsational (but nonradial) storage modes.[10,17,19,22,24] These remain to be explored in detail, along with other storage modes of nonthermal kinetic energy and the whole question of magnetic storage modes.

What is important from all this is that we recognize the star as, to first priority of attention, a storage-concentration of matter and energy, whose *observed* parts—the several regions of the transition zone to the environment, or the atmosphere—have their properties fixed by whatever is the *optimum* set of storage modes in the subatmosphere, the *linear* nonequilibrium region. We *conjecture* that *optimum* means *maximal storage*, maximal concentration. So, in place of asking stability against the onset of some kind of nonthermal phenomenon, we simply ask whether its presence will produce a storage mode that increases the storage of matter and energy. That is, we ask for the greatest possible *organization* of storage-concentration for matter and energy. And we recognize that the time-history of the growth of the star is *toward* such organization, away from the essentially random, nonequilibrium, original state of the matter and energy in the interstellar medium.

V. STELLAR ENVIRONMENT: UNSTABLE MEDIUM

Given the above recognition of what a star actually is, a concentration of matter and energy arising in a nonstable medium, we recognize what *irreversibility*, hence evolution, means, in the stellar case. By contrast to the 19th century statistical mechanics, which was preoccupied with the evolution of concentrations in *equilibrium* media, the actual astronomical case focuses on *nonequilibrium* media. Notions of entropy in the change of state of such media are meaningless. Entropy is simply a sequencing parameter describing the mixing of two equilibrium ensembles, relating initial and final *equilibrium* states. Our concern lies in the evolution of a medium having four characteristics:

1. Strongly nonequilibrium, because it is unbounded.
2. Unstable against concentration, because of its chemical and nuclear composition; any increase of concentration is exothermic.

These characteristics (1) and (2) are *observed*. How the medium reached that state is open to speculation; for our present arguments, it suffices to note that the state is what exists. *Wholly speculatively*, we then add two additional characteristics:

3. The medium is, *globally* over the Universe, in a steady state.

In a sense, this marks a return to the Hoyle-Bondi-Gold steady-state Universe outlook. For us, the essential point is that this configuration be accompanied by the observed characteristics 1 and 2. Then, *locally*, the behavior of the medium is neither steady state nor homogeneous: *concentrations* arise—the question of origin is largely irrelevant, as long as they exceed a size sufficient to introduce a nonadiabatic evolution. Our above detailed considerations focus on stellar concentrations; obviously, very little is required to extend the notion to groups of stars. What is more interesting, and less developed, is the final characteristic:

4. The concentration process consists of establishing a hierarchy of media, *each* with properties 1 and 2 to some degree. Then, the most interesting direction of investigation lies in the comparative development of accreting versus nonaccreting concentrations, obviously as a function of the properties of the medium. The obvious applications of the former lie in the areas of crystal growth and biological systems. Again, we emphasize that the novel aspect of this approach lies in its investigation of irreversible evolution toward greater organization of structure via the characteristics 1 and 2 above rather than any attempt to generalize notions of entropy, which we consider associated with equilibrium systems, not nonequilibrium. Also, we emphasize that our starting point lies with arbitrarily large, not small, departures from equilibrium, evolving toward concentrations having smaller departures. And we believe the utility of the study of a typical *transition zone*, in this case the stellar atmosphere, can provide a useful guide to the relation between concentration and environment.

The matter summarized above is treated in more detail in two series of lectures at the Collège de France (academic years 1973–1974 and 1974–

1975) and will then appear in monograph form *The Stellar Atmosphere as a Transition Region*, with some further applications, as indicated above.

References

1. L. Biermann, *Naturwiss.*, **33**, 118 (1946).
2. R. Christy, *Proc. Symp. Cos. Aerodyn. 5th*, **1967** 105.
3. S. Dumont, N. Heidmann, L. Kuhi, and R. N. Thomas, *Astron. Astrophys.*, **29**, 199 (1973).
4. A. S. Eddington, *Internal Constitution of the Stars*, 1930.
5. R. Emden, *Gaskugeln* Teubner Publ., 1907.
6. K. B. Gebbie and R. N. Thomas, *Natl. Bur. Std. U. S. Spec. Pull.*, **353**, 84 (1971). Proc. Menzel Symposium, K. B. Gebbie, Ed.
7. J. T. Jefferies, *Spectral Line Formation*, Blaisdell Publ., 1970.
8. D. H. Menzel, unpublished submitted in competition for A. Cressy Morisson Prize, 1950.
9. P. Ledoux and T. Walraven, *Handbook of Physics*, Vol. 51, S. S. Flügge, Ed., 1958, p. 353.
10. J. V. Leibacher, Thesis, Harvard University, 1971.
11. D. Mihalas, *Stellar Atmospheres*, Freedman Publ., 1971.
12. E. N. Parker, *Ap. J.*, **128**, 664 (1958).
13. J. C. Pecker and R. N. Thomas, *Proc. Symp. Cos. Aerodyn., 4th*, **1961** 1.
14. J. C. Pecker, F. Praderie, and R. N. Thomas, *Astron. Astrophys.*, **29**, 289 (1973).
15. Proceedings of the IAU-IUTAM Symposia on Cosmical Gas Dynamics, 1st Symposium, G. Kolchogoff, Ed., 1951, Central Air Documents Office U.S. Army Forces: ATI103 347; 2nd Symposium, J. M. Burgers and H. Van de Hulst, Eds., 1953; 3rd Symposium, J. M. Burgers and R. N. Thomas, Eds., *Rev. Mod. Phys.*, **30**, 905 (1958); 4th Symposium R. N. Thomas, Ed., *Nuovo Cimento, Suppl. to XXII* 1968; 5th Symposium R. N. Thomas, Ed., Academic Press, 1967, 6th Symposium H. Habing, Ed., Reidel Publishers, 1970.
16. M. Schwarzschild, *Ap. J.*, **107**, 1 (1948).
17. P. Souffrin, IAU General Assembly 1973: Joint Discussion on Waves in Solar and Stellar Atmospheres (unpublished).
18. R. N. Thomas, *Some Aspects of Non-Equilibrium Thermodynamics in The Presence of a Radiation Field*, University of Colorado Press, Colorado, 1965.
19. R. N. Thomas, *Astron. Astrophys.*, **29**, 297 (1973).
20. R. N. Thomas and R. G. Athay, *Physics of the Solar Chromosphere*, Interscience, New York, 1961.
21. R. N. Thomas and K. B. Gebbie, IAU Comm. 36 report on Theory and Analysis of Stellar Atmospheres, 1973.
22. R. K. Ulrich, *Ap. J.*, **162**, 993 (1970).
23. O. C. Wilson, in *Stellar Atmospheres*, J. H. Greenstein, Ed., Univ. Chicago Press, 1960, p. 463.
24. C. L. Wolff, *Ap. J.*, **177**, 187 (1972).

THE BÉNARD INSTABILITY IN LIQUID MIXTURES

G. THOMAES

Faculty of Applied Sciences, Free University of Brussels, Brussels, Belgium

Abstract

A short survey of the experimental work done in Brussels on the stability of a two-component liquid layer subjected to a temperature gradient is given. The phenomenon was studied simultaneously by a static and a flow method. Comparison of the results obtained by the two methods can explain the anomalies observed in a thermodiffusion flow cell.

Thermal diffusion is an interference phenomenon between diffusion and heat conduction which gives rise to a gradient of concentration formed as a result of a temperature gradient.

For binary liquid mixtures, the only ones we shall consider here, the phenomenon is characterized by the Soret coefficient D'/D where D' is the thermal diffusion coefficient and D the isothermal one. The study is almost always carried out by enclosing the liquid between two plane, accurately horizontal surfaces maintained at two different temperatures, the upper being the hotter. The depth of the liquid layer is usually of the order of 1 to 2 cm. In this case, the separation at the steady state between the top and the bottom is given with a good approximation by

$$\Delta N_1 = -\frac{D'}{D} N_1 N_2 \Delta T$$

where N_1 and N_2 are the weight fractions of the two components and ΔT is the difference of temperature. D'/D can be positive or negative and we shall adopt here the convention that it is positive when the more dense component migrates towards the cold side.

When we look at the experimental results reported in the literature, we notice not only very large discrepancies between results given by different authors but also that even the sign is not always the same.

Moreover, in contradistinction to the predictions of the kinetic, thermodynamic, and statistical theories, only very few systems seem to have a negative Soret coefficient. In other words, for almost every binary mixture, thermal diffusion results in a concentration increase of the more dense component at the cold side. But experimenters who have studied these phenomena have generally considered that the systems were stable as long as the global density gradient increases downwards. As we shall see later, instability can occur even though the density gradient is not adverse.[1]

From another point of view, Prigogine has suggested that measurement of the Bénard critical point could provide a new method to determine the Soret coefficient. The concentration contribution to the density gradient can either reduce or increase that established by the temperature gradient depending on the direction of migration of the more dense component.

This was the reason for a series of experiments in our laboratory on the Bénard effect in mixtures by means of a Schmidt-Milverton type apparatus. This apparatus consists of two gilded horizontal brass plates separated by a distance of the order of 1 mm. The lower plate is heated by an electrical heating element whose power can be regulated with accuracy. The upper one is maintained at constant temperature by circulating water. The difference of temperature between the two plates is measured by two resistance thermometers. For pure liquids, by measuring ΔT as a function of the power supplied (W) the onset of instability at the Ra^{cr} (1708) can be detected thanks to the difference of heat conduction between the state at rest and the convective one.

The linear stability theory[2–4] has shown that for liquid mixtures heated from below, a negative value of D'/D, which indicates that the more dense component migrates towards the hot wall (here the lower plate), increases the stability whereas a positive value of D'/D strongly reduces the stability. In our experiments, we did observe for a negative D'/D the predicted increase of stability[3] (Fig. 1).

These experiments were performed with water–methanol and water–ethanol mixtures which are known to have a negative Soret coefficient in

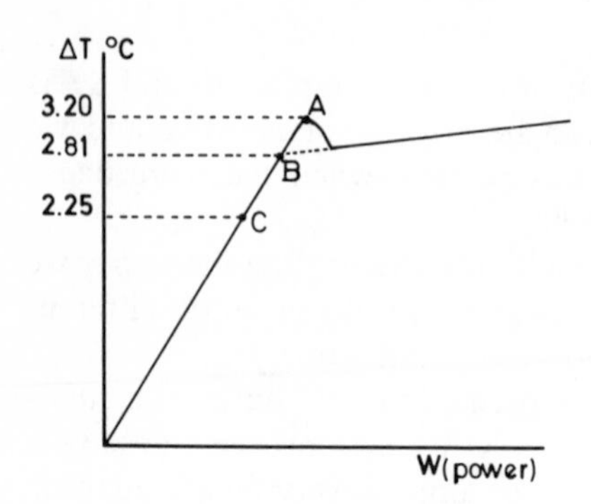

Fig. 1.

the dilute range (more than 70 wt % water). For $CH_3OH–H_2O$, at $N_{H_2O}=$ 0.9

$$\Delta T^{cr}(A)=3.20°C$$

One can show through the thermodynamics of irreversible process that the contribution of thermodiffusion to heat conductivity is of the order of 10^{-2}% and is thus negligible. In this case, the slope of $0A$ should be the same if thermodiffusion does not take place. Calculation of ΔT^{cr} for the *same* solution, disregarding thermodiffusion, gives 2.25°C (C). By extrapolation of the convective state line, point B is obtained at 2.81°C. It seems thus that there remains a certain effect of the concentration distribution in the convective state. In the second case ($D'/D>0$) we did not observe any apparent decrease of stability as indicated by the theory. The diagram (Fig. 2) is the same as for pure liquids.[3,4] The following mixtures were investigated:

$C_6H_6–CCl_4$ $\quad$ $C_6H_5Cl–CCl_4$ $\quad$ $C_6H_{12}–CCl_4$ $\quad$ $C_2H_2Br_4–C_2H_2Cl_4$

for different compositions as well as $CH_3OH–H_2O$ and $C_2H_5OH–H_2O$ in the concentrated range (less than 70 wt % water). All are known to be characterized by a positive Soret coefficient. When data were available in the literature, it appeared that, assuming no thermodiffusion, ΔT^{cr} calculated agrees with the experimental value.

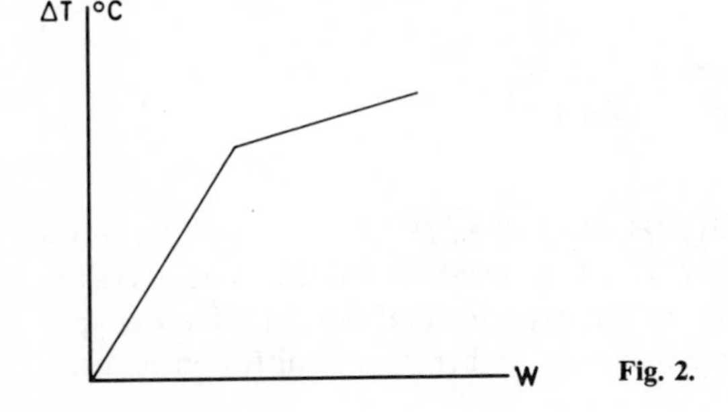

Fig. 2.

For example, for $CH_3OH–H_2O$($N_{H_2O}=0.30$, $D'/D=2.5.10^{-3}$°K^{-1}) one finds ΔT^{cr} calc$=0.99$°C as compared to ΔT^{cr} exptl.$=1.02$°C. The stability seems to be the same as for an equivalent pure liquid. Before explaining these results, I would like to summarize them by giving ΔT^{cr} as a function of the weight fraction. In the case $D'/D>0$, the experimental points lie on the theoretical curve calculated without thermodiffusion (Fig. 3).

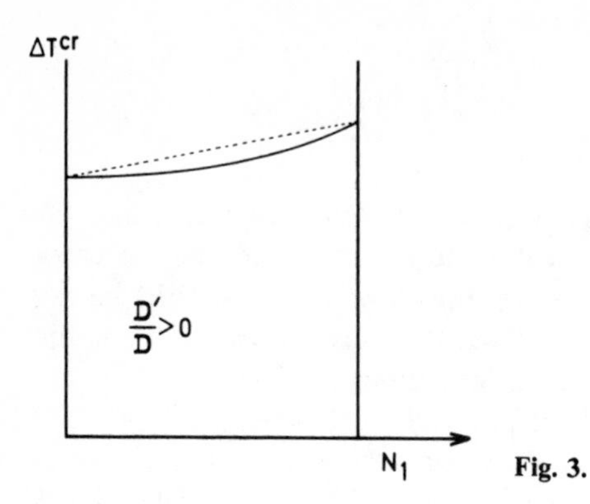

Fig. 3.

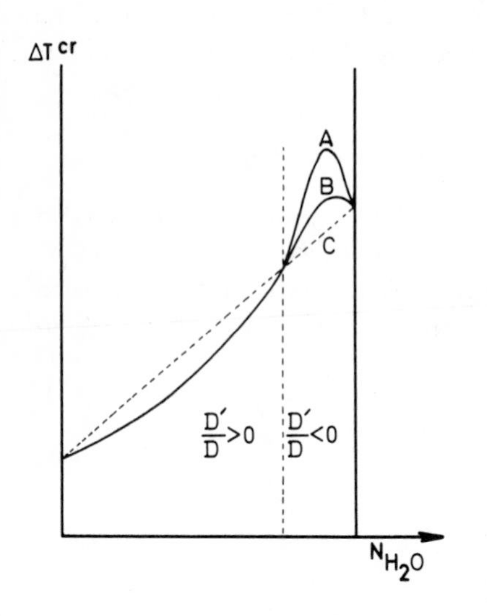

Fig. 4.

The same is observed for CH_3OH–H_2O when D'/D is positive, but for a weight fraction of more than 0.7, D'/D is negative and the three curves in Fig. 4 correspond, respectively, to the experimental one (A), the extrapolated one (B), and the calculated one (C) when thermodiffusion is disregarded.

Now, in the first case ($D'/D > 0$), we tried to determine whether or not there was a concentration gradient before the observed critical point (cf. Fig. 2).

By removing some samples of liquids from the top and the bottom of the cell, we showed that there is no significant change of composition inside the liquid.

Theory indicates that when D'/D is positive, Ra^{cr} decreases steeply as the separation increases and that the critical wave number k^{cr} tends rapidly to zero. This could generate a very broad convection cell remixing the mixture without increasing the heat conduction. The observed critical point very likely is a second instability state. We shall see later that the actual ΔT^{cr} can be estimated to a value of the order of 10^{-2}°C. This is too small to be detected in our apparatus.

The difficulties in observing the thermal diffusion separation in a Schmidt-Milverton apparatus led us to study the stability in a thermal diffusion flow type cell.[5] In this kind of apparatus, the hot and cold plates are separated by a distance of only 0.2 mm. Liquid flows through the cell in a horizontal direction, the rate of flow being adjusted so that the thermal diffusion steady state is established before the liquid leaves the apparatus. At the exit, a knife-edge separator splits the issuing liquid into two samples which can be analyzed.

This apparatus, designed a few years ago for the determination of the Soret coefficient, has the advantage of a very small relaxation time, because of the small thickness of the liquid layer but presents some working anomalies. The dimensions are 20 cm long, 4 cm wide, and 0.2 mm thick. Measurements were carried out for different mixtures with positive or negative Soret coefficient for several ΔT, positive and negative, and for rate of flow from 0 to 20 ml/hour. At these rates, the *Re* number is very small and the flow is laminar. For comparison, let us consider the four quadrants of the stability diagram obtained for a liquid initially at rest (Fig. 5). The two upper quadrants correspond to a negative gradient of T,

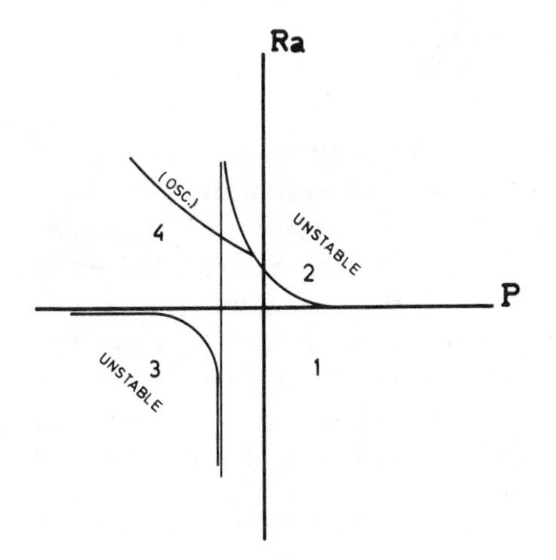

Fig. 5.

the two lower ones to a positive gradient of T. The two quadrants on the right refer to positive D'/D and the two on the left to negative D'/D.

In this figure Ra is given as a function of $P=\nu/\alpha \cdot D'/D \cdot N_1 N_2$* (the ratio of the thermal diffusion contribution to the density gradient to the pure thermal contribution).

1. For grad $T>0$, $D'/D>0$ (Fig. 6), that is the more frequent case in thermodiffusion, we obtain with a flow cell, as usual, separation proportional to ΔT. When extrapolated to $v=0$ this corresponds to quadrant 1. The only anomaly is a slight increase of ΔN_1 for very slow rates we cannot explain. In fact, the separations obtained in a usual static cell correspond to those at higher rates.

2. For grad $T<0$, $D'/D>0$ (Fig. 7), and for small values of ΔT, we obtain the same separation as in 1 above, but in the other direction which is normal since the gradient of T is in the opposite direction. For higher values of ΔT, we observe a rapid decrease of ΔN_1 and even in some cases an inversion of the direction of separation for $v=0$. This corresponds to quadrant 2.

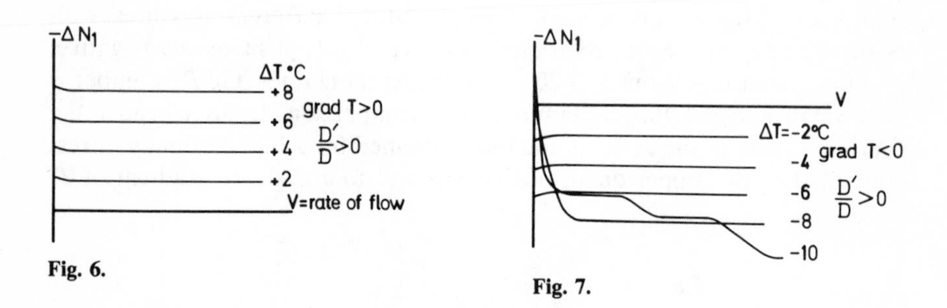

Fig. 6.

Fig. 7.

As mentioned above, we apparently did not observe an instability for the same mixtures in the Schmidt-Milverton type apparatus. But taking account of the respective dimensions of the two apparatus, ΔT^{cr} observed in the flow cell method corresponds to ΔT^{cr} in the Schmidt-Milverton apparatus of 3×10^{-2}°C for a C_6H_{12}–CCl_4 (N=0,5) mixture. The separation curves obtained for $\Delta T<\Delta T^{cr}$ correspond to the stable region. For $\Delta T>T^{cr}$ it seems, by analogy with pure liquids, that the velocity of the fluid increases the stability and that the different plateaus correspond to different convective intermediate states. The inversion of separation for

*$\alpha=-(1/\rho)(\partial\rho/\partial T)$: thermal expansion coefficient; $\nu=(1/\rho)(\partial\rho/\partial N)$: relative variation of the density with component.

$V=0$ in the latter case could be due to the large single convection cell assumed above.

3. For $\text{grad}\,T>0$, $D'/D<0$ (Fig. 8), we have observed a normal separation; for $V=0$: this corresponds to quadrant 3. The fact that the system is stable means that we are either on the right of the asymptote, in which case the system is always stable, or on the left but with a ΔT too small to reach the instability region. For instance, for the C_2H_5OH–H_2O ($N_{H_2O}=0.9$) mixture and $\Delta T=+10°C$, the value of P is -0.320, whereas the asymptote is located at $P(\text{asympt})=-Pr(Pr+Sc)^{-1}=-0.012$, $Ra=-2.0$ and $Ra^{cr}=-69$. In a conventional static thermal diffusion apparatus with a usual liquid height of 2 cm, the Ra number would be -9×10^5, well beyond the critical value. An experimenter working under these conditions would not detect a separation and would deduce that there is no thermodiffusion effect for this mixture. This could explain why so few mixtures are reported as having a negative Soret coefficient.

4. For $\text{grad}\,T<0$, $D'/D<0$ (Fig. 9), we have observed the same separation but in the other direction. In this case, there is no problem, as far as ΔT is not too large.

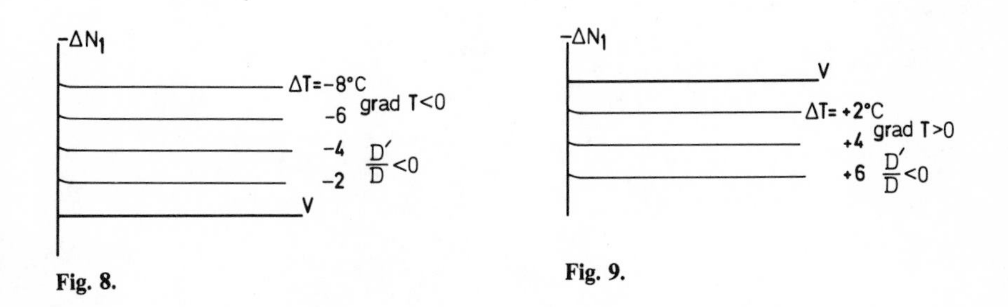

Fig. 8.

Fig. 9.

It is obvious that the stability in a flowing system does not correspond to the problem studied in the Bénard effect where it is assumed that the liquid is at rest. The horizontal flow should be taken into account but nevertheless an extrapolation at $V=0$ can explain some anomalies.

At the present time, we are studying the stability of a flowing system by means of the local potential variational technique.

This report is a summary of the experimental work done in our department by Dr. J.Cl. Legros, Mr. P. Poty, Mr. D. Longrée, and by the author in this field.

Acknowledgments

Our working group is grateful to Professor I. Prigogine and to Professor P. Glansdorff for constant interest and stimulating comments. We also wish to acknowledge helpful discussions with Professor R. S. Schechter from the University of Texas.

References

1. M. G. Velarde and R. S. Schechter, *Chem. Phys. Lett.*, **12** 312 (1971).
2. R. S. Schechter, I. Prigogine and J. R. Hamm, *Phys. Fluids*, **15**, 379 (1972); M. G. Velarde and R. S. Schechter, *Phys. Fluids*, **15**, 1707 (1972); D. T. J. Hurle and E. Jakeman, *J. Fluid Mech.*, **47**, 667 (1971); J. C. Legros, J. K. Platten and P. G. Poty, *Phys. Fluids*, **15**, 1383 (1972); J. C. Legros, P. G. Poty and G. Thomaes, *Physica*, **64**, 481 (1973). J. C. Legros, Bulletin Acad. Roy. Belg. (Cl. Sciences) 5e Serie, Tome LIX, 382 (1973–1974).
3. J. C. Legros, Ph.D. Thesis, Université Libre de Bruxelles, 1971.
4. J. C. Legros, W. A. Van Hook and G. Thomaes, *Phys. Lett.*, **1**, 698 (1968) and **2**, 251 (1968); J. C. Legros, D. Rasse and G. Thomaes, *Phys. Lett.*, **4**, 632 (1970).
5. G. Thomaes, *J. Chim. Phys.*, **53**, 407 (1956); B. D. Butler and J. C. R. Turner, *Trans. Faraday Soc.*, **62**, 3114 (1966).

ON THE NATURE OF OSCILLATORY CONVECTION IN TWO-COMPONENT FLUIDS

D. T. J. HURLE AND E. JAKEMAN

Royal Radar Establishment
Malvern,
Worcs, England

We have predicted[1] and demonstrated[2] overstable temperature oscillations in a two-component fluid resulting from the action of the Soret effect. The oscillations were of a transient nature and it was concluded that they represented the growth of an overstable mode which triggered a finite amplitude stationary mode when some critical amplitude was reached. This behavior is expected[3–5] because of the tendency of the overstable oscillations to destroy the stabilizing density gradient established by the Soret effect.

Platten and Chavepeyer[6,7] have reported steady temperature oscillations in binary fluids which they have ascribed to Soret-driven overstability and have suggested[8] that our oscillations were transient because we were not in a steady-state condition but rather were increasing the heating rate slowly. However we have established that the same transient behavior obtains when the heating rate is constant.

In early Soret-effect experiments with a badly designed cell we obtained steady oscillations, but these were observed for fluids having a Soret coefficient of either sign and, in one experiment, in pure water.[9,10] One expects to obtain overstability only for the case when the sign of the Soret coefficient is such as to produce a stabilizing contribution to the density gradient. Clearly then, these oscillations of constant amplitude were not due directly to the Soret effect, and in subsequent experiments[10] it was demonstrated that a preexisting circulation was a likely prerequsite for their occurrence. Accordingly we have suggested that Platten and Chavepeyer's steady oscillations may also be of this type.[9]

Much light has been thrown on this problem by Caldwell[11] who has performed most careful experiments on Soret-driven convection in saline solutions. He has observed two kinds of oscillation. In Fig. 1, at point A, at which the break in the Schmidt-Milverton plot occurs, exponentially growing overstable oscillations are observed. At some finite, but small, amplitude the oscillations change in character, taking on a longer and somewhat irregular period. At this point the Nusselt number increases significantly, and the Rayleigh number falls corresponding to point *B*.

It seems clear that Platten and Chavepeyer's oscillations are of this latter type since they occur in the region of *B* and have somewhat irregular waveform and period, characteristic of Caldwell's finite amplitude mode.

Two questions arise from these observations:

1. If our transient oscillations are overstable ones, why did we not see the oscillatory finite amplitude mode in our experiments?
2. Do the steady oscillations which we see in the poorly designed cell, where there is thought to be a preexisting circulation, bear any relationship to those of Platten and Chavepeyer and of Caldwell?

The answer to the first question probably lies in the nature of our apparatus which consisted of massive thermodes of high conductivity which prevented the fall in Rayleigh number observed by Caldwell. Hence, in our experiments, upon reaching a critical amplitude the system transferred from the overstable state *A* to the stationary state *C* (where Caldwell did not observe oscillations). Additionally it may be observed that nonlinear numerical computer simulations performed by Platten and Tellier[12,13] with constant Rayleigh number demonstrate a transient rather than a constant-amplitude overstability. The transient overstability is replaced by a finite amplitude stationary mode with increased Nusselt number exactly as we have observed experimentally.[2]

The second question is more difficult to answer. In Caldwell's experiments, the state attained at a given point on the Schmidt-Milverton plot depended on the path by which it was approached. Thus, just to the right of point *D* in Fig. 1 oscillations were obtained for decreasing Nusselt number, whereas, for increasing Nusselt number, the conduction curve was traced towards *A*. Possibly the fact that the Nusselt number is greater than unity in our poorly designed cell with preexisting circulation prevents the system reaching point *A* and instead triggers the finite amplitude mode when point *D* is reached. However, this does not account for the fact that oscillations are observed for both signs of Soret coefficient and further experiments are necessary to resolve the relationship between our steady oscillations and those of Platten and Chavepeyer and of Caldwell.

Finally we wish to reiterate a point made by Shirtcliffe[14] and by ourselves,[2] that we believe that overstable oscillations inevitably grow (at constant Rayleigh number) until they trigger a finite amplitude mode.

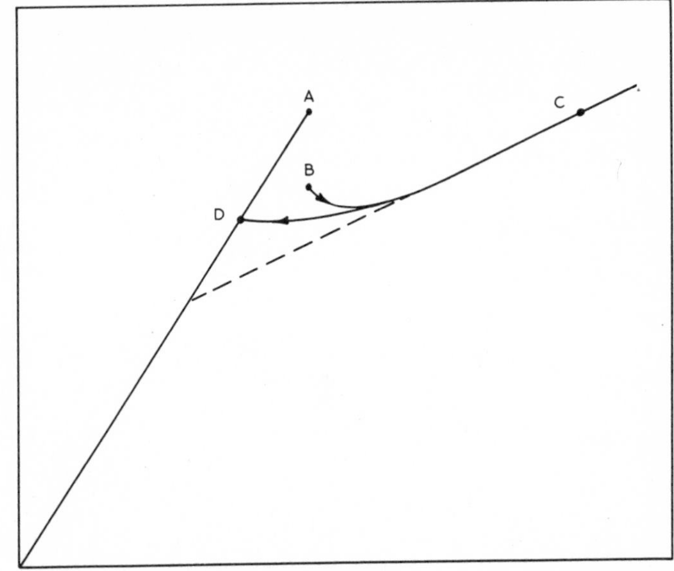

Fig. 1.

Acknowledgments

The authors gratefully acknowledge the receipt of preprints of their work from Dr. Caldwell and Dr. Platten. This paper is contributed by permission of the Director RRE.

References

1. D. T. J. Hurle and E. Jakeman, *Phys. Fluids*, **12**, 2704 (1969).
2. D. T. J. Hurle and E. Jakeman, *J. Fluid Mech.*, **47**, 667 (1971).
3. G. Veronis, *J. Mar. Res.*, **23**, 1 (1965).
4. G. Veronis, *J. Fluid Mech.*, **34**, 315 (1969).
5. R. L. Sani, *A.I.Ch.E.J.*, **11**, 971 (1965).
6. J. K. Platten and G. Chavepeyer, *Phys. Lett.*, **A40**, 287 (1972).
7. J. K. Platten and G. Chavepeyer, *J. Fluid Mech.*, **60**, 305 (1973).
8. J. K. Platten, G. Chavepeyer and J. Tellier, *Phys. Lett.*, **44A**, 479 (1973).
9. D. T. J. Hurle and E. Jakeman, *Phys. Lett.*, **43A**, 127 (1973).
10. D. T. J. Hurle and E. Jakeman, *Phys. Fluids*, **16**, 2056 (1973).
11. D. R. Caldwell, *J. Fluid Mech.*, **64** 347 (1974).
12. J. Tellier, Thesis, Université de l'état a Mons, 1973.
13. J. K. Platten and J. Tellier, paper presented at the Conference on Instability and Dissipative structures in Hydrodynamics, Brussels, 1973.
14. J. G. L. Shirtcliffe, *J. Fluid Mech.*, **35**, 677 (1969).

FINITE AMPLITUDE INSTABILITY IN THE TWO-COMPONENT BÉNARD PROBLEM

J.K. PLATTEN AND G. CHAVEPEYER

University of Mons, Faculty of Sciences
Mons, Belgium

CONTENTS

I. CONCENTRATION GRADIENTS AND HYDRODYNAMIC STABILITY

Onset of convection in a pure liquid layer heated from below is a well-known phenomena and is now quite well understood.[1]

The appearance of strange phenomena when small concentration gradients are present is much more recent. It has been recognized since 1960 that small concentration gradients can have a profound influence on natural convection. This is the so-called "thermohaline convection" and the associated "salt fingers."[2–9]

In 1967, Prof. I. Prigogine suggested that thermal diffusion should be related to the onset of instability in a binary liquid layer heated from below. Research in this field was initiated by the group of the Brussels school, both experimentally and theoretically. The first paper published on this subject is that of Legros, Van Hook, and Thomaes.[10] In that paper, as well as in subsequent publications,[11-13] they presented Schmidt-Milverton plots[14] for binary mixtures heated from below, and evidence is given of anomalous heating curves. Such anomalous heating curves and a hysteresis loop in Schmidt-Milverton plots were also reported by Caldwell[15] in seawater. Cladwell observed in this system very irregular temperature fluctuations. Finally, Platten and Chavepeyer[16-18] reported a hysteresis loop and very regular oscillations in the temperature field of water–ethanol and water–isopropanol systems heated from below and related these oscillations to the Soret effect.[19] However, Hurle and Jakeman[20] expressed some doubt about the origin of the observed oscillations. A controversy still exists on this subject and the paper by Hurle and Jakeman[21] presented at this Conference and published in this volume is indeed devoted to a critical review of experimental work on possible oscillations in the temperature field.

As a matter of fact, sustained oscillations, if they exist, and a hysteresis loop can only be explained in the framework of a nonlinear analysis. The aim of this paper is to give an account of our numerical approach to the solutions of the nonlinear conservation equations.

The linear stability analysis of the two-component Bénard problem, taking into account thermal diffusion, seems now to be firmly established. We would like to mention the work of Hurle and Jakeman,[22-23] Schechter, Prigogine, and Hamm,[24] and that of Platten, Legros, Chavepeyer, and Poty.[17,25,26] A summary of the results of the linear theory is given in section II for convenience.

Finally, a review of the two-component Bénard problem and related topics, such as instabilities due to chemical reactions, has been published recently,[27] but the authors focused their attention mainly on the linear theory. In some sense, this chapter is a continuation of Ref. 27.

II. FORMULATION OF THE PROBLEM

A. The Conservation Equations

The system to be investigated is a transversely infinite, horizontal fluid layer of thickness d, with two components. Initially the system is in mechanical equilibrium and homogeneous in composition. The system is subjected to an inverse temperature gradient (heated from below). Because of this temperature gradient, thermal diffusion (also known as the Soret effect in liquids[19]) takes place. A mass fraction distribution is established in the liquid layer and influences the onset of convection and the supercritical states as well. For a Boussinesq fluid, with constant phenomenological coefficients (such as viscosity, isothermal diffusion, and thermal diffusion coefficients...) and neglecting the so-called Dufour effect[19] (the reverse effect of thermal diffusion), the conservation equations are as follows

- Conservation of mass (continuity equation)

$$\sum_j \frac{\partial u_j}{\partial x_j} = 0 \tag{1}$$

- Conservation of individual species: for a two component system with N_γ being the mass fraction of component γ, we have only one equation, say for N_1, "1" being the more dense component

$$\frac{\partial N_1}{\partial t} = -\sum_j u_j \frac{\partial N_1}{\partial x_j} + D \sum_j \frac{\partial^2 N_1}{\partial x_j^2} + D' \sum_j \frac{\partial}{\partial x_j}\left(N_1 \frac{\partial T}{\partial x_j}\right) \tag{2}$$

This equation is only valid for dilute solutions. The generalization to concentrated solutions is straightforward.[28]

- Conservation of momentum

$$\frac{\partial u_i}{\partial t} = -\sum_j u_j \frac{\partial u_i}{\partial x_j} - \frac{1}{\rho_0}\frac{\partial p}{\partial x_i} + \frac{\rho}{\rho_0}F_i + \nu \sum_j \frac{\partial^2 u_i}{\partial x_j^2} \tag{3}$$

- Conservation of energy

$$\frac{\partial T}{\partial t} = -\sum_j u_j \frac{\partial T}{\partial x_j} + \kappa \sum_j \frac{\partial^2 T}{\partial x_j^2} \tag{4}$$

Moreover a linearized equation of state is adopted

$$\rho = \rho_0[1 - \alpha(T - T_0) + \gamma(N_1 - N_1^{in})] \tag{5}$$

The notation is classical and is that adopted in the papers by Platten, Legros, and Chavepeyer.[17,25,26]

The reference temperature T_0 is the temperature of the lower boundary $z=0$ and the reference mass fraction distribution N_1^{in} is the initial mass fraction of component 1, that is, before thermal diffusion sets in.

An appropriate set of boundary conditions must be specified. Two kinds of boundary conditions are generally adopted: free boundaries and rigid boundaries. The analytical conditions for these boundary conditions are given in the subsequent sections.

B. Summary of Linear Theory

The conservation equations are first linearized in the disturbances. Thus we write

$$N_1(x,z,t) = \overline{N_1(z)} + \varepsilon n(x,z,t)$$

$$T(x,z,t) = \overline{T(z)} + \varepsilon \vartheta(x,z,t)$$

$$u_i(x,z,t) = 0 + \varepsilon u_i(x,z,t) \tag{6}$$

$$p(x,z,t) = \overline{p(z)} + \varepsilon p'(x,z,t)$$

$$\rho(x,z,t) = \overline{\rho(z)} + \varepsilon \delta\rho(x,z,t)$$

and

$$\delta\rho = \rho_0(-\alpha\vartheta + \gamma n) \tag{7}$$

In a linear theory, only terms in ε^1 are retained, and the time evolution of the perturbations behaves like $e^{\sigma t}$. In general σ is a complex quantity ($\sigma = \sigma_R + i\sigma_I$). In a linear theory of hydrodynamic stability, we are primarily interested in neutral stability ($\sigma_R = 0$) corresponding either to the so-called "principle of exchange of stabilities" ($\sigma_I = 0$) or to overstability ($\sigma_I \neq 0$).

When a dimensionless analysis is performed, the following dimensionless number are of interest

- The Prandtl number

$$Pr = \frac{\nu}{\kappa}$$

- The Schmidt number

$$Sc = \frac{\nu}{D}$$

- The usual Rayleigh number

$$Ra = \frac{g\alpha\Delta T d^3}{\kappa\nu}$$

- A Rayleigh number for the concentration field due to thermal diffusion

$$R_{Th} = \frac{g\gamma N_1^{in} d^3}{\kappa D}$$

- The Soret number

$$\mathbb{S} = \frac{D'}{D} \cdot \Delta T$$

The real problem corresponding to experimental situations is that of rigid boundaries, but in that case, exact analytical solutions are not known. The case of free boundaries is more tractable from a theoretical point of view and we will summarize the final results. First we would like to emphasize that an additional assumption must be introduced in order to get analytical solutions for free boundaries: the thermal diffusion term in (2) $(\partial/\partial x_j)(N_1 \partial T/\partial x_j)$ is replaced by $N_1^{in} \partial^2 T/\partial x_j^2$, that is N_1 is replaced by its initial value N_1^{in} which is a constant not subjected to fluctuations.

Two cases must be considered:

1. $\mathbb{S} > 0$: the more dense component migrates towards the cold upper boundary and has a destabilizing effect. In that case it has been shown that the principle of exchange of stabilities is valid and the critical Rayleigh number is given by[29]

$$Ra_{(\mathrm{ex})}^{(\mathrm{crit})} = \frac{27\pi^4}{4} - R_{Th}\mathbb{S}\left(1 + \frac{Sc}{Pr}\right) \tag{8}$$

2. $\mathbb{S} < 0$: this is a stabilizing case. However (8) holds only for small $|\mathbb{S}|$. When $\mathbb{S} < 0$ and

$$|\mathbb{S}| > \frac{27\pi^4}{4} \frac{Pr(Pr+1)}{R_{Th} \cdot Sc^2} \tag{9}$$

overstability prevails and the critical Rayleigh number is then given by[29]

$$Ra_{(\mathrm{over})}^{(\mathrm{crit})} = \frac{27\pi^4}{4} \frac{(1+Sc)(Sc+Pr)}{Sc^2} - R_{Th}\mathbb{S}\left(\frac{Pr}{1+Pr}\right)$$

and the frequency by

$$\sigma_I = \left[-\frac{R_{Th} \cdot \mathbb{S}}{3Pr(Pr+1)} - \frac{9\pi^4}{4Sc^2} \right]^{1/2} \tag{10}$$

Finally the reduced critical wave number is unaffected by thermal diffusion for free boundaries and is

$$k^{\mathrm{crit}} = \frac{\pi}{\sqrt{2}}$$

For rigid boundary conditions only numerical results are available.[25,26] The behavior of the critical Rayleigh numbers is quite similar to that given by (8) and (10). The main difference is that k^{crit} is now a function of $\mathbb{S}$. It was shown[17,25,26] that $k^{\text{crit}} \to 0$ for small positive values of $\mathbb{S}$. That means that the critical wavelength goes to infinity. We expect thus only one convection cell, which cannot markedly affect the total heat flux. For $\mathbb{S} < 0$, k^{crit} grows slowly with $|\mathbb{S}|$, thus the number of cells increases.

III. NONLINEAR ANALYSIS: THE CASE OF FREE BOUNDARIES

A. The Nonlinear Equations

Let us start with equations (1) to (4) and

$$F_i = -g\vec{1}_z$$

where $\vec{1}_z$ is a unit vector in the z direction. For two-dimensional flows we define a stream function ψ such that

$$u_i = \begin{pmatrix} u_z \\ u_x \end{pmatrix} = \begin{pmatrix} -\partial\psi/\partial x \\ +\partial\psi/\partial z \end{pmatrix} \tag{11}$$

The continuity equations is thus satisfied. Conservation of the individual species becomes

$$\frac{\partial N_1}{\partial t} = \frac{\partial(\psi, N_1)}{\partial(x,z)} + D\nabla^2 N_1 + D' N_1^{in} \nabla^2 T \tag{12}$$

with the notation

$$\frac{\partial(A,B)}{\partial(a,b)} = \frac{\partial A}{\partial a}\cdot\frac{\partial B}{\partial b} - \frac{\partial A}{\partial b}\cdot\frac{\partial B}{\partial a}$$

$$\nabla^2 = \frac{\partial^2}{\partial x^2} + \frac{\partial^2}{\partial z^2}$$

The energy equation is now

$$\frac{\partial T}{\partial t}=\frac{\partial(\psi,T)}{\partial(x,z)}+\kappa\nabla^2 T \tag{13}$$

and with the pressure eliminated we get for the momentum equation

$$\frac{\partial}{\partial t}\nabla^2\psi=\frac{\partial(\psi,\nabla^2\psi)}{\partial(x,z)}-\frac{g}{\rho_0}\frac{\partial\rho}{\partial x}+\nu\nabla^2(\nabla^2\psi) \tag{14}$$

We now introduce dimensionless variables

$$\begin{aligned} t^* &= \frac{t\nu}{d^2} \\ x_i^* &= \frac{x_i}{d} \\ u_i^* &= \frac{u_i d}{\nu} \\ N_1^* &= \frac{N_1}{N_1^{in}} \\ \psi^* &= \frac{\psi}{\nu} \\ T^* &= \frac{T-T_0+\Delta T}{\Delta T} \end{aligned} \tag{15}$$

The dimensionless form of (12 to (14) are easily written down. Decomposition (6) is then introduced in the new equations, the steady state being characterized by

$$\begin{aligned} \overline{N_1^*(z^*)} &= 1+\mathcal{S}(z^*-\tfrac{1}{2}) \\ \overline{T^*(z^*)} &= 1-z^* \end{aligned} \tag{16}$$

The final equations to be integrated are (all the * denoting dimensionless variables are omitted):

$$Pr\frac{\partial}{\partial t}(\nabla^2\psi) = Pr\frac{\partial(\psi,\nabla^2\psi)}{\partial(x,z)} - Ra\,\frac{\partial T}{\partial x} + R_{Th}\,\frac{\partial n}{\partial x} + Pr\nabla^2(\nabla^2\psi) \qquad (17)$$

$$Sc\frac{\partial n}{\partial t} = Sc\left[\frac{\partial(\psi,n)}{\partial(x,z)} + \mathcal{S}\,\frac{\partial\psi}{\partial x}\right] + \nabla^2 n + \mathcal{S}\,\nabla^2 T \qquad (18)$$

$$Pr\frac{\partial\vartheta}{\partial t} = Pr\left[\frac{\partial(\psi,\vartheta)}{\partial(x,z)} - \frac{\partial\psi}{\partial x}\right] + \nabla^2\vartheta \qquad (19)$$

We have now to specify a set of boundary conditions. We have in a first step adopted free boundaries

$$\vartheta = 0 \qquad \text{for} \qquad z = 0, 1$$

$$\frac{\partial\vartheta}{\partial x} = 0 \quad \text{for} \quad x = 0, \frac{L}{d}\,\text{(aspect ratio)}$$

$$\frac{\partial\psi}{\partial z} = \frac{\partial^2}{\partial x^2}\left(\frac{\partial\psi}{\partial z}\right) = 0 \qquad \text{for } x = 0, \frac{L}{d}$$

$$\frac{\partial\psi}{\partial x} = \frac{\partial^2}{\partial z^2}\left(\frac{\partial\psi}{\partial x}\right) = 0 \qquad \text{for} \qquad z = 0, 1$$

$$n = \frac{\partial^2 n}{\partial z^2} = 0 \qquad \text{for } z = 0; 1$$

In fact the boundaries are free and permeable. The general method of solution consists of expanding ψ, ϑ, and n in Fourier components with time-dependent coefficients and following the time evolution of each coefficient by direct numerical integration.

B. Analysis with Five Fourier Coefficients

The main advantage of a seriously truncated Fourier representation is that it allows an analytical treatment of the equations, which in turn provides a guide for further numerical experimentation.

Once convection has set in, the mean temperature and concentration field are distorted by the convective motions. Veronis[8,30] has discussed the minimal representation which takes account of the finite amplitude motion. Following the arguments of Veronis, we have adopted a severe truncated representation

$$\begin{aligned}\psi(x,z,t)&=A(t)\sin\pi rx\cdot\sin\pi z\\ \vartheta(x,z,t)&=B(t)\cos\pi rx\cdot\sin\pi z+C(t)\sin 2\pi z\\ n(x,z,t)&=D(t)\cos\pi rx\cdot\sin\pi z+E(t)\sin 2\pi z\end{aligned} \tag{20}$$

where r is a wavelength in the horizontal direction. This representation is substituted into the nonlinear equations (17) to (19) and the time evolution of each Fourier coefficient is obtained using a Galerkin procedure. The following set of ordinary nonlinear and coupled differential equations is obtained:

$$-\pi^2(r^2+1)Pr\dot{A}=Ra\,\pi rB-R_{Th}\pi rD+\pi^4(r^2+1)^2A \tag{21}$$

$$Pr\dot{B}=-Pr\,\pi^2 rAC-Pr\,\pi rA-\pi^2(r^2+1)B \tag{22}$$

$$Pr\dot{C}=-4\pi^2C+Pr\,\pi^2 r\frac{AB}{2} \tag{23}$$

$$Sc\dot{D}=-Sc\,\pi^2 rAE+Sc\,\mathbb{S}\,\pi rA-\pi^2(r^2+1)D+\mathbb{S}\,\pi^2(r^2+1)B \tag{24}$$

$$Sc\dot{E}=-4\pi^2E-4\pi^2\mathbb{S}\,C+Sc\,\pi^2 r\frac{AD}{2} \tag{25}$$

At the steady state ($\dot{A}=\dot{B}=\cdots=0$) the algebraic equations are solved for one of the Fourier coefficients, for example, A; the following equation is found

$$\begin{aligned}A\cdot\Bigg\{&\left(\frac{A^2}{8}\right)^2 Sc^2Pr^2\pi^4(r^2+1)^2r^4\\ &+\left(\frac{A^2}{8}\right)\left[-r^4ScRa+\pi^4(r^2+1)^3r^2(Pr^2+Sc^2)\right]\\ &+\left[-(r^2+1)r^2Ra-R_{Th}\mathbb{S}(r^2+1)r^2\left(\frac{Sc+Pr}{Pr}\right)+\pi^4(r^2+1)^4\right]\Bigg\}=0\end{aligned} \tag{26}$$

The solution $A = 0$ corresponds to the state of rest and real values of A correspond to finite amplitude convection. Real values of A means real *positive* values of $(A^2/8)$ and we have to study the roots of an equation of the second degree in $(A^2/8)$. Once A is known, the other Fourier coefficients are evaluated

$$
\begin{aligned}
B &= \frac{-rPrA}{\pi(r^2+1) + \pi r^2 Pr^2 (A^2/8)} \\
C &= \tfrac{1}{8} Pr \cdot rAB \\
D &= \frac{Ra\, rB + \pi^3 (r^2+1)^2 PrA}{rR_{Th}} \\
E &= \frac{Sc}{8} rAD - \mathfrak{S} C
\end{aligned}
\tag{27}
$$

The Nusselt number is defined as the ratio of the total heat flux to the conductive heat flux. Usually the Nusselt number is averaged horizontally and as it may vary with Z, in time-dependent processes, we decided to take its value at $z = 0$. We have found

$$
\begin{aligned}
\langle Nu \rangle_{z=0} &= 1 - 2\pi C \\
&= 1 + 2\frac{Pr^2(A^2/8)}{3 + Pr^2(A^2/8)}
\end{aligned}
\tag{28}
$$

where we have substituted r by its critical value

$$
k^{\text{crit}} = \pi r^{\text{crit}} = \frac{\pi}{\sqrt{2}} \tag{29}
$$

It is convenient to rearrange (26)

$$
A \cdot \left\{ \left(\frac{A^2}{8} \right)^2 PrSc^2 Ra^{(0)} - \left(\frac{A^2}{8} \right) 3Sc^2 \left[Ra - Ra^{(0)} \left(1 + \frac{Pr^2}{Sc^2} \right) \right] - 9\left(Ra - Ra^{(\text{crit})}_{(\text{ex})} \right) \right\} = 0 \tag{30}
$$

where $Ra^{(0)}$ stands for $27\pi^4/4$. This is possible thanks to (29) and (8). The solution of (30) is

$$\frac{A^2}{8}=\frac{3}{2}\frac{1}{Pr^2}\left[\frac{Ra}{Ra^{(0)}}-\left(1+\frac{Pr^2}{Sc^2}\right)\right]$$

$$\pm\left\{\frac{9}{4}\frac{1}{Pr^4}\left[\frac{Ra}{Ra^{(0)}}-\left(1+\frac{Pr^2}{Sc^2}\right)\right]^2+\left(\frac{Ra-Ra^{(\text{crit})}_{(\text{ex})}}{Ra^{(0)}}\right)\frac{9}{Sc^2Pr^2}\right\}^{1/2} \tag{31}$$

We see immediately that for $\mathbb{S}>0$ no finite amplitude instability exists below the critical point. Indeed, in that case $Ra^{(\text{crit})}_{(\text{ex})}<Ra^{(0)}$ and as we are interested in Rayleigh numbers such that $Ra<Ra^{(\text{crit})}_{(\text{ex})}$, we have

$$\frac{Ra}{Ra^{(0)}}<1$$

and thus the first term in (31) is negative. Even with a plus sign in front of the square root, positive values of $A^2/8$ cannot be obtained in the range

$$Ra<Ra^{(\text{crit})}_{(\text{ex})}<Ra^{(0)}$$

However the situation is completely different if $\mathbb{S}<0$ because $Ra^{(\text{crit})} > Ra^{(0)}$. In the range

$$Ra^{(0)}<Ra<Ra^{(\text{crit})}_{(\text{over})}$$

where $Ra^{(\text{crit})}_{(\text{over})}$ is itself smaller than $Ra^{(\text{crit})}_{(\text{ex})}$, positive values of $A^2/8$ are indeed possible. From (28) and (31), the Nusselt number has been computed for $Pr=10$, $Sc=10^3$, $R_{Th}=4.10^4$, $\mathbb{S}=10^{-2}$ (thus $Ra^{(\text{crit})}_{(\text{over})}=1028.3$) and for various values of Ra. The results are shown in Fig. 1.

Curve (a) corresponds to $Nu=1$, or to $A=0$, thus to the state of rest.

Curve (b) corresponds to finite amplitude motion, thus to the two real positive values of $A^2/8$ [see (31)] when they exist. In fact curve (b) cuts curve (a) at $Ra=Ra^{(\text{crit})}_{(\text{ex})}$. This is thus a bifurcating point. There is a striking analogy with recent results of Nicolis and Auchmuty for instability with respect to diffusion in a chemical reation.[31] The minimum of curve (b) on the Ra-axis is called $Ra_{\text{f.a.}}$.

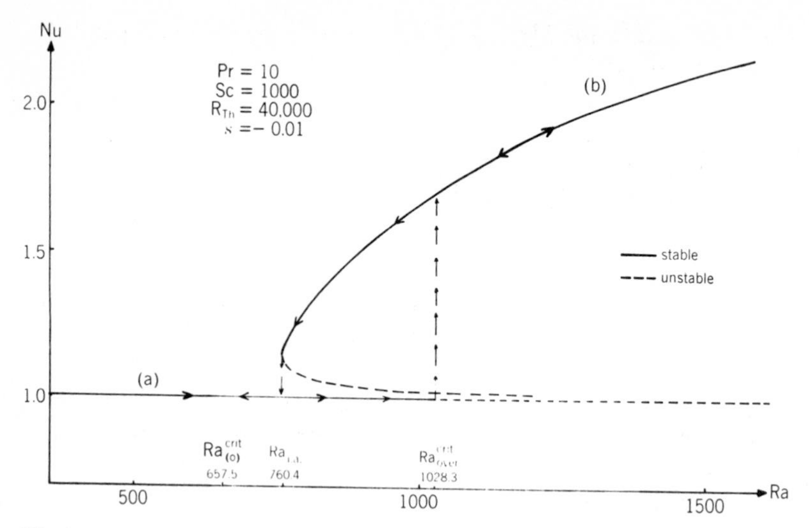

Fig. 1.

If $Ra < Ra_{\text{f.a.}}$ only one steady state is possible, the state of rest, and this state is stable against all perturbations. If $Ra > Ra_{\text{f.a.}}$, three steady states are possible and we have thus to investigate the stability of each steady state. Equations (21) to (25) are linearized in the disturbances and we write

$$A = A^{\text{st}} + a$$

$$B = B^{\text{st}} + b$$

etc.

where A^{st}, $B^{\text{st}}\ldots$ are the possible steady states deduced from (30) or (31) and (27). This is a standard matter: we have to find the eigenvalues of a 5×5 real matrix.

The stability of the state of rest was already studied previously and again we find that this state becomes unstable if $Ra > Ra^{\text{crit}} = 1028.3$. The associated eigenvalue is complex and overstability prevails.

The lower branch of curve (b) with a negative slope is always unstable. On the contrary, the upper branch is stable. Thus we see that in the range

$$Ra_{\text{f.a.}} < Ra < Ra^{\text{(crit)}}_{\text{(over)}}$$

two stable steady states are possible and the state reached by the system depends on the initial conditions. Thus a hysteresis loop is possible and is described in more details in Section III.E.

$Ra_{\text{f.a.}}$ is the smallest value of the Rayleigh number such that finite amplitude instability exists ($A^2/8$ is real and positive). In other words, $Ra_{\text{f.a.}}$ is the value of Ra which makes the quantity under the square root sign in (31) equal to zero. Thus we have

$$\frac{9}{4}\frac{1}{Pr^4}\left[\frac{Ra_{\text{f.a.}}}{Ra^{(0)}}-\left(1+\frac{Pr^2}{Sc^2}\right)\right]^2+\left(\frac{Ra_{\text{f.a.}}-Ra_{(\text{ex})}^{(\text{crit})}}{Ra^{(0)}}\right)\cdot\frac{9}{Sc^2Pr^2}=0 \quad (32)$$

The variation of $Ra_{\text{f.a.}}$ with $\mathbb{S}$ can be given thanks to (8). By combining (32) and (8) we have

$$\frac{Ra_{\text{f.a.}}}{Ra^{(0)}}=1-\left(\frac{Pr}{Sc}\right)^2+2\frac{Pr}{Sc}\cdot\left\{-\frac{R_{Th}\cdot\mathbb{S}}{Ra^{(0)}}\left[1+\left(\frac{Pr}{Sc}\right)^{-1}\right]\right\}^{1/2} \quad (33)$$

This last equation shows clearly that finite amplitude convection cannot exist for positive Soret numbers.

Table I gives the three Rayleigh numbers of interest [$Ra_{(\text{ex})}^{(\text{crit})}$; $Ra_{(\text{over})}^{(\text{crit})}$; $Ra_{\text{f.a.}}$] for various negative $\mathbb{S}$.

TABLE I.
Variation of the critical Rayleigh numbers with the Soret number

$\mathbb{S}$	$Ra_{\text{ex}}^{\text{crit}}$	$Ra_{\text{over}}^{\text{crit}}$	$Ra_{\text{f.a.}}$
-10^{-7}	657.9	664.7	657.8
-10^{-6}	661.5	664.8	658.5
-10^{-5}	697.9	665.1	660.7
-10^{-4}	1061.5	668.4	667.8
-10^{-3}	4697.5	701.1	690.0
-5×10^{-3}	20857.5	846.6	730.3
-10^{-2}	41057.5	1028.4	760.5
-2×10^{-2}	81457.5	1392.0	803.2
-4×10^{-2}	162257.5	2119.3	863.6
-6×10^{-2}	243057.5	2846.6	909.6
-8×10^{-2}	323857.5	3573.8	949.0
-10^{-1}	404657.5	4301.1	983.4

We have proven in this paragraph that convective motions may exist below the critical point given by the linear theory, for finite amplitude perturbations. In some range of the Rayleigh number, two stable steady states may exist and a hysteresis loop in the *Nu*–*Ra* plane is possible. Of course, with such a severe truncated Fourier development we cannot expect "quantitative results." On the other hand the term "quantitative results" is more or less nonsense for the case of two free boundaries, where experimental results are missing.

However the results with five coefficients can be used as a guide in further numerical experiments.

Finally we have not studied possible oscillations with five coefficients. This is left for the next section.

C. Analysis with 16 Fourier Coefficients

In fact we expect expansion (20) to be valid only for small $(Ra - Ra^{\text{crit}})$. A complete representation is

$$\psi = \sum_{m=1}^{M} \sum_{n=1}^{N} A_{mn}(t) \sin\left(m\pi \frac{d}{L} x\right) \sin(n\pi z)$$

$$\vartheta = \sum_{m=0}^{M} \sum_{n=1}^{N} B_{mn}(t) \cos\left(m\pi \frac{d}{L} x\right) \sin(n\pi z) \tag{34}$$

$$n = \sum_{m=0}^{M} \sum_{n=1}^{N} C_{mn}(t) \cos\left(m\pi \frac{d}{L} x\right) \sin(n\pi z)$$

We give in this section a summary of the results that we have obtained[32] with $M = N = 2$, that is, 4 Fourier coefficients for ψ, 6 for θ or n, thus 16 coefficients in total. The development (34) is substituted into equations (17) to (19). Equation (17) is then multiplied by $\sin(p\pi(d/L)x)\sin(q\pi z)$, and (18) and (19) by $\cos(p\pi(d/L)x)\cdot\sin(q\pi z)$. Integration is performed on x from 0 to L/d and on z from 0 to 1. The orthogonality conditions are then used. In this way the time evolution of each Fourier coefficient is obtained

$$Pr\frac{d}{dt}A_{pq}\cdot\pi^2(p^2r^2+q^2)=-PrA_{pq}\pi^4(p^2r^2+q^2)^2$$

$$-\pi rp(RaB_{pq}-R_{Th}C_{pq})$$

$$-Pr\cdot 4\pi^2r\sum_{m=1}^{M}\sum_{n=1}^{N}\sum_{k=1}^{M}\sum_{l=1}^{N}A_{mn}\cdot A_{kl}(k^2r^2+l^2)$$

$$\times[nkI_{mpk}^{ssc}I_{lqn}^{ssc}-ml\,I_{kpn}^{ssc}I_{nql}^{ssc}] \tag{35}$$

$$(p=1,2,3,\ldots,M \qquad q=1,2,3,\ldots,N)$$

$$Pr\frac{d}{dt}B_{pq}=-B_{pq}\pi^2(p^2r^2+q^2)-Pr\cdot\pi rpA_{pq}$$

$$+(1-\tfrac{1}{2}\delta_{0,p})Pr\cdot 4r\cdot\sum_{m=1}^{M}\sum_{n=1}^{N}\sum_{k=0}^{M}\sum_{l=1}^{N}A_{mn}B_{kl}$$

$$\times[nkI_{mkp}^{ssc}I_{lqn}^{ssc}+ml\,I_{mkp}^{ccc}I_{nql}^{ssc}] \tag{36}$$

$$(p=0,1,2,3,\ldots,M \qquad q=1,2,3,\ldots,N)$$

$$Sc\frac{d}{dt}C_{pq}=-\pi^2(p^2r^2+q^2)(C_{pq}+\mathbb{S}\,B_{pq})+Sc\cdot\mathbb{S}\,\pi rp\cdot A_{pq}$$

$$+(1-\tfrac{1}{2}\delta_{0,p})Sc\cdot 4r\cdot\sum_{m=1}^{M}\sum_{n=1}^{N}\sum_{k=0}^{M}\sum_{l=1}^{N}A_{mn}\cdot C_{kl}$$

$$\times[nkI_{mkp}^{ssc}I_{lqn}^{ssc}+ml\,I_{mkp}^{ccc}I_{nql}^{ssc}] \tag{37}$$

$$(p=0,1,2,3,\ldots,M \qquad q=1,2,3,\ldots,N)$$

with

$$r=\frac{d}{L}$$

and

$$I_{ijk}^{ccc}=\int_0^{\pi}\cos ix\cos jx\cos kx\,dx$$

$$=\frac{\pi}{4}\quad \text{if}\quad k=|i-j|\quad \text{or}\quad k=i+j$$

$$=\frac{\pi}{2}\quad \text{if}\quad k=0\quad \text{and}\quad i=j$$

$$=0\quad \text{otherwise}$$

$$I_{ijk}^{ssc}=\int_0^{\pi}\sin ix\sin jx\cos kx\,dx$$

$$=+\frac{\pi}{4}\quad \text{if}\quad k=|i-j|$$

$$=-\frac{\pi}{4}\quad \text{if}\quad k=i+j$$

$$=\frac{\pi}{2}\quad \text{if}\quad k=0\quad \text{and}\quad i=j$$

$$=0\quad \text{otherwise}$$

In order to proceed with numerical integration of (35) to (37), initial conditions must be specified. Generally the initial state is the state of rest ($A_{pq}=B_{pq}=C_{pp}=0$) with an arbitrarily imposed perturbation on ϑ. (The four first coefficients B_{pq} were set equal to 10^{-6} and sometimes to -10^{-6}.) An account of the complete numerical work that we have performed will be published in the future.[32] We give in this section only a few examples, but treat them in more detail than in the original publication. Let us start immediately with the most illustrative example:

$$R_{Th}=40{,}000;\qquad Pr=10;\qquad Sc=1000;\qquad \mathfrak{S}=-10^{-2};\qquad Ra=1100$$

For this set of parameters, the critical Rayleigh number is $Ra_{\text{(over)}}^{\text{(crit)}}=1028.35$. Moreover, overstability prevails and instability should arise as oscillations of increasing amplitude with a period given by [cf. (10)]:

$$\mathbf{T}=\frac{2\pi}{\sigma_I}$$

thus for this example $\mathbf{T}=5.71$.

Figure 2 shows the time evolution of Nu for $20 < t < 40$ for a very small initial perturbation. This is a typical example of overstability. The same behavior is seen for $t < 160$, the only difference being the maximum value of the Nusselt number during each time interval. The period of oscillations of the Nusselt number, called $\mathbf{T}_{Nu}$ is

$$\mathbf{T}_{Nu} = 2.84$$

in reduced units.

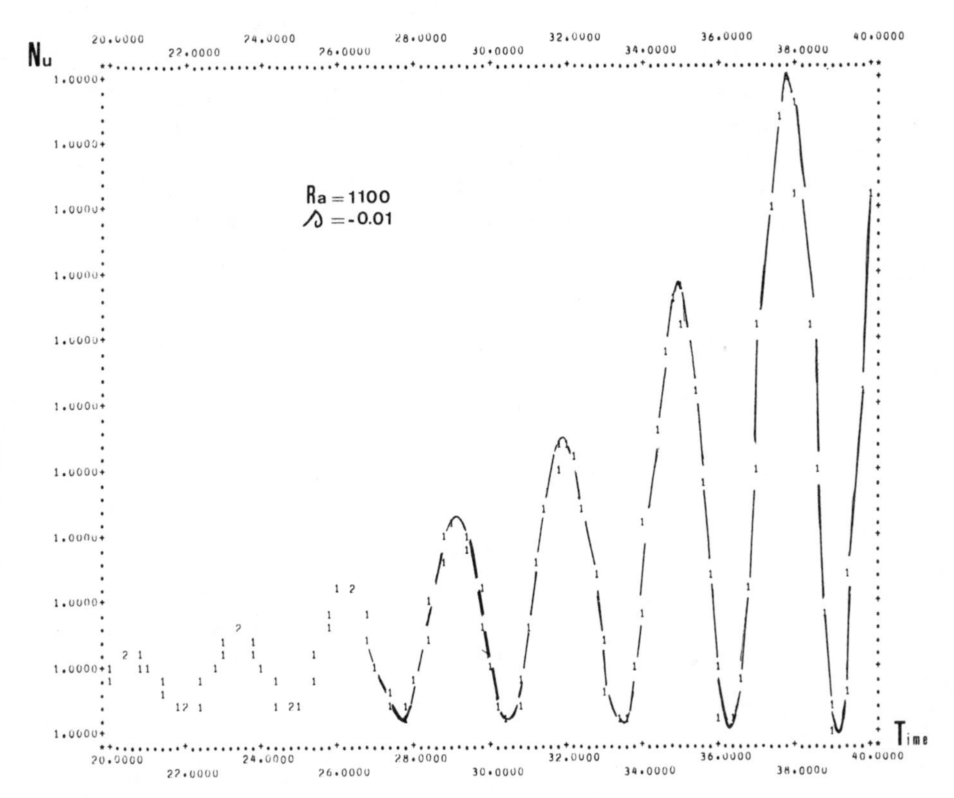

Fig. 2.

The stream function ψ at every point $x_i z_i$, oscillates with a period twice that of Nu. Thus $\mathbf{T}_\psi = 5.68$, which agrees with the linear theory. This is followed without difficulty because there is only one Fourier coefficient of ψ (namely, $A_{1,1}$) which is amplified, and this particular coefficient oscillates from positive to negative values. It is easily understood that either a maximum or a minimum (maximum in modulus) in ψ corresponds to a maximum in the heat transfer (in Nu). When ψ is equal to zero, the Nusselt number becomes equal to unity (there is a small difference in phase probably due to numerical integration).

When $t > 160$, the nonlinear contributions become important and there is a modification in the behavior of $Nu(t)$. There is a change in the period and simultaneously the mean value of the Nusselt number no longer increases exponentially, but much more slowly (Fig. 3). Figures 4 and 5 show $Nu(t)$ for $250 < t < 270$. The main point is that the two minima of Nu at $t \cong 254$ and 258 are of the order of 1, corresponding to a change of sign in the stream function, but during the next oscillation the Nusselt number drops to a value close to 1.16 at $t \cong 262$. The next minima at $t \cong 266$ and 270 increase too, whereas the successive maxima decrease. At the same time the stream function no longer changes sign and now oscillates with exactly the same period as the Nusselt number. There is a kind of acceleration followed by a retardation of the convective motion.

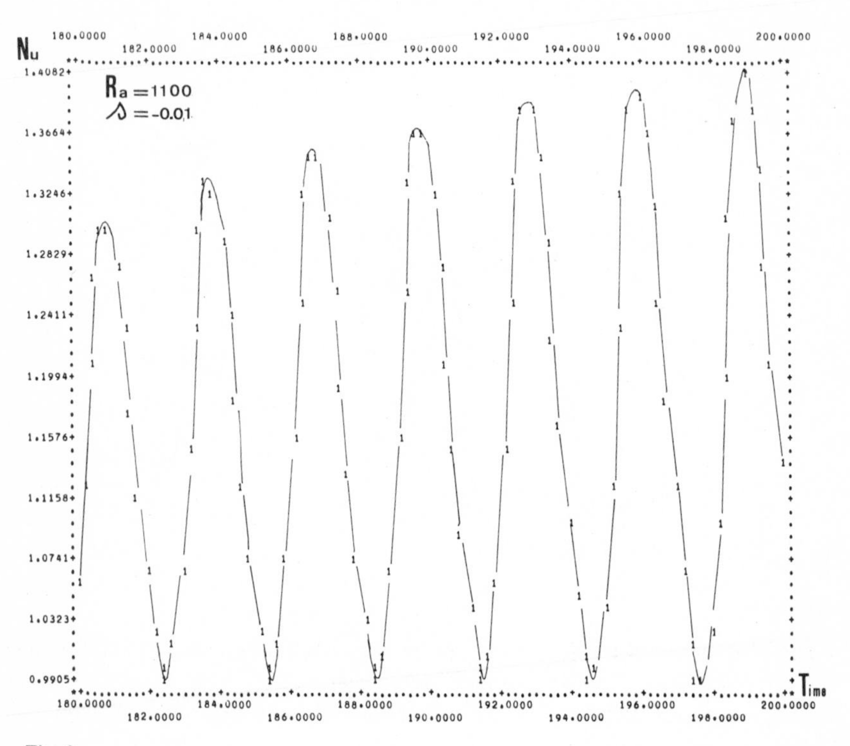

Fig. 3.

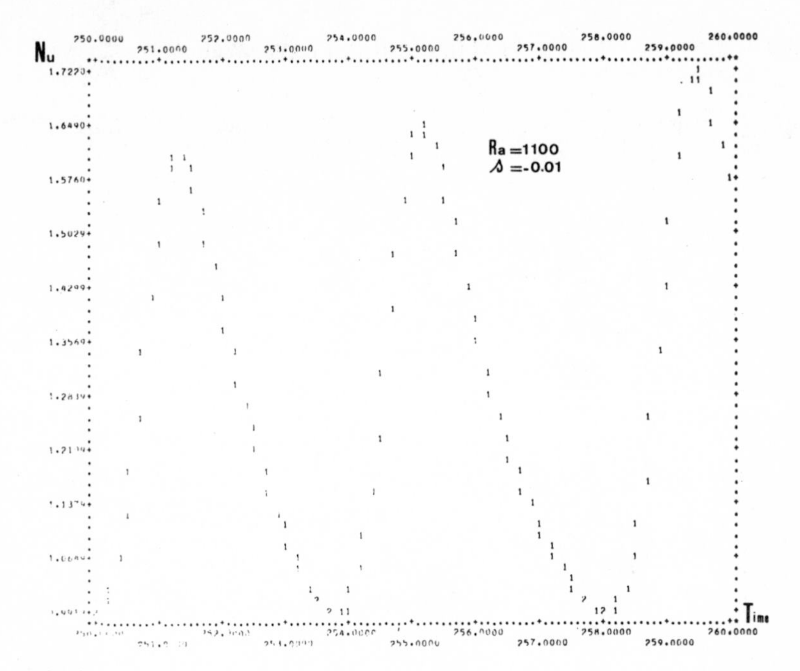

Fig. 4.

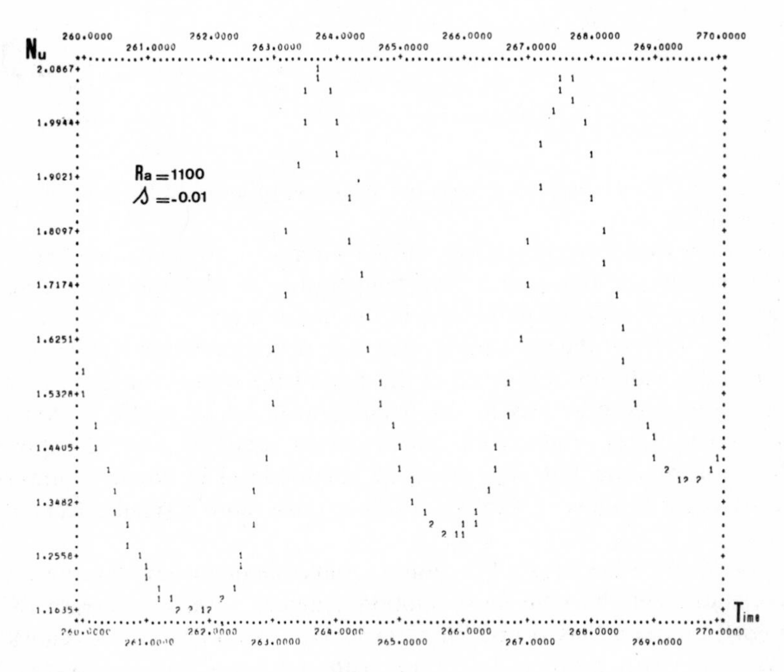

Fig. 5.

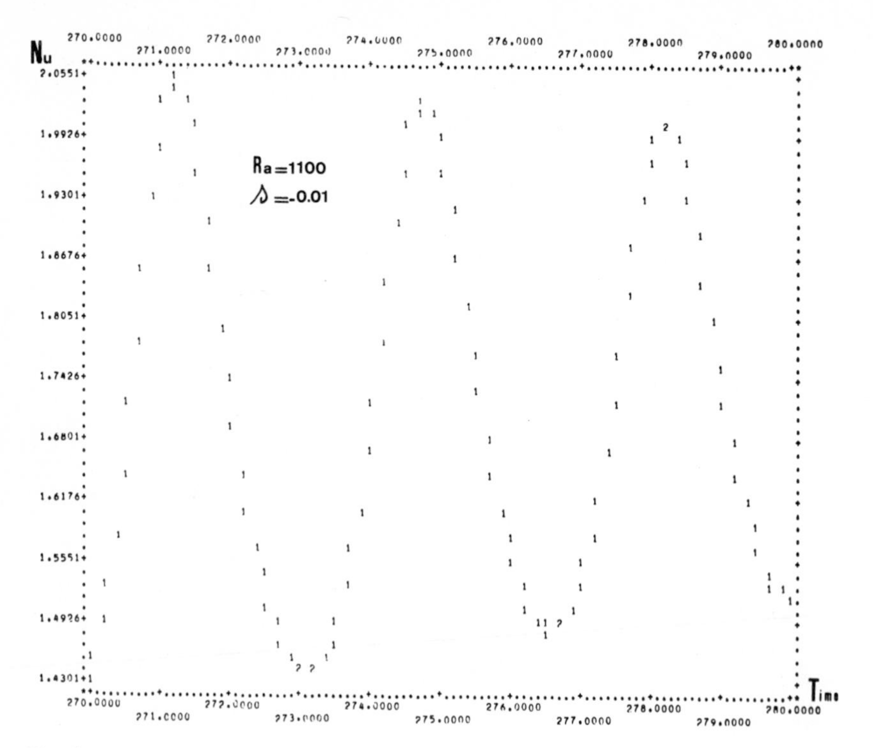

Fig. 6.

For $t > 270$, oscillations persist but decrease in amplitude. An example is given in Fig. 6.

At each time step, possessing all the Fourier coefficients, we can compute the stream function ψ, the temperature $\mathbf{T}$ and the mass fraction distribution N_1 at each point x_i, z_i in the liquid layer.

Figure 7 shows the streamlines (iso $-\ \psi$) at three different reduced times and clearly indicates a reversal in the sign of the stream function. Part (*a*) corresponds more or less to the maximum of Nu at $t = 255.2$, that is, a maximum in $|\psi|$, and to an anticlockwise rotation. Part (*b*) shows a clockwise rotation but very small in amplitude. The Nusselt number is nearly equal to unity ($t = 257.6$). There is once more a small difference in phase.

For $t > 257.6$ the stream function no longer changes sign, but there is an acceleration of the convective motion together with an increase of the Nusselt number. Part (*c*) of Fig. 7 ($t = 259.1$; $Nu \cong 1.6$) indeed shows the same clockwise rotation this time with an important amplitude. The successive extremum values of ψ are listed in Table II.

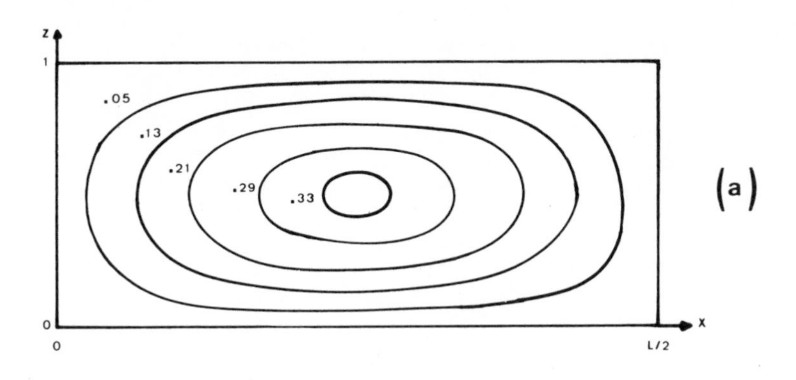
z
1
.05
.13
.21
.29
.33
0
0
L/2
x
(a)

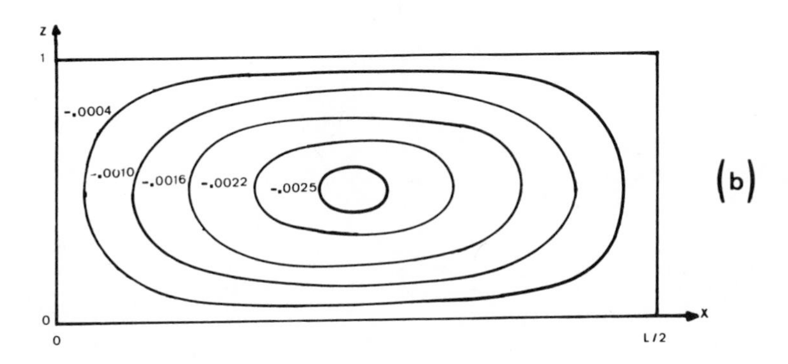
z
1
-.0004
-.0010
-.0016
-.0022
-.0025
0
0
L/2
x
(b)

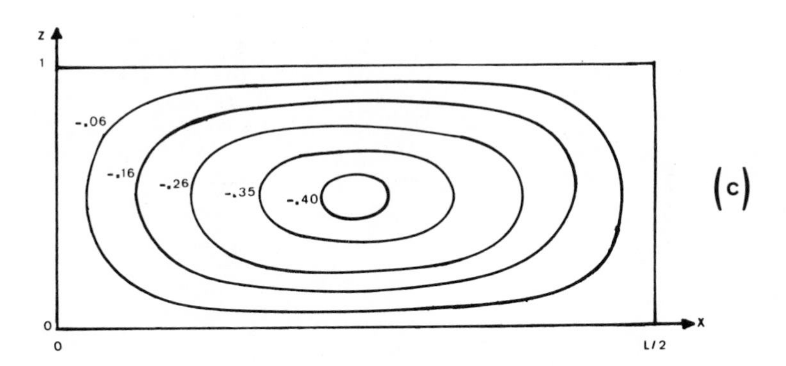
z
1
-.06
-.16
-.26
-.35
-.40
0
0
L/2
x
(c)

Fig. 7.

TABLE II
Relation between the nature of the extrema of ψ and Nu at $Ra = 1100$ and $\mathbb{S} = -0.01$

t	Extremum of ψ	Nature of the extremum of Nu
255.3	$+0.3398$ max	Max
257.6	$-0.0026 \cong 0$	Min
259.1	-0.4158 min	Max
261.5	-0.1330 max	Min
263.5	-0.5412 min	Max
265.5	-0.1932 max	Min
267.4	-0.5292 min	Max
269.3	-0.2278 max	Min

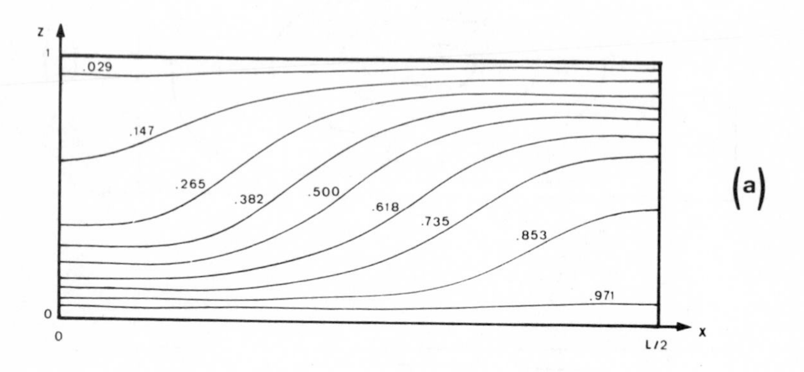

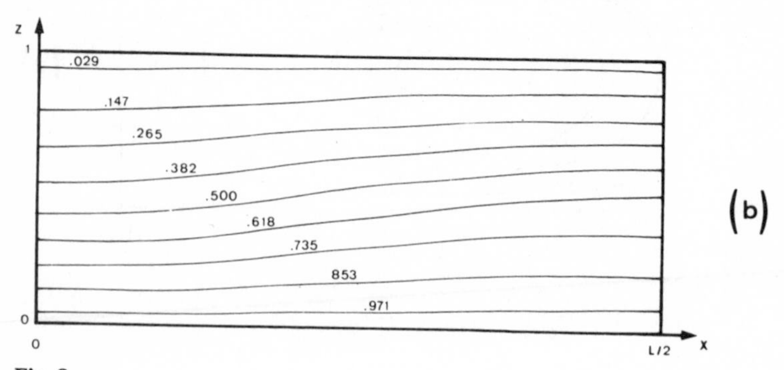

Fig. 8.

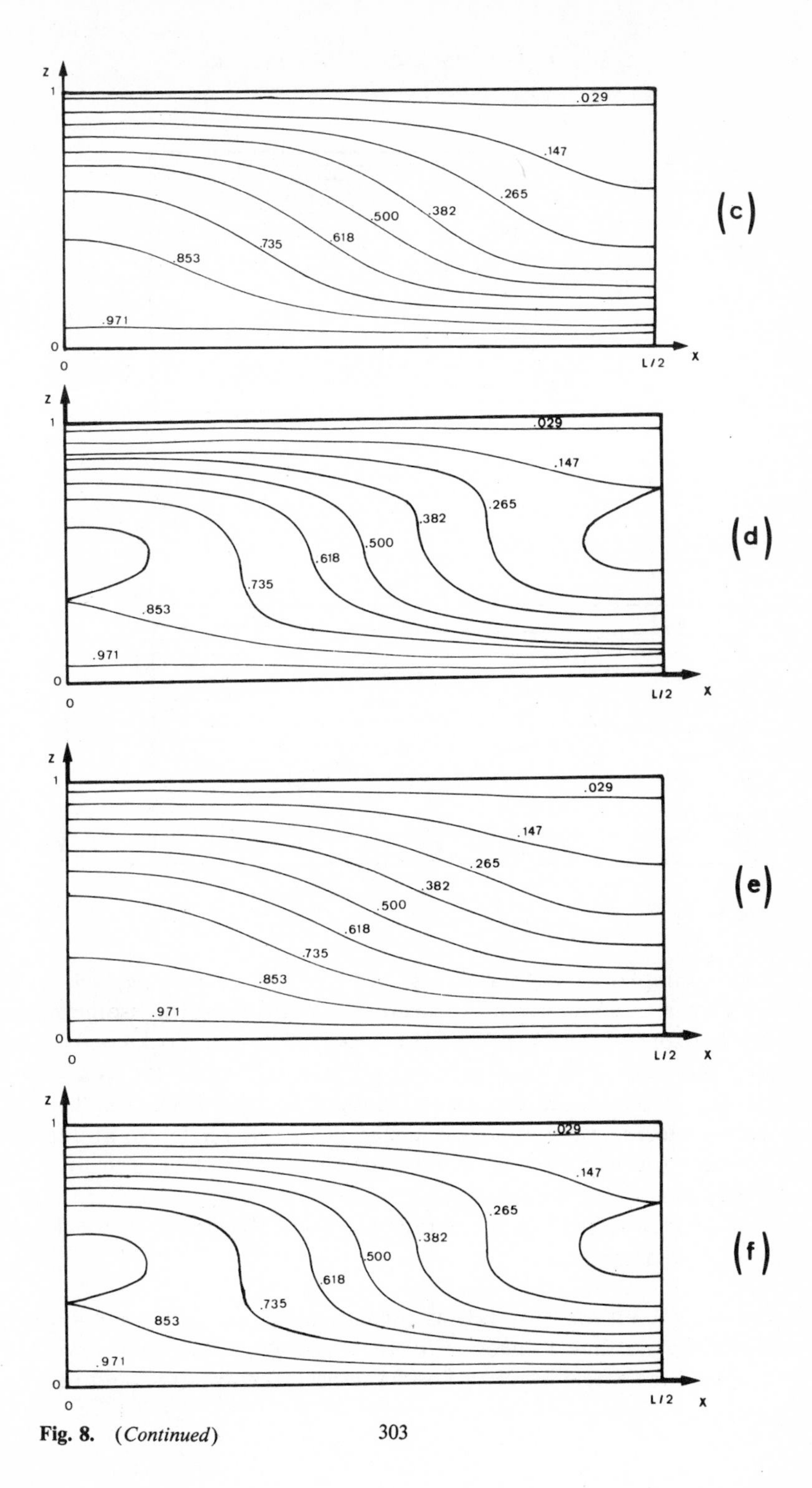

Fig. 8. (*Continued*)

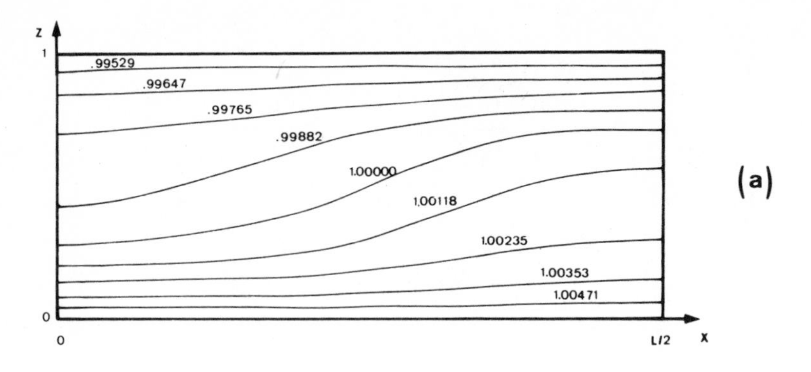

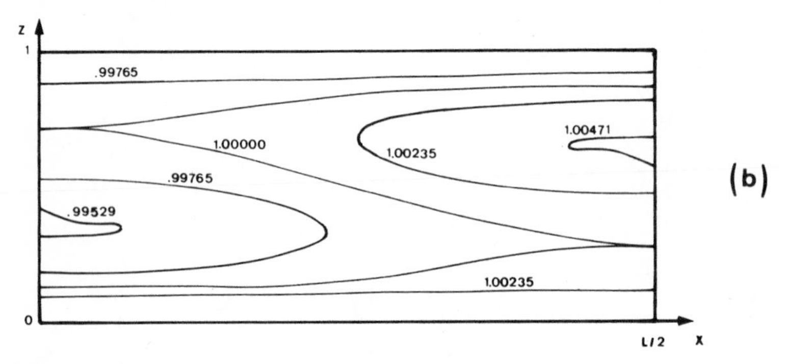

Fig. 9.

Some examples of isotherms are given in Fig. 8. Part (*a*) shows the anticlockwise rotation; part (*b*) is nearly the state of rest (the isotherms are nearly horizontal); parts (*c*) to (*f*) correspond to the clockwise rotation and to the successive accelerations and retardations of the convection. When the amplitude of the stream function increases (in modulus), the isotherms become more and more distorted, clearly related to an increase of the heat transfer, that is, of the Nusselt number. Figure 9 gives two examples of isotiters. In Fig. 9*a*, the isotiters are slightly distorted due to slow motions, whereas Fig. 9*b* corresponds to heavy mixing. In conclusion, there are two domains

1. For $t < 160$, instability arises as oscillations of increasing amplitude as predicted by the linear theory. In other computer experiments we have investigated the whole range of Soret numbers, both positive and negative

and near the neutral stability curve, and the predictions of the linear analysis were always confirmed for "small t": for $\mathbb{S} < 0$, the period of oscillations confirms, in our nonlinear analysis, the value of the linear theory ($\mathbf{T} \div 1/(\mathbb{S})^{1/2}$), whereas for $\mathbb{S} > 0$, a monotonous growth of the initial perturbation is observed according to the principle of exchange of stabilities.

2. $t > 160$ is the pure nonlinear domain where no relationship exists with the prediction of the linear analysis. The lifetime of the linear behavior of the system ($0 < t < 160$ at $Ra = 1100$) is shortened when we increase ($Ra - Ra^{crit}$). For example at $\mathbb{S} = -10^{-2}$ and $Ra = 2630$ the linear behavior is restricted to $t < 5$ (see Fig. 10).

At $Ra = 2630$, different runs were performed at $\mathbb{S} = \pm 10^{-2}$, $\mathbb{S} = \pm 5 \times 10^{-4}$, and $\mathbb{S} = \pm 1.2 \times 10^{-4}$. It is worth mentioning that oscillations are observed both for positive and negative values of the Soret number and

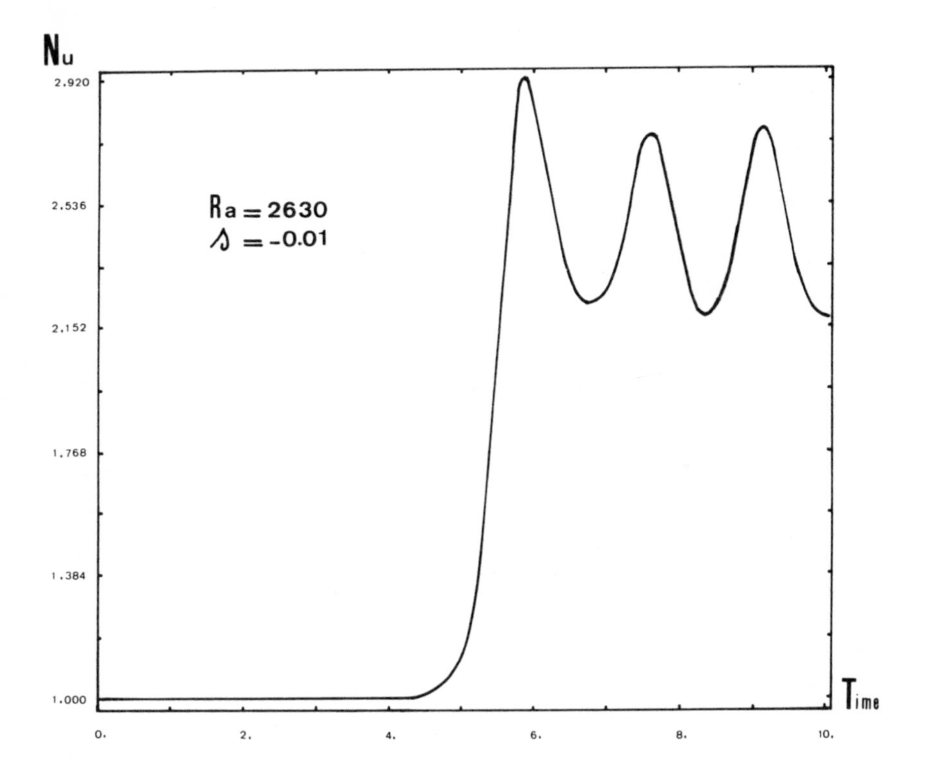

Fig. 10.

that the principle of exchange of stabilities, valid for $\mathbb{S} > 0$, cannot be extrapolated far from the critical Rayleigh number. In each case, oscillations start after a rapid growth of the Nusselt number (see Fig. 10) and immediately decrease in amplitude. The system seems to reach a steady state as $t \to \infty$, but in fact oscillations are always present. Figure 11 shows the behavior of Nu at $\mathbb{S} = -10^{-2}$ and $Ra = 2630$ for $190 < t < 200$. Regular oscillations are still observed but their amplitude is now of the order of 0.003! As already stated the period of oscillation is not predicted by (10) $(\mathbf{T} \div 1/(\mathbb{S})^{1/2})$, but we have observed that $\mathbf{T}$ was independent of $\mathbb{S}$. However the initial amplitude is proportional to $\mathbb{S}$.

As pointed out by Veronis[33] we have observed that in development (34) for ψ, θ, and n, all the contributions A_{nm}, B_{nm}, and C_{nm}, with $(n+m)$ odd are never amplified. This is also true for any value of N and M. Moreover, for $N = M = 2$ (16 coefficients) 3 of the 8 coefficients with $(n+m)$ even, do not contribute to the Nusselt number. Thus only 5 coefficients are ampli-

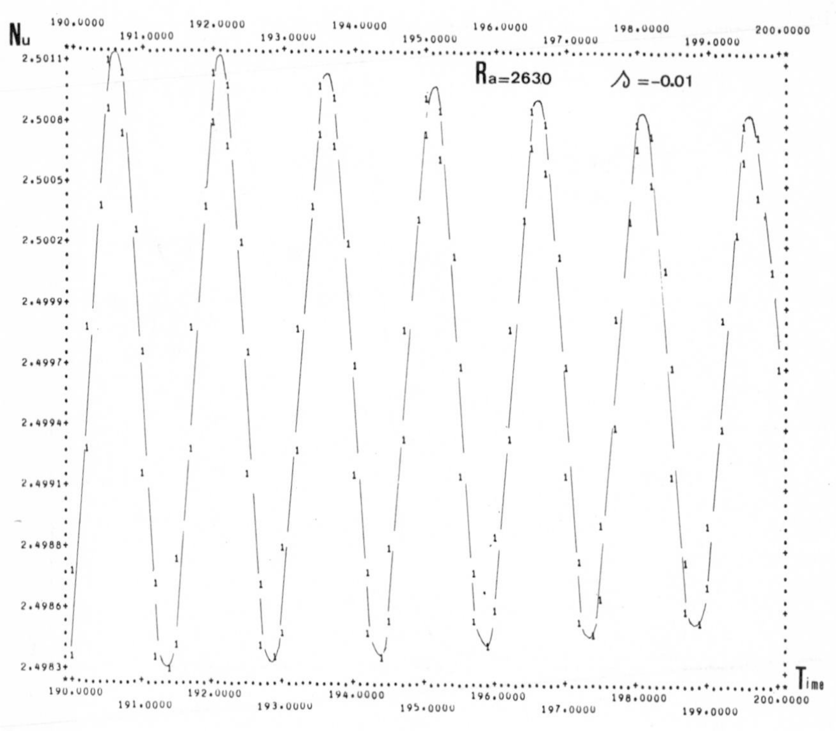

Fig. 11.

fied, and these 5 coefficients are exactly those considered in Section III.B. This shows the validity of the truncated representation (20), which is confirmed in Section III.D where 56 and sometimes 120 coefficients are considered. In other words, the numerical results with 16 coefficients are necessarily identical to those obtained with 5 coefficients. Thus a hysteresis loop must be described with 16 coefficients (see Section III.E).

D. Analysis with 56 Coefficients

Considering the identity of the results between 5 and 16 coefficients, we decided to increase the number of Fourier coefficients and to take $N = M = 4$ (that is, 56 coefficients). First of all, it was verified that the 28 contributions with $(n+m)$ odd are not amplified. In contradistinction with the previous section, *all* the coefficients with $(n+m)$ even *do* contribute to the time evolution of the system. In order to save computer time, the noncontributing terms are neglected. There are two main differences with the lower expansion:

1. The mean value of the Nusselt number is slightly higher than in the former case. In fact this mean value during oscillations at a given Rayleigh number is the same as the steady value of the usual one-component problem, where no oscillations are seen. Moreover, this mean value does not depend on the Soret number and agrees with previous findings[33] (see Table III).

TABLE III $Ra = 2630$

Comparison of the mean value of Nu at different $\mathbb{S}$ and different truncation levels in the Fourier development

$\mathbb{S}$	N	M	$\langle Nu \rangle^a$	Nu from Ref. 33
$+10^{-2}$	2	2	2.50 (os)	—
-10^{-2}	2	2	2.50 (os)	—
$+5\times10^{-4}$	2	2	2.50 (os)	—
-5×10^{-4}	2	2	2.50 (os)	—
$+1.2\times10^{-2}$	2	2	2.50 (os)	—
-1.2×10^{-2}	2	2	2.50 (os)	—
0	2	2	2.50 (st)	2.50
0	4	4	3.04 (st)	3.04
-10^{-2}	4	4	3.04 (os)	—

[a]os: oscillations in $Nu(t)$; st: steady value.

2. The oscillations of Nu with t are quite different from those shown, for example, in Fig. 5 or 6. Indeed beats are observed. The general expression for the Nusselt number is

$$Nu = 1 - \pi \sum_{n}^{N} nB_{0n} \tag{38a}$$

In fact only the contributions with n even must be considered. With 16 coefficients, Nu is given by

$$Nu = 1 - 2\pi B_{02} \tag{38b}$$

whereas with 56 coefficients we have

$$Nu = 1 - 2\pi B_{02} - 4\pi B_{04} \tag{38c}$$

Beats in Nu can be obtained if B_{02} and B_{04} oscillate each with a single but different frequency. In fact this is not the case and beats are observed both in B_{02} and B_{04}. With a higher expansion ($N = M = 6$) involving 120 coefficients (in fact 60 $(n+m)$ even contributions) we have

$$Nu = 1 - 2\pi B_{02} - 4\pi B_{04} - 6\pi B_{06} \tag{38d}$$

In this case too, each of the three contributing B_{0n}, presents the same behavior, their ratios B_{04}/B_{02} and B_{06}/B_{02} being, respectively, $\sim 1/10$ and $\sim 1/100$.

Figure 12 shows for $\mathcal{S} = -10^{-2}$ and $Ra = 1800$ the behavior of $B_{02}(t)$ in equation (38d), and thus of $Nu(t)$, except a scaling on the y-axis. By curve fitting numerical results, we have found that B_{02} obeys the law (within an error less than 1%)

$$-B_{02}(t) = 0.2035 + 0.003\ (\sin \tfrac{7}{6}\pi t + \sin \tfrac{5}{6}\pi t) e^{-0.09t} \tag{39}$$

Two frequencies are thus present and are the same as with 56 coefficients; the only difference is a faster decay of the amplitude in the present case.

E. A Hysteresis Loop

Analytical calculations involving five Fourier coefficients have shown that two stable steady states exist below the critical Rayleigh number, provided that

$$Ra > Ra_{\text{f.a.}}$$

where $Ra_{\text{f.a.}}$ is a function of the Soret number tabulated in Table I. As previously stated the numerical results with 16 coefficients produce for

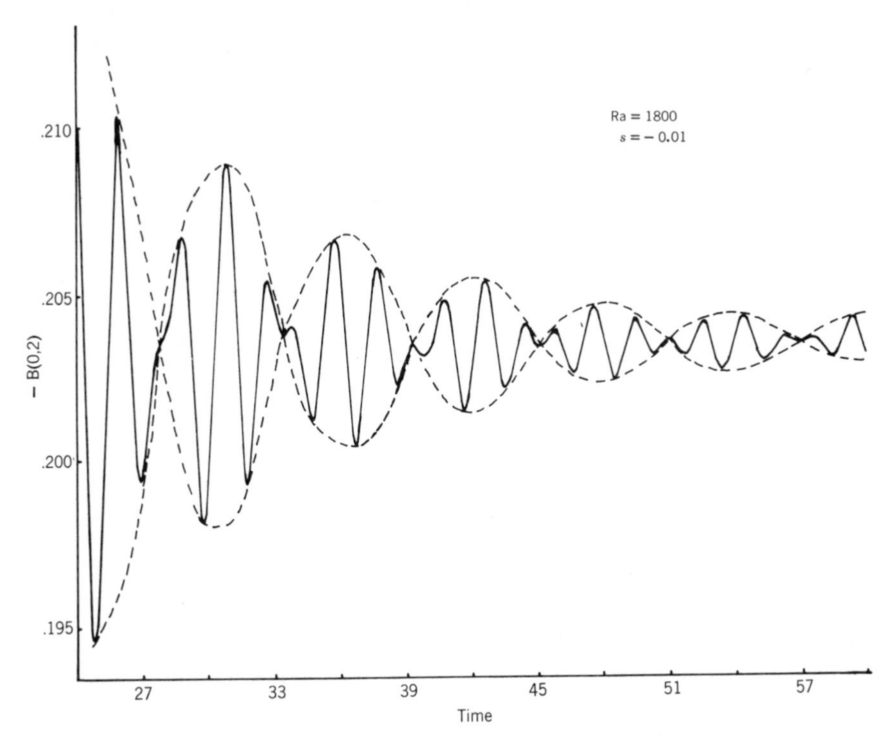

Fig. 12.

large t an identical mean value of Nu, and for $t \to \infty$ an identical steady state. Thus the hysteresis loop shown in Fig. 1 must be exactly reproduced with 16 coefficients. This is verified in numerical experiments, summarized in Table IV. Clearly from Table IV, if $Ra = 1000$, that is, below the critical value 1028.35, the system does not return to the state of rest, if the initial state is a finite amplitude convective state, in our example, the convective state at $Ra = 1100$. From runs 7 and 8 and from a supplementary run at $Ra = 750$ we deduce that

$$750 < Ra_{\text{f.a.}} < 780$$

It is pertinent to question if the existence of the hysteresis loop is not linked to the poorness of too low a Fourier expansion. Numerical experiments were performed with 56 coefficients and are summarized in Table V. The hysteresis loop still exists, with this representation, but we notice that

$$913.35 < Ra_{\text{f.a.}} < 928.35$$

The hysteresis loop with the two expansions is shown in Fig. 13.

The boundary conditions adopted in our model and described in Section III must be modified in a more realistic numerical simulation of laboratory experiments. A comparison between the results of the linear theory for "free" and "rigid" boundaries shows that all the physical information (e.g., overstability) can be obtained from the more tractable analysis with free boundaries. Of course the numerical values of critical numbers are quite different. Moreover, the nonlinear analysis with free boundaries presented in Section III was sufficient to describe the existence of a hysteresis loop, experimentally observed. However a comparison between numerical values of computer simulations and experimental findings requires the incorporation of the realistic rigid boundary conditions in the nonlinear theory. It seems to us that the boundary conditions on the stream function are less important than those on the mass fraction. Indeed thermal diffusion requires that the mass flux vanishes at the boundaries (the boundaries are impermeable to chemical species) and the analytical conditions are

$$\frac{\partial n}{\partial z} + \mathbb{S}\,\frac{\partial \vartheta}{\partial z} = 0 \qquad \text{at} \quad z = 0, 1 \tag{40}$$

instead of

$$\begin{aligned} n &= 0 \\ \frac{\partial^2 n}{\partial z^2} &= 0 \end{aligned} \qquad \text{at} \quad z = 0, 1 \tag{41}$$

corresponding to permeable boundaries. Even for "free" but "impermeable" boundary conditions, the double Fourier development (34) is not the proper solution. Rigid or impermeable boundary conditions impose another mathematical tool.

B. Discretization of the Equations

A classical finite differences scheme is adopted to approximate the spatial derivatives. The z-and the x-axises are divided, respectively, into $(N_z - 1)$ and $(N_x - 1)$ intervals. The spacings

$$DZ = \frac{1}{N_z - 1} \qquad \text{and} \qquad DX = \frac{L}{N_x - 1}$$

are related for convenience by

$$DX = K \cdot DZ$$

The numerical value of K is assigned by the number of convective cells that we wish to see. Indeed other researchers focused their attention on only one cell of fixed horizontal length, given by a principle of "maximum Nusselt number." On the contrary, in order to verify that the critical wave number (or the size of a cell) *does* depend on the Soret coefficient, we impose the length **L** of a Soret apparatus, into which a free number of cells can develop, and this number will be a function of the Soret coefficient.

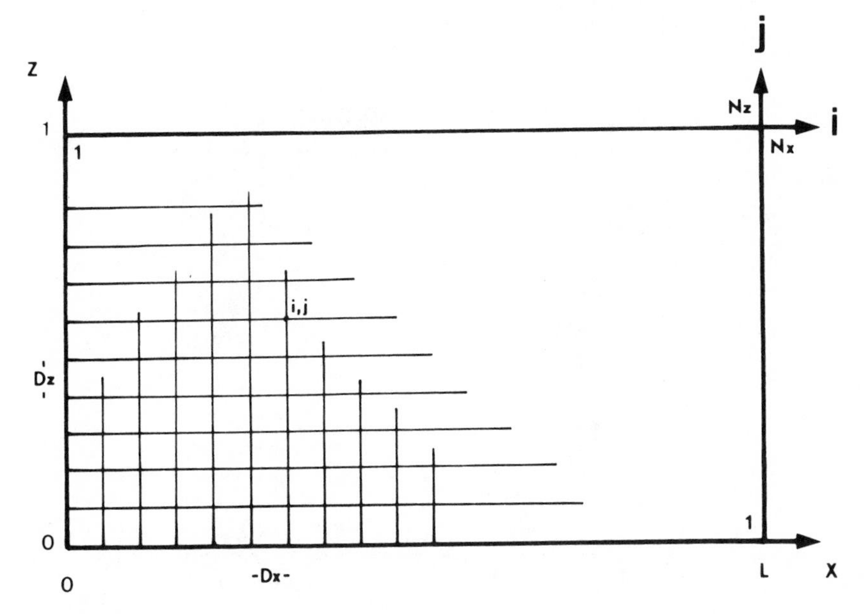

Fig. 14.

At a current grid point i,j (see Fig. 14) the differential system (17) to (19) is equivalent to a larger system of ordinary differential equations, using for the space derivatives of a variable f the following expressions

$$\left(\frac{\partial f}{\partial r}\right)_k = \frac{f_{k+1} - f_{k-1}}{2Dr}; \qquad \left(\frac{\partial^2 f}{\partial r^2}\right)_k = \frac{f_{k+1} - 2f_k + f_{k-1}}{(Dr)^2} \tag{42}$$

Defining

$$\phi = \nabla^2 \psi \tag{43}$$

we get

$$\frac{d\phi_{i,j}}{dt} = \frac{\psi_{i+1,j} - \psi_{i-1,j}}{2DX} \cdot \frac{\phi_{i,j+1} - \phi_{i,j-1}}{2DZ} - \frac{\psi_{i,j+1} - \psi_{i,j-1}}{2DZ} \cdot \frac{\phi_{i+1,j} - \phi_{i-1,j}}{2DX} - \frac{Ra}{Pr} \frac{(\vartheta_{i+1,j} - \vartheta_{i-1,j})}{2DX} + \frac{R_{Th}}{Pr} \frac{(n_{i+1,j} - n_{i-1,j})}{2DX} + \frac{(\phi_{i+1,j} - 2\phi_{i,j} + \phi_{i-1,j})}{(DX)^2} + \frac{(\phi_{i,j+1} - 2\phi_{i,j} + \phi_{i,j-1})}{(DZ)^2} \tag{44}$$

$$\frac{dn_{i,j}}{dt} = \frac{\psi_{i+1,j} - \psi_{i-1,j}}{2DX} \cdot \frac{n_{i,j+1} - n_{i,j-1}}{2DZ} - \frac{\psi_{i,j+1} - \psi_{i,j-1}}{2DZ} \cdot \frac{n_{i+1,j} - n_{i-1,j}}{2DX} + \mathbb{S} \frac{(\psi_{i+1,j} - \psi_{i-1,j})}{2DX} + \frac{\mathbb{S}}{Sc} \cdot \left(\frac{\vartheta_{i+1,j} - 2\vartheta_{i,j} + \vartheta_{i-1,j}}{(DX)^2} + \frac{\vartheta_{i,j+1} - 2\vartheta_{i,j} + \vartheta_{i,j-1}}{(DZ)^2} \right) + \frac{1}{Sc} \left(\frac{n_{i+1,j} - 2n_{i,j} + n_{i-1,j}}{(DX)^2} + \frac{n_{i,j+1} - 2n_{i,j} + n_{i,j-1}}{(DZ)^2} \right) \tag{45}$$

$$\frac{d\vartheta_{i,j}}{dt} = \frac{\psi_{i+1,j} - \psi_{i-1,j}}{2DX} \cdot \frac{\vartheta_{i,j+1} - \vartheta_{i,j-1}}{2DZ} - \frac{\psi_{i,j+1} - \psi_{i,j-1}}{2DZ} \cdot \frac{\vartheta_{i+1,j} - \vartheta_{i-1,j}}{2DX} - \frac{\psi_{i+1,j} - \psi_{i-1,j}}{2DX} + \frac{1}{Pr} \left(\frac{\vartheta_{i+1,j} - 2\vartheta_{i,j} + \vartheta_{i-1,j}}{(DX)^2} + \frac{\vartheta_{i,j+1} - 2\vartheta_{i,j} + \vartheta_{i,j-1}}{(DZ)^2} \right) \tag{46}$$

Equation (43) becomes

$$\phi_{i,j} = \frac{\psi_{i+1,j} - 2\psi_{i,j} + \psi_{i-1,j}}{(DX)^2} + \frac{\psi_{i,j+1} - 2\psi_{i,j} + \psi_{i,j-1}}{(DZ)^2} \tag{47}$$

Boundary and initial conditions must be given before proceeding with numerical integration of (44) to (47). The boundary conditions for ψ are given, for example, by Veronis[33]

$$\psi = 0; \qquad \frac{\partial^2 \psi}{\partial z^2} = 0; \qquad \theta = 0 \quad \text{on} \quad z = 0, 1 \qquad \text{(free)}$$

or

$$\psi = 0; \qquad \frac{\partial \psi}{\partial z} = 0; \qquad \theta = 0 \quad \text{on} \quad z = 0, 1 \qquad \text{(rigid)}$$

The mass fraction perturbation must satisfy condition (40) or (41). The lateral boundary conditions are

$$\frac{\partial n}{\partial x} = \frac{\partial \vartheta}{\partial x} = 0 \qquad \text{on} \qquad x = 0, L$$

(no heat or mass flux through the lateral boundaries).

As pointed out recently by Moore and Weiss,[34] it is essential to establish the accuracy of any difference scheme and the maximum Rayleigh number that can be studied on a given mesh. They believe that with $N_z = 18$, accurate results are restricted to $Ra \leqslant 4Ra^{\text{crit}}$, and with $N_z = 34$ to $Ra \leqslant 25Ra^{\text{crit}}$. For $N_z = 18$, the error in the Nusselt number rises from 1% at $Nu = 3$ to 2% at $Nu = 6$. We are primarily interested in oscillations, cell size variation due to Soret effect around the critical point, and subcritical instabilities. Thus in this first study, the Rayleigh numbers of interest for those purposes are restricted to

$$Ra \leqslant 2Ra^{\text{crit}}$$

It is reasonable to believe that $N_z = 14$ will produce sufficiently accurate results. With the choice $N_x = 66$, we hope to resolve, at least four cells with $DX \cong DZ$, that is, $N_x \cong N_z$ for one cell. The number of equations to be integrated is therefore of the order of 2700, and at each time step, the stream function ψ is obtained by solving Poisson's equation (43) or (47).

We have now to make a choice for the integration routine of (44) to (46). A Runge-Kutta scheme, associated with a predictor–corrector process (Runge-Kutta-Hammings), as used in Section III to integrate around 100 equations, fails for the 2700 equations considered here, mainly because the processor time in machine calculations becomes too long. We have thus to use finite differences in time. An explicit scheme of the Euler type can be used, namely,

$$f_{i,j}^{(k+1)} = f_{i,j}^{k} + \left(\frac{df_{i,j}}{dt}\right)^{k} \cdot \Delta t \tag{48}$$

where $f_{i,j}^{k}$ means the value of the function f at the grid point i,j at a time $\tau = k \cdot \Delta t$. The derivatives df/dt are directly given by (44) to (46). Peaceman and Rachford[35] have studied the two-dimensional heat flow equation with regard to the stability and accuracy of finite differences schemes. It is shown that with $DX = DZ$, the condition required to avoid numerical instabilities is

$$\Delta t \leqslant \frac{(DX)^2}{4}$$

that is, in our case $\Delta t \cong 10^{-3}$. For the differential system considered one iteration is performed in 0.5 sec on a CDC 6600 computer. We need at least 2 hours of central processor time in order to see a few oscillations around the mean value of Nu. It is thus useful to look at other integration schemes taking a greater time step, which moreover do not lead to numerical instabilities. The alternating-direction implicit method described by Peaceman and Rachford,[35] which is always stable (numerically), was used. In this method, only the derivatives with respect to one variable (say $\partial/\partial x$ and $\partial^2/\partial x^2$) are replaced by differences evaluated in terms of the *unknown* functions, while the other derivatives ($\partial/\partial z$ and $\partial^2/\partial z^2$) are replaced by differences evaluated in terms of the *known* functions; in this way sets of simultaneous equations are formed that can be solved easily without iterations. These equations are implicit in the x-direction and explicit in the z-direction. When the procedure was repeated for a second time step of equal size, with the difference equations implicit in the z-direction and explicit in the x-direction, it was shown *on the heat equation* that the overall procedure for the two time steps is stable for any size time step. This is of course not proven for the nonlinear and coupled equations [(44) to (46)], but we expect that numerical instability should arise at a much larger Δt than in a usual explicit method. This is confirmed by testing the size of the time step and we have found that for the same accuracy the computer time may be reduced by a factor of 20. Moreover the method proposed by Peaceman and Rachford was slightly modified in order to be applicable to our nonlinear and coupled equations: the implicit procedure in a particular space direction is used in the right-hand side *only* on the variables appearing in the left-hand side. Equation (44) becomes then

$$\frac{\phi_{i,j}^{2n+1}-\phi_{i,j}^{2n}}{\Delta t}=\frac{\psi_{i+1,j}^{2n}-\psi_{i-1,j}^{2n}}{2DX}\cdot\frac{\phi_{i,j+1}^{2n}-\phi_{i,j-1}^{2n}}{2DZ}$$

$$-\frac{\psi_{i,j+1}^{2n}-\psi_{i,j-1}^{2n}}{2DZ}\cdot\frac{\phi_{i+1,j}^{2n+1}-\phi_{i-1,j}^{2n+1}}{2DX}$$

$$-\frac{Ra}{Pr}\frac{\left(\vartheta_{i+1,j}^{2n}-\vartheta_{i-1,j}^{2n}\right)}{2DX}+\frac{R_{\mathrm{Th}}}{Pr}\frac{\left(n_{i+1,j}^{2n}-n_{i-1,j}^{2n}\right)}{2DX}$$

$$+\frac{\phi_{i+1,j}^{2n+1}-2\phi_{i,j}^{2n+1}+\phi_{i-1,j}^{2n+1}}{(DX)^2}+\frac{\phi_{i,j+1}^{2n}-2\phi_{i,j}^{2n}+\phi_{i,j-1}^{2n}}{(DZ)^2} \tag{49a}$$

(at an odd time step implicit on $\phi_{i,j}$ in the x-direction, otherwise explicit).

$$\frac{\phi_{i,j}^{2n+2}-\phi_{i,j}^{2n+1}}{\Delta t}=\frac{\psi_{i+1,j}^{2n+1}-\psi_{i-1,j}^{2n+1}}{2DX}\cdot\frac{\phi_{i,j+1}^{2n+2}-\phi_{i,j-1}^{2n+2}}{2DZ}$$

$$-\frac{\psi_{i,j+1}^{2n+1}-\psi_{i,j-1}^{2n+1}}{2DZ}\cdot\frac{\phi_{i+1,j}^{2n+1}-\phi_{i-1,j}^{2n+1}}{2DX}$$

$$-\frac{Ra}{Pr}\frac{\left(\vartheta_{i+1,j}^{2n+1}-\vartheta_{i-1,j}^{2n+1}\right)}{2DX}+\frac{R_{Th}}{Pr}\frac{\left(n_{i+1,j}^{2n+1}-n_{i-1,j}^{2n+1}\right)}{2DX}$$

$$+\frac{\phi_{i+1,j}^{2n+1}-2\phi_{i,j}^{2n+1}+\phi_{i-1,j}^{2n+1}}{(DX)^2}$$

$$+\frac{\phi_{i,j+1}^{2n+2}-2\phi_{i,j}^{2n+2}+\phi_{i,j-1}^{2n+2}}{(DZ)^2} \tag{49b}$$

(at an even time step implicit on $\phi_{i,j}$ in the z-direction, otherwise explicit). In the same way, we get for $n_{i,j}$ and $\vartheta_{i,j}$ at odd or even number of time steps the following sets of linear algebraic equations:

$$
\begin{aligned}
\frac{n_{i,j}^{2n+1}-n_{i,j}^{2n}}{\Delta t} = {} & \frac{\psi_{i+1,j}^{2n}-\psi_{i-1,j}^{2n}}{2DX}\cdot\frac{n_{i,j+1}^{2n}-n_{i,j-1}^{2n}}{2DZ} \\
& - \frac{\psi_{i,j+1}^{2n}-\psi_{i,j-1}^{2n}}{2DZ}\cdot\frac{n_{i+1,j}^{2n+1}-n_{i-1,j}^{2n+1}}{2DX} \\
& + \mathbb{S}\,\frac{\psi_{i+1,j}^{2n}-\psi_{i-1,j}^{2n}}{2DX} \\
& + \frac{\mathbb{S}}{Sc}\left(\frac{\vartheta_{i+1,j}^{2n}-2\vartheta_{i,j}^{2n}+\vartheta_{i-1,j}^{2n}}{(DX)^2}+\frac{\vartheta_{i,j+1}^{2n}-2\vartheta_{i,j}^{2n}+\vartheta_{i,j-1}^{2n}}{(DZ)^2}\right) \\
& + \frac{1}{Sc}\left(\frac{n_{i+1,j}^{2n+1}-2n_{i,j}^{2n+1}+n_{i-1,j}^{2n+1}}{(DX)^2}+\frac{n_{i,j+1}^{2n}-2n_{i,j}^{2n}+n_{i,j-1}^{2n}}{(DZ)^2}\right)
\end{aligned} \tag{50a}
$$

$$
\begin{aligned}
\frac{n_{i,j}^{2n+2}-n_{i,j}^{2n+1}}{\Delta t} = {} & \frac{\psi_{i+1,j}^{2n+1}-\psi_{i-1,j}^{2n+1}}{2DX}\cdot\frac{n_{i,j+1}^{2n+2}-n_{i,j-1}^{2n+2}}{2DZ} \\
& - \frac{\psi_{i,j+1}^{2n+1}-\psi_{i,j-1}^{2n+1}}{2DZ}\cdot\frac{n_{i+1,j}^{2n+1}-n_{i-1,j}^{2n+1}}{2DX} \\
& + \mathbb{S}\,\frac{(\psi_{i+1,j}^{2n+1}-\psi_{i-1,j}^{2n+1})}{2DX} \\
& + \frac{\mathbb{S}}{Sc}\left(\frac{\vartheta_{i+1,j}^{2n+1}-2\vartheta_{i,j}^{2n+1}+\vartheta_{i-1,j}^{2n+1}}{(DX)^2}+\frac{\vartheta_{i,j+1}^{2n+1}-2\vartheta_{i,j}^{2n+1}+\vartheta_{i,j-1}^{2n+1}}{(DZ)^2}\right) \\
& + \frac{1}{Sc}\left(\frac{n_{i+1,j}^{2n+1}-2n_{i,j}^{2n+1}+n_{i-1,j}^{2n+1}}{(DX)^2}+\frac{n_{i,j+1}^{2n+2}-2n_{i,j}^{2n+2}+n_{i,j-1}^{2n+2}}{(DZ)^2}\right)
\end{aligned} \tag{50b}
$$

$$\frac{\vartheta_{i,j}^{2n+1}-\vartheta_{i,j}^{2n}}{\Delta t}=\frac{\psi_{i+1,j}^{2n}-\psi_{i-1,j}^{2n}}{2DX}\cdot\frac{\vartheta_{i,j+1}^{2n}-\vartheta_{i,j-1}^{2n}}{2DZ}$$

$$-\frac{\psi_{i,j+1}^{2n}-\psi_{i,j-1}^{2n}}{2DZ}\cdot\frac{\vartheta_{i+1,j}^{2n+1}-\vartheta_{i-1,j}^{2n+1}}{2DX}$$

$$-\frac{\psi_{i+1,j}^{2n}-\psi_{i-1,j}^{2n}}{2DX}$$

$$+\frac{1}{Pr}\left(\frac{\vartheta_{i+1,j}^{2n+1}-2\vartheta_{i,j}^{2n+1}+\vartheta_{i-1,j}^{2n+1}}{(DX)^2}+\frac{\vartheta_{i,j+1}^{2n}-2\vartheta_{i,j}^{2n}+\vartheta_{i,j-1}^{2n}}{(DZ)^2}\right) \tag{51a}$$

$$\frac{\vartheta_{i,j}^{2n+2}-\vartheta_{i,j}^{2n+1}}{\Delta t}=\frac{\psi_{i+1,j}^{2n+1}-\psi_{i-1,j}^{2n+1}}{2DX}\cdot\frac{\vartheta_{i,j+1}^{2n+2}-\vartheta_{i,j-1}^{2n+2}}{2DZ}$$

$$-\frac{\psi_{i,j+1}^{2n+1}-\psi_{i,j-1}^{2n+1}}{2DZ}\cdot\frac{\vartheta_{i+1,j}^{2n+1}-\vartheta_{i-1,j}^{2n+1}}{2DX}$$

$$-\frac{\psi_{i+1,j}^{2n+1}-\psi_{i-1,j}^{2n+1}}{2DX}$$

$$+\frac{1}{Pr}\left(\frac{\vartheta_{i+1,j}^{2n+1}-2\vartheta_{i,j}^{2n+1}+\vartheta_{i-1,j}^{2n+1}}{(DX)^2}+\frac{\vartheta_{i,j+1}^{2n+2}-2\vartheta_{i,j}^{2n+2}+\vartheta_{i,j-1}^{2n+2}}{(DZ)^2}\right) \tag{51b}$$

The details on the numerical solution of the algebraic systems (49) to (51), including the boundary conditions, can be found elsewhere.[36] It is worth mentioning that every kind of boundary condition (free, rigid, permeable...) can be included in the core of the computer program.[36]

The numerical procedure described in this section was tested for the case of free and permeable boundaries at $Ra=1800$ and $\mathfrak{S}=-10^{-2}$. The results of Section III are recovered. K was set equal to 1, thus $DX=DZ$ or $\mathbf{L}=5d$ ($d=1$ arbitrary unit). In fact, due to symmetry requirements, the value of $\mathbf{L}$ of interest is half the total length of the Soret apparatus. With

$K=1$, it is thus possible for free boundaries to describe four convection cells. An example of streamlines and isotherms is given in Fig. 15.

The study of rigid boundaries is now in progress.

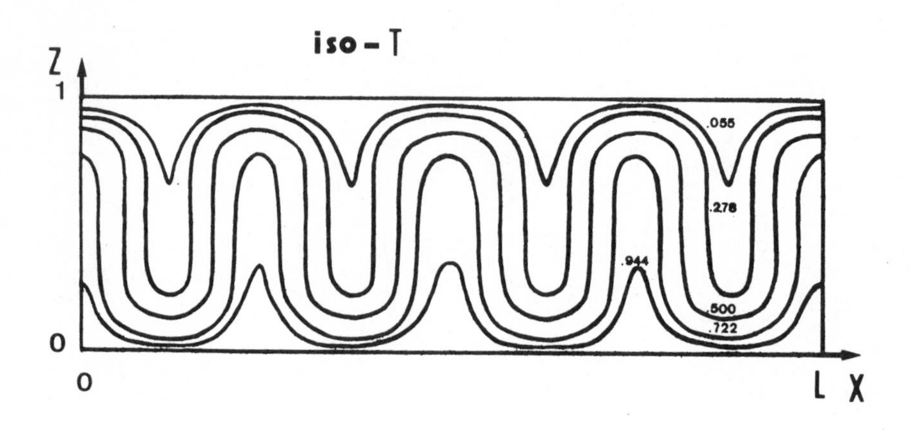

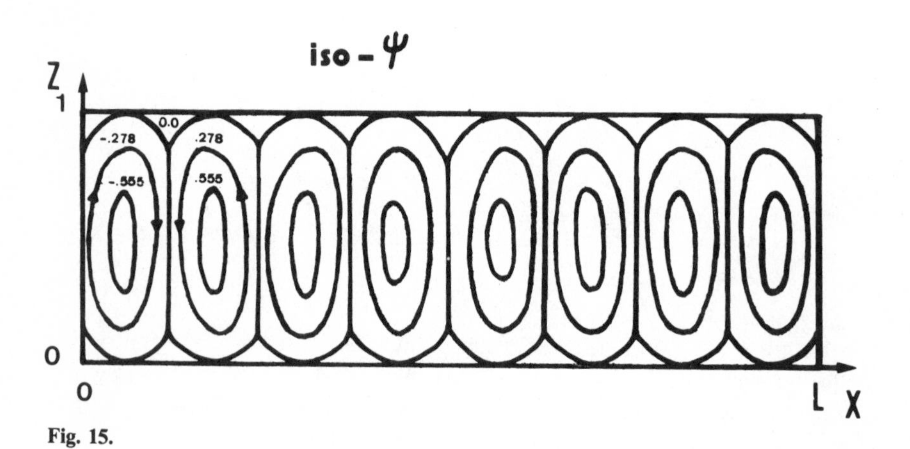

Fig. 15.

Acknowledgments

We are deeply indebted to Professors I. Prigogine, P. Glansdorff, and G. Nicolis whose continuous interest and stimulating comments were indispensable for the realization of this work. One of us (G.C.) wishes to thank the Institut pour l'Encouragement de la Recherche Scientifique dans l'Industrie et l'Agriculture (I.R.S.I.A., Brussels) for a grant.

References

1. S. Chandrasekhar, *Hydrodynamics and Hydrodynamic Stability*, Clarendon, Oxford, 1961, Chap. 2.
2. D. A. Nield, "The Thermohaline Rayleigh-Jeffreys Problem," *J. Fluid Mech.*, **29**, 545 (1967).
3. M. E. Stern, "The Salt-Fountain and Thermohaline Convection," *Tellus*, **12**, 172 (1960).
4. M. E. Stern, "Collective Instability of Salt Fingers," *J. Fluid Mech.*, **35**, 209 (1969).
5. M. E. Stern and J. S. Turner, "Salt Fingers and Convecting Layers," *Deep Sea Res.*, **16**, 497 (1969).
6. H. Stommel, A. B. Arons, and D. Blanchard, "An Ocean Curiousity: The Perpetual Salt Fountain," *Deep Sea Res.*, **3**, 152 (1956).
7. J. S. Turner, "The Behavior of a Stable Salinity Gradient Heated from Below," *J. Fluid Mech.*, **33**, 183 (1968).
8. G. Veronis, "On Finite Amplitude Instability in Thermohaline Convection," *J. Marine Res.*, **23**, 1 (1965).
9. G. Veronis, "Effect of a Stabilizing Gradient of Solute on Thermal Convection," *J. Fluid Mech.*, **34**, 315 (1968).
10. J. C. Legros, W. A. Van Hook, and G. Thomaes, "Convection and Thermal Diffusion in a Solution Heated from Below," *Chem. Phys. Lett.*, **1**, 696 (1968).
11. J. C. Legros, W. A. Van Hook, and G. Thomaes, "Convection and Thermal Diffusion in a Solution Heated from Below II: The System $CHBr_2 \cdot CHBr_2$–$CHCl_2 \cdot CHCl_2$." *Chem. Phys. Lett.*, **2**, 249 (1968).
12. J. C. Legros, W. A. Van Hook, and G. Thomaes, "Thermal Diffusion in $CHBr_2 \cdot CHBr_2$–$CHCl_2$ Solutions," *Chem. Phys. Lett.*, **2**, 251 (1968).
13. J. C. Legros, D. Rasse, and G. Thomaes, "Convection and Thermal Diffusion in a Solution Heated from Below," *Chem. Phys. Lett.*, **4**, 632 (1970).
14. R. S. Schmidt and S. W. Milverton, "On the Instability of a Fluid when Heated from Below," *Proc. Roy. Soc. London, Ser. A*, **152**, 589 (1935).
15. D. R. Caldwell, "Non-linear Effects in a Rayleigh-Bénard Experiment," *J. Fluid Mech.*, **42**, 161 (1970).
16. J. K. Platten and G. Chavepeyer, "Oscillations in a Water-Ethanol Liquid Heated from Below," *Phys. Lett.*, **40**, 287 (1972).
17. J. K. Platten and G. Chavepeyer, "Oscillatory Motion in a Bénard Cell Due to the Soret Effect," *J. Fluid Mech.*, **60**, 305 (1973).
18. I. Prigogine, "Time, Irreversibility and Structure," in *The Physicist's Conception of Nature* J. Mehra, Ed., Reidel Pub. Co., 1973, p. 584.
19. S. R. De Groot and P. Mazur, "Nonequilibrium Thermodynamics," North Holland Pub. Co., Amsterdam, 1962, pp. 273–284.
20. D. T. J. Hurle and E. Jakeman, "Natural Oscillations in Heated Fluid Layers," *Phys. Lett.*, **43A**, 127 (1973).
21. D. T. J. Hurle and E. Jakeman, "On the Nature of Oscillatory Convection in Two-Component Fluids" in *Advances in Chemical Physics*, Vol. 32, I. Prigogine and S. Rice, Eds. Interscience, New York, 1974, pp. 277–279.
22. D. T. J. Hurle and E. Jakeman, "Significance of the Soret Effect in the Rayleigh-Jeffreys Problem," *Phys. Fluids*, **12**, 2704 (1969).
23. D. T. J. Hurle and E. Jakeman, "Soret-driven Thermosolutal Convection," *J. Fluid Mech.*, **47**, 667 (1971).
24. R. S. Schechter, I. Prigogine, and J. R. Hamm, "Thermal Diffusion and Convective Stability," *Phys. Fluids*, **15**, 379 (1972).

25. J. C. Legros, J. K. Platten, and P. G. Poty, "Stability of a Two-Component Fluid Layer Heated from Below," *Phys. Fluids*, **15**, 1383 (1972).
26. J. K. Platten and G. Chavepeyer, "Soret Driven Instability," *Phys. Fluids*, **15**, 1555 (1972).
27. R. S. Schechter, M. G. Velarde, and J. K. Platten, "The Two Component Bénard Problem," *Advances in Chemical Physics*, Vol. 26, I. Prigogine and S. A. Rice, Eds., Interscience New York, 1974, p. 265.
28. J. C. Legros, P. G. Poty, and G. Thomaes, "Thermal diffusion in a two-component fluid Bénard problem," *Physica*, **64**, 481 (1973).
29. J. K. Platten, "Le problème de Bénard dans les mélanges: cas de surfaces libres," *Bull. Acad. Roy. Belg. Cl. Sci.*, **57**, 669 (1971).
30. G. Veronis, "Motions at Subcritical Values of the Rayleigh Number in Rotating Fluid," *J. Fluid Mech.*, **24**, 545 (1966).
31. G. Nicolis and J. F. Auchmuty, "Dissipative Structures, Catastrophes and Pattern Formation: a Bifurcation Theory," *Proc. Natl. Acad. Sci. U.S.*, **71**(7), 2748 (1974).
32. J. K. Platten and G. Chavepeyer, "Nonlinear Two Dimensional Bénard Convection with Soret Effect: Free Boundaries," submitted to the *Int. J. Heat Mass Transfer*.
33. G. Veronis, "Large amplitude Bénard convection," *J. Fluid Mech.*, **26**, 49 (1966).
34. D. R. Moore and N. O. Weiss, "To-Dimensional Rayleigh-Bénard Convection," *J. Fluid Mech.*, **58**, 289 (1972).
35. D. W. Peaceman and H. H. Rachford Jr., "The Numerical Solution of Parabolic and Elliptic Differential Equations," *J. Soc. Ind. Appl. Math.*, **3**, 28 (1955).
36. G. Chavepeyer, "Application de la méthode des différences finies à l'étude non linéaire du problème de Bénard dans les mélanges," *Bull. Acad. Roy. Belg. Cl. Sci.*, in press (1975).

AUTHOR INDEX

Numbers in parentheses are reference numbers and show that an author's work is referred to although his name is not mentioned in the text. Numbers in *italics* indicate the pages on which the full reference appears.

SUBJECT INDEX